Complex Systems Design & Management

Marc Aiguier · Frédéric Boulanger
Daniel Krob · Clotilde Marchal
Editors

Complex Systems Design & Management

Proceedings of the Fourth International Conference on Complex Systems Design & Management CSD&M 2013

 Springer

Editors

Marc Aiguier
Ecole Centrale Paris
Grande Voie des Vignes
Châtenay-Malabry
France

Daniel Krob
Ecole Polytechnique
LIX/DIX
Palaiseau Cedex
France

Frédéric Boulanger
Supélec
Plateau Moulon
Gif-sur-Yvette Cedex
France

Clotilde Marchal
EADS
Suresnes Cedex
France

ISBN 978-3-319-37763-6   ISBN 978-3-319-02812-5   (eBook)
DOI 10.1007/978-3-319-02812-5
Springer Cham Heidelberg New York Dordrecht London

Preface

Introduction

This volume contains the proceedings of the Fourth International Conference on "Complex System Design & Management" (CSD&M 2013; see the conference website: http://www.csdm2013.csdm.fr for more details).

The CSD&M 2013 conference was jointly organized, the December 4–6, 2013 at the Cité Internationale Universitaire of Paris (France), by the two following founding partners:

1. the research & training Ecole Polytechnique – ENSTA ParisTech – Télécom ParisTech – Dassault Aviation – DCNS – DGA – Thales chair "Engineering of Complex Systems",
2. the non profit organization C.E.S.A.M.E.S. (Center of Excellence on Systems Architecture, Management, Economy and Strategy).

The conference benefited of the permanent support of many academic and institutional organizations such as Conservatoire National des Arts et Métiers (CNAM), Digiteo labs, Ecole Centrale de Paris, Ecole Nationale Supérieure des Techniques Avancées (ENSTA ParisTech), Ecole Polytechnique, Ecole Supérieure d'Electricité (Supélec), Institut des Systèmes Complexes – Ile de France, Ministère de l'Enseignement Supérieur et de la Recherche and Télécom ParisTech which were deeply involved in its organization.

A special thank also goes to Dassault Aviation, DCNS, Direction Générale de l'Armement (DGA), EADS, EDF, Faurecia, Institut de Recherche Technologique IRT-SystemX, MEGA International, Nexter Systems and Thales which were the main industrial sponsors of the conference. The generous specific support of EADS and IRT-SystemX shall be especially pointed out here.

We are also grateful to many non-profit organizations such as Association Française d'Ingénierie Système (AFIS), International Council on Systems Engineering (INCOSE), Institut pour la Maîtrise des Risques (IMdR) and the Systematic cluster which supported strongly our communication.

All these institutions helped us a lot through their constant participation to the organizing committee during the one-year preparation of CSD&M 2013. Many thanks therefore to all of them.

Why a CSD&M Conference?

Mastering complex systems requires an integrated understanding of industrial practices as well as sophisticated theoretical techniques and tools. This explains the creation of an annual *go-between* forum at European level (which did not existed yet) dedicated both to academic researchers and industrial actors working on complex industrial systems architecture and engineering in order to facilitate their *meeting*. It was actually for us a *sine qua non* condition in order to nurture and develop in Europe this complex industrial systems science which is now emergent.

The purpose of the "Complex Systems Design & Management" (CSD&M) conference is exactly to be such a forum, in order to become, in time, *the* European academic-industrial conference of reference in the field of complex industrial systems architecture and engineering, which is a quite ambitious objective. The last three CSD&M 2010, CSD&M 2011 and CSD&M 2012 conferences – which were held in October 2010, December 2011 and December 2012 in Paris – were the first steps in this direction with respectively more than 200, 250 and 280 participants coming from 20 different countries with an almost perfect balance between academia and industry.

The CSD&M Academic–Industrial Integrated Dimension

To make the CSD&M conference this convergence point of the academic and industrial communities in complex industrial systems, we based our organization on a principle of *complete parity* between academics and industrialists (see the conference organization sections in the next pages). This principle was first implemented as follows:

- the Programme Committee consisted of 50% academics and 50% industrialists,
- the Invited Speakers are coming in a balanced way from numerous professional environments.

The set of activities of the conference followed the same principle. They indeed consist of a mixture of research seminars and experience sharing, academic articles and industrial presentations, software and training offers presentations, etc. The conference topics cover in the same way the most recent trends in the emerging field of complex systems sciences and practices from an industrial and academic perspective, including the main industrial domains (transport, defense & security, electronics & robotics, energy & environment, health & welfare services, media & communications, e-services), scientific and technical topics (systems fundamentals, systems architecture & engineering, systems metrics & quality, systemic tools) and system types (transportation systems, embedded systems, software & information systems, systems of systems, artificial ecosystems).

The CSD&M 2013 Edition

The CSD&M 2013 edition received 76 submitted papers, out of which the program committee selected 22 regular papers to be published in these proceedings, which corresponds to a 29% acceptance ratio which is fundamental for us to guarantee the high quality of the presentations. The program committee also selected 41 papers for a collective presentation in the poster workshop of the conference.

Each submission was assigned to at least two program committee members, who carefully reviewed the papers, in many cases with the help of external referees. These reviews were discussed by the program committee during a physical meeting held in C.E.S.A.M.E.S. office in Paris by June 11, 2013 and via the EasyChair conference management system.

We also chose 17 outstanding speakers with various industrial and scientific expertise who gave a series of invited talks covering all the spectrum of the conference, mainly during the two first days of CSD&M 2013, the last day being dedicated to a special "vision session" and the presentations of all accepted papers in parallel with three system-focused tutorials. The first day of the conference was especially organized around a common topic – Systems of systems – that gave a coherence to all the initial invited talks.

Futhermore, we had a poster workshop, for encouraging presentation and discussion on interesting but "not-yet-polished" ideas, and a software tools presentation session, in order to provide to each participant a good vision on the present status of the engineering tools market offer.

Acknowledgements

We would like finally to thank all members of the program and organizing committees for their time, effort, and contributions to make CSD&M 2013 a top quality conference. A special thank is addressed to the C.E.S.A.M.E.S. non-profit organization team which managed permanently with a huge efficiency all the administration, logistics and communication of the CSD&M 2013 conference (see http://www.cesames.net).

The organizers of the conference are also greatly grateful to the following sponsors and partners without whom the CSD&M 2013 event would just not exist:

Founding Partners

Center of Excellence on Systems Architecture, Management, Economy and Strategy (C.E.S.A.M.E.S.)
Ecole Polytechnique – ENSTA ParisTech – Télécom ParisTech – Dassault Aviation – DCNS – DGA – Thales chair "Engineering of Complex Systems"

Academic Sponsors

Conservatoire National des Arts et Métiers (CNAM)
Ecole Centrale de Paris
Ecole Polytechnique

Ecole Supérieure d'Electricité (Supélec)
Ecole Nationale Supérieure des Techniques Avancées (ENSTA ParisTech)
Télécom ParisTech

Industrial Sponsors

Dassault Aviation
DCNS
Direction Générale de l'Armement (DGA)
EADS
EDF
Faurecia
MEGA International
Nexter Systems
Thales

Institutional Sponsors

Digiteo labs
Institut de Recherche Technologique (IRT) SystemX
Ministère de l'Enseignement Supérieur et de la Recherche
Institut des Systèmes Complexes – Ile de France

Supporting Partners

Association Française d'Ingénierie Système (AFIS)
Institut pour la Maîtrise des Risques (IMdR)
International Council on Systems Engineering (INCOSE)
Systematic cluster

Participating Partners

Atego
IBM Rational Software
Knowledge Inside
Obeo
PragmaDev
Project Performance International
PTC
The CosMo Company
The MathWorks

Paris, August 02, 2013   Marc Aiguier – Ecole Centrale de Paris
                    Frédéric Boulanger – Ecole Supérieure d'Electricité (Supélec)
                        Daniel Krob – C.E.S.A.M.E.S. & Ecole Polytechnique
                                      Clotilde Marchal – EADS

Conference Organization

Conference Chairs

- **General Chair**
  - Daniel Krob, institute professor, Ecole Polytechnique - France
- **Organizing Committee Chair**
  - Marc Aiguier, professor, Ecole Centrale de Paris - France
- **Program Committee Chairs**
  - Frédéric Boulanger, professor, Supélec - France (academic co-chair)
  - Clotilde Marchal, head of EADS Systems Engineering Group Capabilities Corporate Technical Office, EADS - France (industrial co-chair)

Program Committee

The PC consists of 50 members (25 academic and 25 industrial): all are personalities of high international visibility. Their expertise spectrum covers all of the conference topics.

Academic Members

- **Co-Chair**
  - Frédéric Boulanger, Supélec, France
- **Other Members**
    - Erik Aslaksen, Gumbooya Pty. - Australia
    - Christian Attiogbé, Université de Nantes - France
    - Julien Bernet, Trusted Labs - France
    - Manfred Broy, TUM - Germany
    - Michel-Alexandre Cardin, National University of Singapore - Singapore
    - Vincent Chapurlat, Ecole des Mines d'Alès - France
    - Robert De Simone, INRIA Sophia-Antipolis - France

- o   Olivier De Weck, MIT - USA
- o   Holger Giese, Hasso Plattner Institut - Germany
- o   Patrick Godfrey, University of Bristol - Great Britain
- o   Omar Hammami, ENSTA ParisTech - France
- o   Paulien Herder, University of Delft - Netherlands
- o   Mike Hinchey, Irish Software Engineering Research Centre - Ireland
- o   Marjan Mernik, University of Maribor - Slovenia
- o   Gérard Morel, Université de Nancy - France
- o   Pieter J. Mosterman, MacGill University - Canada
- o   Antoine Rauzy, Ecole Polytechnique - France
- o   Arend Rensink, University of Twente - Netherlands
- o   Adam Ross, MIT - USA
- o   Bernhard Rumpe, Aachen University - Germany
- o   Pierre-Yves Schobbens, Facultés Universitaires Notre-Dame de la Paix - Belgium
- o   Stavros Tripakis, University of California at Berkeley - USA
- o   Paul Valckenaers, Catholic University of Leuven - Belgium
- o   John Wade, Stevens Institute of Technology - USA

Industrial Members

- •   **Co-Chair**
  - -   Clotilde Marchal, EADS - France

- •   **Other Members**
  - o   Gérard Auvray, Astrium ST - France
  - o   André Ayoun, Cassidian - France
  - o   Martine Callot, EADS Innovation Works - France
  - o   Jean-Pierre Daniel, AREVA - France
  - o   Brigitte Daniel-Allegro, Brigitte Daniel-Allegro - France
  - o   Alain Dauron, Renault - France
  - o   François-Xavier Fornari, Esterel Technologies - France
  - o   Greg Gorman, IBM - USA
  - o   Alan Harding, BAE SYSTEMS - Great Britain
  - o   Erik Herzog, SAAB - Sweden
  - o   Carl Landrum, Honeywell - USA
  - o   Emmanuel Ledinot, Dassault Aviation - France
  - o   Juan Llorens, The Reuse Company - Spain
  - o   Tim Lochow, EADS Innovation Works - Germany
  - o   David Long, VITECH - USA
  - o   Padman Nagenthiram, BOEING - USA
  - o   Andrew Pickard, ROLLS ROYCE - Great Britain
  - o   Jean-Claude Roussel, EADS Innovation Works – France

- o   Dominique Seguela, CNES - France
- o   Hillary Sillitto, Thales - Great Britain
- o   Robert Swarz, MITRE - USA
- o   Philippe Thuillier, Altran - France
- o   Xavier Warzee, Palo IT - USA
- o   Mike Wilkinson, ATKINS - Great Britain

Organizing Committee

- **Chair**

- Marc Aiguier, Ecole Centrale de Paris - France

- **Other Members**

  - o   Anas Alfaris, CCES & MIT - Saudi Arabia
  - o   Emmanuel Arbaretier, EADS - France
  - o   Eric Bonjour, Université de Lorraine ENSGSI - France
  - o   Karim Azoum, System@tic - France
  - o   Guy Boy, Florida Institute of Technology - USA
  - o   Michel-Alexandre Cardin, National University of Singapore - Singapore
  - o   Gilles Fleury, Supélec - France
  - o   Pascal Foix, Thales - France
  - o   Vassilis Giakoumakis, Université d'Amiens - France
  - o   Eric Goubault, CEA – France
  - o   Paul Labrogère, IRT-SystemX - France
  - o   Isabelle Perseil, INSERM - France
  - o   Garry Roedler, Lockheed Martin Corporate Engineering - USA
  - o   François Stephan, IRT-SystemX - France
  - o   Nicolas Treves, CNAM - France
  - o   John Wade, Stevens Institute of Technology - USA
  - o   David Walden, Sysnovation & INCOSE - USA
  - o   Fatiha Zaïdi, Université Paris Sud 11 - France

Invited Speakers

Societal Challenges: Systems of Systems

- Pao-Chuen Lui, former professor at the National University of Singapore, ministries adviser - Singapore
- Claude Feliot, project manager, Conseil Régional de Martinique - France
- Jean-François Janin, senior engineer, French Ministry of Sustainable Development - France
- Michael Henshaw, professor, University of Loughborough - Great Britain

Industrial Challenges: Systems of Systems

- Emmanuel Teboul, head of the methods department of Transilien, SNCF - France
- Pascal Pezzani, head of COEIE1 & COEIE4, Cassidian - France
- Alain Bovis, director, DCNS - France
- Isabelle Buret, head of the project Iridium, Thales TAS - France

Scientific State-of-the-art

- John Fitzgerald, professor, Newcastle University - Great Britain
- Edward Lee, professor, Berkeley University - USA
- Kirstie Bellman, head of the AISC, Aerospace Corp. - USA
- Michel Morvan, associate professor, the Santa Fe Institute - USA

Methodological State-of-the-art

- Judith Dahmann, principal senior scientist, Mitre Corp. - USA
- Dominique Luzeaux, director of Land Systems acquisition, Direction Générale de l'Armement - France
- Eric Honour, past president, INCOSE - USA
- Olivier De Weck, professor, MIT - USA

AGeSYS Workshop

- Eric Bantegnie, CEO, Esterel Technologies/System@tic - France

Contents

8   A Hybrid Event-B Study of Lane Centering.........................   97
Richard Banach, Michael Butler

**9   A Method for Managing Uncertainty Levels in Design Variables
during Complex Product Development 113**
João Fernandes, Elsa Henriques, Arlindo Silva

21 Handling Complexity in System of Systems Projects – Lessons Learned from MBSE Efforts in Border Security Projects 281
Emrah Asan, Oliver Albrecht, Semih Bilgen

22 The Multidimensional Hierarchically Integrated Framework (MHIF) for Modeling Complex Engineering Systems 301
Ahmad Alabdulkareem, Anas Alfaris, Vivek Sakhrani, Adnan Alsaati, Olivier de Weck

Foundations for Model-Based Engineering of Systems of Systems

John Fitzgerald, Peter Gorm Larsen, and Jim Woodcock

Abstract. The engineering of Systems of Systems presents daunting challenges. In this paper it is argued that rigorous semantic foundations for model-based techniques are potentially beneficial in addressing these. Three priorities for current research are identified: contractual interface definition, the verification of emergent behaviour, and the need to deal with semantic heterogeneity of constituent systems and their models. We describe the foundations of an approach in which architectural models in SysML may be analysed in a formal modelling language that has an extensible semantics given using the Unifying Theories of Programming (UTP).

1 Introduction

A *System of Systems* (SoS) is a group of independently owned and managed systems that can be treated as an integrated system offering an emergent service [21, 18]. SoSs are typically distributed in character and may evolve as a result of changes in their operating environment and in the goals of the owners of the autonomous constituent systems [5]. As networking capabilities improve, the range of SoSs increases. Examples include infrastructure such as smart grids and transport networks, as well as applications in healthcare, emergency response, and defence. In many cases, reliance comes to be placed on the emergent behaviour offered by SoSs, and so it is imperative that they are engineered in such a way that this reliance is justified.

John Fitzgerald
Newcastle University, UK
e-mail: John.Fitzgerald@ncl.ac.uk

Peter Gorm Larsen
Aarhus University, DK
e-mail: pgl@iha.dk

Jim Woodcock
University of York, UK
e-mail: Jim.Woodcock@york.ac.uk

M. Aiguier et al. (eds.), *Complex Systems Design & Management 2013*,
DOI: 10.1007/978-3-319-02812-5_1, © Springer International Publishing Switzerland 2014

Systems of Systems Engineering (SoSE) is a branch of Systems Engineering that addresses the challenges inherent in developing SoSs and maintaining them as they evolve. SoSE emphasises cross-disciplinary thinking, the role of socio-technical aspects such as governance, business aspects such as procurement and the management of multiple diverse stakeholders. There is a lively SoSE community[1] and the European Commission has developed the field [12], funding two projects on model-based methods (COMPASS[2] and DANSE[3]), an agenda-setting project on transatlantic cooperation(T-AREA-SoS[4]), a roadmapping project (ROAD2SOS[5]), and four further projects expected to be launched in 2013. Although lively, SoSE is a young discipline, and general principles and patterns remain to be discovered.

The developer or maintainer of an SoS faces challenges that stem from the lack of centralised control over constituents, the complexity of verifying properties at the SoS level, and the diversity of stakeholders in SoS development. Model-based methods are increasingly seen as a way of addressing these aspects [8]. In the COMPASS project, our approach is further informed by formal methods that exploit rigorous mathematical semantics for computer-based languages [28], allowing models to be constructed at levels of abstraction appropriate to each stage in the development process and subjected to machine-assisted analysis, or to enable the exploration of design alternatives by permitting trade-off analysis. Models at differing levels of abstraction may be linked by a formal *refinement* relation which ensures that properties verified in the more abstract model will still hold in the more concrete refined version.

While model-based and formal techniques are gradually being exploited in software engineering, their realisation in SoSE faces several challenges. Our work in the COMPASS project focuses on three problems that we believe must be addressed in order to realise fully the benefits of model-based SoSE technology:

Independence of constituent systems: Constituent systems are provided and operated by stakeholders independent of the SoS as a whole. These stakeholders have the freedom to choose what information to reveal and what to withhold from their peers in the SoS. In particular, they may change the goals and operation of constituent systems during the life of the SoS. As a consequence, the SoS engineer may not expect to be able to model constituent system behaviour completely. Instead, the constituent system must enter into a *contract* with its fellows. Modelling techniques must therefore describe *ranges* of permitted behaviours (the things each constituent can rely on and guarantee).

Verification of emergent behaviour: A system shows emergent behaviour when observations can be made of the overall system that cannot be described by ob-

---

[1] See, for example, the IEEE Conferences on Systems Engineering
(www.ieeesyscon.org) and SoSE (www.sose2013.org);
INCOSE (www.incose.org), which has an active SoS working group; the IEEE Systems Engineering Journal and the Journal of System of Systems Engineering.

[2] www.compass-research.eu

[3] www.danse-ip.eu

[4] www.tareasos.eu

[5] www.road2sos-project.eu

serving the behaviour of any of its components. For example, consider an SoS in which several banks produce smart card electronic cash systems that allow amounts of money to be paid between cards. Developers might wish to verify or refute the emergent property that the total amount of money in the SoS does not increase. The verification of emergent properties requires the composition of the verified properties of constituent systems.

Semantic heterogeneity: SoSs and their models have many facets, including functionality, mobility, cyber and physical elements, human and socio-economic aspects. Engineering for the interactions between these facets requires a multi-disciplinary approach [10], but the models themselves, being rooted in different disciplines, may not share a common semantic framework, making it impossible to perform analyses, such as trade-offs, that cross the boundaries. For example, we may have constituent systems that have components that are synchronous and others that are asynchronous; some may have discrete state or time and others continuous state or time. There may also be different levels of abstraction. For example, a socio-technical medical SoS may have patients undergoing courses of treatment that last for months, while their medicines must be taken daily and their adaptive pacemaker must be accurate to 100ms. The semantic heterogeneity of this range of models must be addressed if emergent properties are to be verified with a level of automated support.

It should be noted that socio-technical, business and human aspects of SoSs are vitally important, our current work focuses on theories, methods and tools for computational aspects of SoSE.

In this paper we describe a formal framework for SoS modelling that aims to address the challenges of SoSE outlined above. We present the COMPASS framework and tools in outline (Sect. 2) before discussing the approach taken to the three issues of collaborative development through contracts (Sect. 3), the verification of emergence (Sect. 4), and the apparatus required to deal with semantic heterogeneity (Sect. 5). We conclude by identifying challenges for future research (Sect. 6).

2 The COMPASS Framework

Our goal in COMPASS is to develop methods and tools for SoSE based on architectural modelling and formal languages that are rich enough to describe SoS abstractions, and are able to support the verification of key properties, automating tracability with test and management of interfaces between constituent systems. In this section, we first review the basic concepts of COMPASS, and then describe the tools framework.

2.1 Basic Concepts

We start from a view of system as an entity that interacts with other entities that form its *environment*. The common frontier between a system and its environment is the *system boundary* [3]. A system has an *interface* that defines the interactions of interest between the system and its environment. The *behaviour* of a system is the collection of traces or sequences of events that may occur at the system boundary. An SoS is a system composed of units (termed *constituents*) which are themselves systems. In an SoS, the environment of each constituent is composed at least in part of other constituents, and interfaces define the interactions between constituents.

We address the independence of constituent systems by permitting relatively loose specification of interfaces in a *contractual* style whereby the behaviour of a constituent is specified in terms of *assumptions* and *guarantees*. An assumption is a statement of a property that the constituent relies on being true of its environment, and a guarantee is a statement of a property that constituent promises will hold provided the assumptions are met. The contractual style is intended to document precisely the requirements on a constituent without unduly compromising its autonomy. Support for the development of contractual specifications is discussed in Sect. 3.

COMPASS aims to support SoS modelling in the well established modelling language SysML [17], although most of the techniques and tools technology developed in the project are notation-neutral. Pragmatic guidelines and patterns are being developed for SoS requirements modelling, architectural modelling, and systems integration, based on SysML[6]. In order to help address the verification problem, we have proposed a formal COMPASS Modelling Language CML [27]. SoS models developed in SysML can be translated to a semantic representation in CML (we define this translation systematically [22]). The formal semantics of CML is realised in a suite of tools supporting simulation, test case generation model checking, verification and simulation. Emergent SoS-level properties are verified by combining models of constituents' interfaces. Verification using CML is discussed further in Sect. 4.

COMPASS approaches semantic heterogeneity by using the Unifying Theories of Programming (UTP) [16], to enable the consistent description of many facets of the semantics: for example allowing the modular but consistent description of semantics for features such as object-orientation and real-time. The approach to semantic heterogeneity is described further in Sect. 5.

2.2 Tools Framework

The independence of constituent systems and the breadth of stakeholder groups involved mean that SoS development is necessarily collaborative. There may not be a single organisation in control of the development or maintenance process, and the organisations developing constituent systems may work remotely, with only limited contact. Tools for model-based SoSE therefore need to be Collaborative Development Environments (CDEs) that support the joint production of models that can still be systematically analysed, even though a development team might not want (or be able) to disclose every part of their model [6]. The tools therefore need to support

---

[6] Available from www.compass-research.eu/deliverables.html

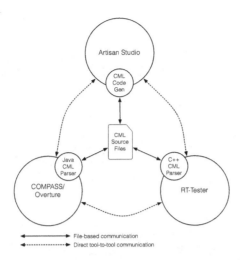

Fig. 1 Overview of the COMPASS Toolset

private and public parts of the model and a mechanism for extracting the public parts of the model which still should be usable for other organisations.

In COMPASS, we aim to develop a CDE supporting the proper engineering of SoSs by independent stakeholders using a model-based approach [9]. Current frameworks for SoS modelling and analysis support simulation, e.g. [25, 13] or static network analysis, e.g. to support dependability assessment [14]. We believe that complementary analysis techniques are essential for being able to properly deal with the problems mentioned above. At the high level, a tool that supports an architectural notation like SysML is necessary to describe the abstract structure of the SoS and to clarify the requirements, and capabilities of the constituent systems. At a detailed level, we require a tool that supports formal modelling of constituent systems in order to verify constituent system behaviour, emergent SoS behaviour, the generation and verification of implementations. At this detailed level, we need to build confidence and insight into an implementation's behaviour by providing test automation for the constituent systems and their interaction in the overall SoS.

The COMPASS platform links three tools packages addressing SySML modelling, formal CML model-based analysis, and test automation. Figure 1 shows the three packages. Communication between the packages may be by the use of flat files (all three packages will be able to read and write CML source files), or by direct communication to allow for synchronisation of their user interfaces. Architectural modelling in SysML is supported by Artisan Studio[7], extended to allow the generation of CML from SysML, and synchronising changes between the two. Construction, maintenance and analysis of CML models is supported by the COMPASS tools implemented on the Eclipse-based Overture Tool platform [20]. Test automation is supported by the RT-Tester tool which identifies test cases, traces them to

---

[7] www.atego.com/products/artisan-studio

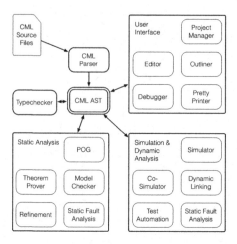

Fig. 2 Overview of the COMPASS/Overture components

requirements and generates concrete test data by means of an integrated constraint solver [23].

Figure 2 indicates the components planned for the COMPASS tool. CML source files of the detailed SoS model are parsed, the Abstract Syntax Tree (AST) is type-checked, ensuring a basic level of validity of the model. At this point the AST is available for use by plugins supporting a range of analytic techniques.

3 Supporting Collaboration and Negotiation through Contracts

When a consortium of partners decides to work together in order to engineer a well-functioning SoS it is essential that they have proper support for collaboration and negotiation of what the different constituent systems shall deliver to the SoS. The basis of this can be established using formally defined contracts that can be analysed efficiently by tools. The stakeholders responsible for constituent systems in a SoS needs to work collaboratively on reaching a common agreement that ensures that a given SoS design is feasible and has the desired functionality.

In practice today such agreements are either not made at all, or are made in the form of textual contracts by lawyers, or are made directly at the coding level e.g. with a service-oriented architecture. In the first case, developers have to work with whatever description they might have for the constituent systems. In the second case, the expense and inflexibility of legal contracts mean that the approach may only be expected to succeed for highly directed SoSs with limited evolution. In the third case, the SoS-level emergent behaviour may not be expressed at all, and the low abstraction level means that analysis capabilities may be limited.

In order to have a firm agreement on the collaboration there is a need to estab-lish a type of contract that describes the connections and interactions between the constituent systems. Such a contract describes the interface a constituent offers to its environment in terms of the externally accessible services that can be called by other constituents, and the types for data that can be passed over the interface. The contract could be evolved to define more details and constraints on the services, data types and the sequence of events that the specific constituent system accepts. This must be augmented by forms in which to express naturally the constraints on the protocol for interacting with constituent systems. These features are present in CML, and are illustrated in the example in Sect. 4.

Managing the information hiding aspects of collaborative development requires both a contractual modelling language that supplies abstraction from proprietary detail as well as specific hiding constructs, as well as tool support to manage the collaboration during development. We expect to include features in CML for managing multiple models in the same SoS context, and specifically handling the explicit hiding of model elements. Given that the environment supports a rich group of analytic tools, it is not trivial to provide facilities such as verification support that retains the confidentiality of parts of models of constituents.

4 Verifying Emergent Behaviour

Emergent behaviour can be observed in systems at all scales, including both complex systems and SoSs. A simple example can be found in Conway's Game of Life, discussed in [26]. The game is played out as a two-dimensional cellular automaton with simple rules involving only individual cells and their eight immediate neighbours, e.g. that a live cell with fewer than two live neighbours dies of isolation. The rules are purely local, but non-local patterns emerge. The most famous such pattern is the glider: a group of five live cells that reconfigure according to the game's rules, and in doing so appear to move as a unit in a particular direction. The glider, its direction, and its movement together constitute a global behaviour that emerges from the local rules: this observed behaviour has not been programmed anywhere. The glider is an example of *weak emergence* [4], where an emergent property can be reduced to its constituents; if we apply the local rules to a particular configuration of cells, the glider pattern emerges. This is in contrast to *strong emergence*, where an emergent property cannot be traced to any direct cause; an example would be consciousness emerging as a property of the brain. Like [26], we confine our attention to weak emergence.

The question that we address here is: how can emergence be verified in an SoS modelled in CML? We divide the problem into two cases. First, how do we engineer an emergent property? Second, how do we analyse a system to check that an emergent property exists?

4.1 Engineering Emergence

Our hypothesis is that certain emergent properties of a distributed system can be specified as global invariants. Here we have in mind global invariants that are purely specification artefacts: they do not exist in the implemented system. Our hypothesis applies equally to component-based systems and SoSs.

Our view is supported by [26], where they use the following motivational example. Suppose that we are want to simulate the behaviour of a flock of birds implemented as autonomous agents. As in the Game of Life, the behaviour of the flock emerges from much simpler behaviour: each bird reacts autonomously to the behaviour of those birds in its immediate environment. Studying the flock as a unit, we can observe global behaviours to do with the position, volume and surface of the flock. These behaviours can be specified as abstract qualities. The correctness of the implementation (the model of the flock of individual birds) can be demonstrated by proving that the birds collectively refine the abstract specification [19].

To illustrate this approach in our framework, consider a simple model in CML. Mondex is an electronic purse hosted on a smart card and developed in the mid-1990s to the high-assurance standard ITSEC Level E6 by a consortium led by NatWest, a UK bank. Purses interact using a communications device and strong guarantees are needed that transactions are secure in spite of failures and attacks. These guarantees ensure that electronic cash cannot be counterfeited, although transactions are distributed. There is no centralised control: all security measures are locally implemented with no real-time external audit-logging or monitoring.

Once again, we notice a similarity with the Game of Life: autonomous smart cards play the role of cells and rules are purely local, with transactions involving only two cards out of the many millions in the overall SoS. The property that we want to emerge from the Mondex SoS is a global invariant, which in its simplest, idealised form can be expressed as constancy of value[8].

Mondex was originally verified as follows. First the protocol for exchanging value between cards was modelled formally. Second an abstract specification was constructed with an Olympian view of the state of every card, with a global invariant expressing the constancy of value. The network of smart cards, with their purely local protocol, was then shown to be a formal refinement of the abstract specification. The global invariant has been completely refined away and no direct evidence of it is left in the network of smart cards. But we know, because we have formally verified it, that the global invariant still holds: the total value does not change in spite of transactions between cards[9].

Suppose that there are N cards, each initialised with V pounds sterling. The money supply, M is then $N*V$. Cards are indexed from the set `Index`, whose values are in the set $\{1,\ldots,N\}$ and money comes from the set `Money`, whose values are in the set $\{0,\ldots,M\}$. We model communication between cards using channels, as follows:

---

[8] In reality, the electronic cash in the system can change via approved routes as cards are destroyed and money enters and leaves the system through banks.

[9] Notice that the Mondex refinement can also be viewed as guaranteeing the absence of an emergent property: the system must not allow counterfeiting to emerge as a behaviour. Our technique of refinement allows us to confirm the presence or absence of an emergent property.

channels
```
  pay, transfer: Index * Index * Money
  accept, reject: Index
```

The first channel, `pay` is connected to the user, who can instruct a card using the communication `pay.i.j.n`: card i should pay j the sum of n pounds. The second channel, `transfer` connects card i to card j and tries to transfer the sum of n pounds. The third and fourth channels are used to signal to the user whether the transaction was accepted or rejected.

A contractual description of the interface to a card would be given as a CML *process* contains a state-based part and a reactive part. The card process is described in Figure 3. The state-based part contains the data and functionality alluded to in Sect. 3. The process is indexed by i and begins by declaring its encapsulated

```
process Card = val i: Index @
  begin
    state value: nat
    operations
      Init: () ==> ()
      Init() == value := V

      Credit: nat ==> ()
      Credit(n) == value := value + n

      Debit: nat ==> ()
      Debit(n) == value := value - n

    actions
      Transfer =
        pay.i?j?n ->
          ( [n > value] & reject!i -> Skip
            []
            [n <= value] &
              transfer.i.j!n -> accept!i -> Debit(n) )

      Receive = transfer?j.i?n -> Credit(n)

      Cycle = ( Transfer [] Receive ); Cycle
  @
    Init(); Cycle
  end
```

Fig. 3 CML model of the Mondex card process

state: the natural number `value`. It defines three operations on the state: `Init`, `Credit`, and `Debit`[10].

The protocol part of the contractual description is given by the definition of two reactive actions, `Transfer` and `Receive` that use the operations to handle outgoing and incoming payments, respectively. `Transfer` is triggered by the user communicating on the `pay` channel: `pay.i?j?n`. This is a communication to card `i` to make a payment; the value of the receiving card is bound to the variable `j`; and the value of the payment is bound to the variable `n`. The subsequent behaviour of the card is described by an external choice guarded so that, if the payment exceeds the funds available ($n > value$), it is rejected and the action terminates. If the payment can proceed ($n <= value$), then the transfer is made, and the action terminates.

The cards are brought together as constituent systems in a network, and channels are connected as described above:

```
process Cards =
  || i: Index @
    [ {| pay.i, transfer.i, accept.i, reject.i |}
      union
        { transfer.j.i.n | j <- Index, n <- Money } ] Card(i)

process Network = Cards \\ {|transfer|}
```

This is a model of a rather homogeneous SoS: although cards are independently owned and managed, they all follow the same rules. The real Mondex system is much more elaborate. For example, the transport medium is modelled; messages between cards can be corrupted, lost, or forged; and the power may go down at any point. In spite of these threats and faults, the property must emerge that all value is accounted for and so there can be no change in the value in the system.

We have a models of a simple SoS. How should we verify a specified emergent property of such a model? The specification for the Mondex system can also be given in CML, as in Figure 4. The specification describes the network of cards as a single, centralised process. It has an Olympian view of the state of every card, enabling it to assert global invariants. This is no way to implement the requirements for Mondex, which, roughly speaking state that Mondex should behave like electronic cash: we don't take our cash to a single common place in the world in order to pay

---

[10] The purpose of this example is to illustrate an approach to verifying emergent behaviour, rather than showing all the features of CML. The operations are simple and are defined explicitly as single assignments. CML provides program-like constructs for defining operations explicitly. However, it also supports an implicit style in which operations are specified contractually by means of postconditions that express the properties of the state after the operation has executed. For the `Credit` operation, rather than giving the assignment, we might define a postcondition that `value = value~ n|`, where `value~` is the state variable before execution of the operation. Whether explicitly or implicitly defined, operations may also be constrained by preconditions that record the conditions assumed to hold before the operation is invoked.

```
process Spec =
  begin
    state
      valueseq: seq of nat
    inv
      len(valueseq) = N
      sum(valueseq) = M
    operations
      Init: () ==> ()
      Init() == valueseq := initseq(N)
    actions
      Pay = i,j: Index, n: Money @
        pay.i.j.n ->
          if n > valueseq(i) then
            reject.i -> Skip
          else
            ( valueseq := subtseq(valueseq,i,n);
              valueseq := addseq(valueseq,j,n);
              accept.i -> Skip
            )

      Cycle =
        ( |~| i,j: Index, n: Money @ Pay(i,j,n) );
        Cycle
  @
    Cycle
  end
```

Fig. 4 CML specification for the Mondex system

for our groceries. But it is not meant to be an implementation, but rather the specification of an emergent property agreed upon by all the SoS stakeholders: constancy of value.

The state of each individual card is recorded as an element of the sequence `valueseq`, accessed by the card's index number. With all the state in one place, it is easy to specify our property: that all cards are accounted for and their total sum is the entire money supply. There is a single action, `Pay`, that plays the part of the simple protocol in the implementation, transferring value atomically with assignments rather than `transfer` operations.

We can now assert that the network of cards formally refines this specification. In the COMPASS tool environment, this assertion will be discharged either using the theorem prover or using the model checker.

4.2 Analysing Existing Systems

The previous section describes the use of formal refinement in verifying the presence or absence of emergent properties. In this section, we explore a complementary problem. Suppose that we have an existing system or system of systems; how do we discover and verify emergent properties? We describe some future work that addresses exactly this point.

Some authors describe the element of surprise that is often associated with emergent properties[11]. Simulation is often used as a technique to explore emergence; but of course, it is not a verification technique. In our future work in this area, we propose to use simulation of CML models to explore behaviours and to try to discover global system invariants.

Our inspiration is the Daikon tool for detection of likely program invariants [11]. Daikon discovers invariants that hold for particular test-data sets, but which are not asserted in the particular program under consideration; in that sense, the invariant emerges from a dynamic analysis. Daikon has applications in documenting and understanding programs, generating further test cases, automating theorem proving and model checking, assisting component evolution and integration, repairing inconsistent data structures, and checking data validity.

Daikon tackles two separate concerns: the choice of which invariants to infer; and the inference of invariants. Daikon can infer only certain types of invariants that match choices from a library of patterns. The user usually specifies the invariant patterns of interest, leaving key parameters as symbolic constants. Daikon's task then is to identify suitable values for these constants, discarding falsified putative invariants. It does this with a simple machine-learning technique.

The major research challenge is to scale up the basic Daikon technique to large and multi-paradigm models. Currently, there is no Daikon-like system for inferring invariants in concurrent programs, object-oriented programs, pointer-rich programs, or for mobile or dynamically reconfigurable systems. But this challenge could be met by CML's notions of contracts and invariants that cover these different paradigms, and by taking advantage of the natural parallelism of the invariant-inferencing technique.

---

[11] For example, many years ago UK telecommunications provider introduced a new feature on its telephone service, which is an extremely complex SoS. The new feature allowed, for the first time, conference calls whereby a caller could call and be connected to a recipient and then add additional recipients one by one. The way this was implemented allowed any of the group to leave the call, including the original call initiator. This feature interacted badly with the call processing subsystem, which is responsible for, amongst other things, billing for the call. Without an initiator, the billing stopped. News soon got around and the feature was used to set up day-long transatlantic calls for which no one was billed. Free telephone calling was an emergent feature popular with customers but not with the telephone company.

5 Semantic Heterogeneity

Models for constituent systems will be semantically heterogeneous, involving a range of paradigms and levels of abstraction. Our approach in COMPASS is to use Unifying Theories of Programming (UTP) to formalise these paradigms and the relationships between them, in order to provide a sound basis for model construction and analysis. In the terms of [15] we use a translational approach between SysML and CML and a unified semantics approach in CML itself. In this section we discuss how UTP provides a framework for managing the extensible description of the semantics of CML.

5.1 Unifying Theories of Programming

UTP is originally the work of Hoare & He [16]. Its long-term research agenda can be summarised as follows: researchers have proposed many different theories of system development and practitioners have proposed many different pragmatic paradigms for industrial application; how do we understand the relationship between all of these? UTP proposes three principal ways:

1. **Computational Paradigms:** UTP groups programming languages according to a classification by computational model; for example, structured, object-oriented, concurrent, synchronous, real-time, and discrete or continuous values. The technique used to give semantics to each computational model is to identify common concepts and deal separately with additions and variations. UTP uses two fundamental and familiar scientific principles: simplicity of presentation and separation of concerns.

2. **Abstraction:** The description of each computational paradigm can also be categorised by different levels of abstraction to capture the development process leading from requirements through to code designed to run using the platform-specific technology of an implementation. Interfaces are specified using contracts to guarantee the correctness of moving a model from one level to another. This mapping between levels is based on a formal notion of refinement that provides guarantees of correctness all the way from requirements to code.

3. **Presentation:** The third classification is by the method chosen to present a language definition. There are three well-known scientific methods.

   a. *Denotational*, given by a function from syntax to a single mathematical meaning: its *denotation*. A specification is then just a set of denotations: the permitted behaviours of a system. Refinement is simply inclusion: every behaviour of the program must also be a behaviour permitted by the specification.

   b. *Algebraic*, given by a collection of equations relating descriptions in the language. Interestingly, no direct meaning is given to the language at all.

   c. *Operational*, given by a set of rules describing how descriptions in the language are executed on an idealised abstract mathematical machine.

As Hoare & He point out, a comprehensive account of constructing systems in any theory needs all three kinds of presentation. The UTP technique allows us to study differences and mutual embeddings, and to derive each from the others by mathematical definition, calculation, and proof.

The UTP Research Agenda has as its ultimate goal to cover all the interesting paradigms of computing, including declarative and procedural, hardware and software. It presents a theoretical foundation for understanding software and systems engineering, and has already been exploited in areas such as hardware ([24]), hardware/software co-design ([7]), and component-based systems ([29]). It also presents an opportunity in constructing new languages, especially ones with heterogeneous paradigms and techniques. Having studied the variety of existing programming languages and identified the major components of programming languages and theories, we can select theories for new, perhaps special-purpose languages. The analogy here is of a theory supermarket, where you shop for exactly those features you need, while being confident that the theories plug-and-play together.

5.2   The COMPASS Modelling Language

Currently, CML contains several paradigms, the first two of which have already been seen in the Mondex example:

1. **State-based description.** The theory of *designs* in UTP provides a nondeterministic programming language with pre- and postcondition specifications as contracts.
2. **Concurrency and communication.** The theory of reactive processes in UTP provides a way of constructing networks of processes that communicate by passing messages.
3. **Object orientation.** The theory of object orientation in UTP is built on the theory of designs and provides a way of structuring state-based descriptions through sub-typing, inheritance, and dynamic binding, with mechanisms for object creation, type testing, type casting, and state-component access.
4. **Pointers.** The theory of pointers in UTP provides a way of modelling heap storage and its manipulations, as found in implementations of object orientation, for which we have a reference semantics. Crucially, it supports modular reasoning about the heap.
5. **Time.** The theory of timed traces in UTP supports the observation of events in discrete time.

5.3   Galois Connections

The semantic models mentioned in the last section are each formalised as sets of relations. For example, state-based descriptions are represented as pre- and postconditions, which are familiar from languages such as VDM and B. In UTP, an operation to decrement a variable x, which must invariantly be positive, would be written as $(x > 0 \vdash x' = x - 1)$. The precondition requires that $x > 0$ and the postcondition ensures that the after-value of x, written as x', is exactly one less than the before-value of x, written $x - 1$. This pair of predicates is modelled as a single predicate (a relation) with two observational variables: $(ok \wedge x > 0 \Rightarrow ok' \wedge x' = x - 1)$. This is read as "if the operation is started (the observation ok is true) and $x > 0$, then the operation must terminate (the observation ok' is true) and when it does, $x' = x - 1$ must be true". Designs are organised into a lattice of relations ordered by refinement. At the bottom of the lattice is the aborting operation and at the top is the infeasible operation that can never be started. All other designs are somewhere in between. The process of correctness by construction starts by specifying the requirements for an operation as a design, moving upwards through the lattice in a series of refinement steps, until an implementation is reached. As usual, the specification is chosen to make the formalisation of requirements as easy and clear as possible, whilst the implementation is chosen to be executable on the chosen technology platform. Choices between alternative, correct refinements in this process are usually determined by non-functional requirements.

Mappings exist between the different semantic lattices, and some of these are shown in Figure 5. These mappings can be used to translate a model in one lattice into a corresponding model in another lattice. For example, the lattice of designs is completely disjoint from the lattice of reactive processes, but the mapping **R** maps

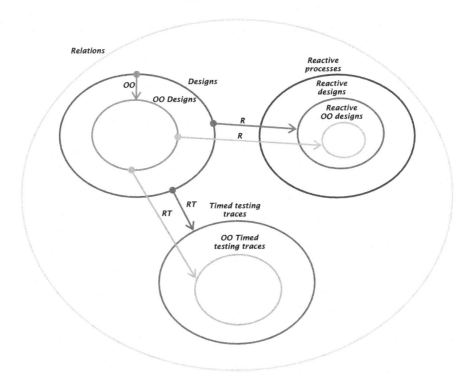

Fig. 5 CML semantic models

every design into a corresponding reactive process. Intuitively, the mapping equips the design with the crucial properties of a reactive process: that it has a trace variable that records the history of interactions with its environment and that it can wait for such interactions. A vital healthiness condition is that this trace increases monoton- ically: this ensures that once an event has taken place, it cannot be retracted—even when the process aborts.

But there is another mapping that can undo the effect of **R**: it is called **H**, and it is the function that characterises what it is to be a design. **H** puts requirements on the use of *ok* and *ok'*, and it is the former that concerns us here. It states that, until the operation has started properly (*ok* is true), no observation can be made of the operation's behaviour. So, if the operation's predecessor has aborted, nothing can be said about any of the operation's variables, not even the trace observational variable. This destroys the requirement of **R** that says that the trace increases monotonically.

This pair of mappings form a structure known as a Galois connection. Galois connections exist between all the semantic domains mentioned in the last section. One purpose of a Galois connection is to embed one theory within another, and this is what gives the compositional nature of UTP and CML. This is the foundation of the contractual approach used in COMPASS: preconditions and postconditions (designs) can be embedded in each of the semantic domains and brings unifor- mity through a familiar reasoning technique. In summary, semantic heterogeneity is achieved through using UTP to include new semantic domains within the COM- PASS framework. New domains are built as lattices of relations and equipped with Galois connections to compose and map models.

6   Conclusions

We have identified three fundamental issues that we believe must be addressed if model-based methods are to achieve their potential in the engineering of Systems of Systems: contractual specification, verification of emergence, and semantic het- erogeneity. We have described the foundations of an approach, implemented in the COMPASS project, that seeks to address these by combining structured modelling in SysML with formal methods. Contractual description is enabled by a formal mod- elling language that combines the description of data and functionality with com- munication and concurrency, object-orientation and real time features. A notion of refinement underpins the verification of emergent behaviour. Semantic heterogene- ity is addressed by using UTP for the systematic integration of semantic domains.

We have focussed on foundations rather than pragmatics in this paper. However, it is important to stress the provision of guidelines and patterns for managing require- ments, architectural modelling and integration in SoS. We expect that the provision of a semantically sound framework for SoS models expressed using SysML can lead to patterns that are amenable to machine-assisted verification. The link from SysML to CML has been outlined [22]. This has allowed us to describe, for ex- ample a profile for fault modelling in SysML models [1], leading to verification of fault tolerance on the CML models derived from them [2]. At the time of writing, the focus in COMPASS is on linking the tools that automate parts of this process to provide a semantically sound chain. Within the project, case studies are being undertaken in networks of home audio/video devices and content streaming network, emergency response, traffic management and smart grid.

Several aspects of SoS engineering provide promising areas of future research. *Cyber-physical* systems combine facets of embedded systems and SoS, integrating continuous-time or continuous-state descriptions of physical phenomena with the discrete models of cyber elements. The integration of semantics for co-analysis and co-simulation of models of cyber-physical systems brings with it opportunities for rapid exploration of the design space, making informed trade-offs across the cyber/physical boundary. Similarly, we have not addressed socio-technical and human factors. Incorporating of such features with a model-based framework would provide an opportunity to integrate radically different models. In both areas, we conjecture that the framework being established in COMPASS could provide a sound basis for engineering in these demanding environments.

Acknowledgements. The work described in the paper is supported by the Commission of the European Communities project Comprehensive Modelling for Advanced Systems of Systems (COMPASS, grant agreement 287829). The authors are grateful to their many collaborators in the project.

References

1. Andrews, Z.H., Fitzgerald, J.S., Payne, R., Romanovsky, A.: Fault modelling for systems of systems. In: 11th IEEE International Symposium on Autonomous Decentralized System (ISADS), pp. 1–8. IEEE Computer Society (March 2013)
2. Andrews, Z.H., Payne, R., Romanovsky, A., Didier, A.L.R., Mota, A.: Model-based development of fault tolerant systems of systems. In: 7th International Systems Conference, IEEE SysCon. IEEE (April 2013)
3. Avizienis, A., Laprie, J.-C., Randell, B., Landwehr, C.: Basic Concepts and Taxonomy of Dependable and Secure Computing. IEEE Transactions on Dependable and Secure Computing 1, 11–33 (2004)
4. Bedau, M.A.: Downward causation and autonomy in weak emergence. Principia Revista Internacional de Epistemologica 6, 5–50 (2003)
5. Boardman, J., Sauser, B.: System of Systems – the meaning of "of". In: Proceedings of the 2006 IEEE/SMC International Conference on System of Systems Engineering, Los Angeles, CA. IEEE (April 2006)
6. Booch, G., Brown, A.W.: Collaborative development environments. Advances in Computers 59, 1–27 (2003)
7. Butterfield, A., Gancarski, P., Woodcock, J.: State visibility and communication in unifying theories of programming. In: Theoretical Aspects of Software Engineering, pp. 47–54. IEEE Computer Society (2009)
8. Cantot, P., Luzeaux, D.: Simulation and Modeling of Systems of Systems. Wiley (2011)

9. Coleman, J.W., Malmos, A.K., Larsen, P.G., Peleska, J., Hains, R., Andrews, Z., Payne, R., Foster, S., Miyazawa, A., Bertolini, C., Didier, A.: COMPASS Tool Vision for a System of Systems Collaborative Development Environment. In: Proceedings of the 7th International Conference on System of System Engineering, IEEE SoSE 2012, pp. 451–456 (July 2012)

10. DeLaurentis, D.A., Crossley, W.A.: A taxonomy-based Perspective for Systems of Systems Design Methods. In: IEEE International Conference on Systems, Man and Cybernetics, vol. 1, pp. 86–91. IEEE (October 2005)

11. Ernst, M.D., Perkins, J.H., Guo, P.J., McCamant, S., Pacheco, C., Tschantz, M.S., Xiao, C.: The daikon system for dynamic detection of likely invariants. Sci. Comput. Program. 69(1-3), 35–45 (2007)

12. European Commission. Directions in Systems of Systems Engineering. Technical report, European Commission, Communications Networks, Content and Technology Directorate- General Unit A3-DG CONNECT (July 2012)

13. Fang, Z., DeLaurentis, D.A., Davendralingam, N.: An Approach to Facilitate Decision Making on Architecture Evolution Strategies. In: 2013 Conference on Systems Engineering Research, Atlanta, Georgia. Procedia Computer Science, vol. 16, pp. 275–282 (2013)

14. Han, S.Y., DeLaurentis, D.A.: Development Interdependency Modeling for System-of-Systems (SoS) using Bayesian Networks: SoS Management Strategy Planning. In: 2013 Conference on Systems Engineering Research, Atlanta, Georgia. Procedia Computer Science, vol. 16, pp. 698–707 (March 2013)

15. Hardebolle, C., Boulanger, F.: Multi-formalism modelling and model execution. International Journal of Computers and their Applications 31(3), 193–203 (2009); Special Issue on the International Summer School on Software Engineering

16. Hoare, T., He, J.: Unifying Theories of Programming. Prentice Hall (April 1998)

17. Holt, J., Perry, S.: SysML for Systems Engineering. IET (2008)

18. Jamshidi, M.: System of Systems Engineering: Innovations for the Twenty-First Century, 1st edn. Wiley (November 2008)

19. Jun, H., Liu, Z., Reed, G.M., Sanders, J.W.: Ensemble engineering and emergence. In: Wirsing, M., Banâtre, J.-P., Hölzl, M., Rauschmayer, A. (eds.) Soft-Ware Intensive Systems. LNCS, vol. 5380, pp. 162–178. Springer, Heidelberg (2008)

20. Larsen, P.G., Battle, N., Ferreira, M., Fitzgerald, J., Lausdahl, K., Verhoef, M.: The Overture Initiative – Integrating Tools for VDM. SIGSOFT Softw. Eng. Notes 35(1), 1–6 (2010)

21. Maier, M.W.: Architecting Principles for Systems-of-Systems. In: Sixth International Symposium of the International Council on Systems Engineering, INCOSE (1996)

22. Miyazawa, A., Lima, L., Cavalcanti, A.: Formal models of sysml blocks. In: Groves, L., Sun, J. (eds.) Accepted for publication at ICFEM 2013, Queenstown, New Zealand. Springer (October 2013)

23. Peleska, J., Vorobev, E., Lapschies, F.: Automated Test Case Generation with SMT-Solving and Abstract Interpretation. In: Bobaru, M., Havelund, K., Holzmann, G.J., Joshi, R. (eds.) NFM 2011. LNCS, vol. 6617, pp. 298–312. Springer, Heidelberg (2011)

24. Perna, J.I., Woodcock, J.: UTP Semantics for Handel-C. In: Butterfield, A. (ed.) UTP 2008. LNCS, vol. 5713, pp. 142–160. Springer, Heidelberg (2010)

25. Sahin, F., Jamshidi, M., Sridhar, P.: A Discrete Event XML based Simulation Framework for System of Systems Architectures. In: IEEE International Conference on System of Systems Engineering, SoSE 2007 (April 2007)

26. Sanders, J.W., Smith, G.: Emergence and Refinement. Formal Aspects of Computing 24(1), 45–65 (2012)
27. Woodcock, J., Cavalcanti, A., Fitzgerald, J., Larsen, P., Miyazawa, A., Perry, S.: Features of CML: a Formal Modelling Language for Systems of Systems. In: Proceedings of the 7th International Conference on System of System Engineering. IEEE (July 2012)
28. Woodcock, J., Larsen, P.G., Bicarregui, J., Fitzgerald, J.: Formal Methods: Practice and Experience. ACM Computing Surveys 41(4), 1–36 (2009)
29. Zhan, N., Kang, E.Y., Liu, Z.: Component publications and compositions. In: Butterfield, A. (ed.) UTP 2008. LNCS, vol. 5713, pp. 238–257. Springer, Heidelberg (2010)

Industrial Cyber-Physical Systems – iCyPhy

Amit Fisher, Clas A. Jacobson, Edward A. Lee, Richard M. Murray,
Alberto Sangiovanni-Vincentelli, and Eelco Scholte

Abstract. ICyPhy is a pre-competitive industry-academic partnership focused on architectures, abstractions, technologies, methodologies, and supporting tools for the design, modeling, and analysis of large-scale complex systems. The purpose of this partnership is to promote research that applies broadly across industries, providing the intellectual foundation for next generation systems engineering. The focus is on cyber-physical systems, which combine a cyber side (computing and networking) with a physical side (e.g., mechanical, electrical, and chemical processes). Such systems present the biggest challenges and biggest opportunities in several critical industrial segments such as electronics, energy, automotive, defense and aerospace, telecommunications, instrumentation, and industrial automation. The approach leverages considerable experience designing complex artifacts in the semiconductor, embedded systems, and software industries, and major recent advances in algorithmic techniques for dealing with complexity. This consortium adapts and extends these techniques to handle the fundamentally different challenges in large-scale cyber-physical systems.

Amit Fisher
IBM, Sunnyvale, CA, USA
e-mail: amitf@il.ibm.com

Clas A. Jacobson
UTRC, East Hartford, CT, USA
e-mail: jacobsCA@utrc.utc.com

Edward A. Lee · Alberto Sangiovanni-Vincentelli
UC Berkeley, Berkeley, CA, USA
e-mail: {eal,alberto}@eecs.berkeley.edu

Richard M. Murray
California Institute of Technology, Pasadena, CA, USA
e-mail: murray@cds.caltech.edu

Eelco Scholte
UTC Aerospace Systems, Windsor Locks, CT, USA
e-mail: Eelco.Scholte@utas.utc.com

M. Aiguier et al. (eds.), *Complex Systems Design & Management 2013*,   21
DOI: 10.1007/978-3-319-02812-5_2, © Springer International Publishing Switzerland 2014

1   Industrial Motivation

Efficient and effective design of distributed multi-scale complex systems is largely an unsolved problem. Complex systems are compositions of heterogeneous components, which for cyber-physical systems often include electromechanical, chemical, thermal, computing, and communication elements. These subsystems are interconnected, often uncertain in specification, and encompassing environmental effects. The dynamics of all the elements — both cyber and physical — are critical to the performance of the overall system. Such systems are software and network enabled, and there is significant cost and schedule pressure during development. The technology drivers causing the change in delivery are the pervasive use of electronic control units, and consequently of communication networks, and the blurring of distinctions between software, firmware, hardware and multi-physics systems. These drivers are creating the possibility for placing vastly more functionality into products, but at the same time increase interconnectivity at the risk of unwanted system interactions found late in the development process.

To solve this problem we need a rigorous approach to systems engineering, specifically a methodology for product system level design, optimization and verification that:

- Provides guarantees of performance and reliability against customer requirements while achieving cost and time-to-market objectives;
- Produces modular, extensible architectures for products incorporating electromechanical components, embedded electronic systems, wired and wireless communication networks and application software;
- Exploits analytical tools and techniques to determine design choices and ensure robust system performance despite variations caused by product manufacturing, integration with other products and customer operation; and
- Achieves these objectives through the coordinated execution of a prescriptive, repeatable and measurable process.

Yet industry is still far from developing and using such a systems engineering methodology. Indeed there are no rigorous foundations in systems engineering that can address the issues of the overall design flow, and no analysis and synthesis tools for the design and verification of highly distributed systems. Consequently systems engineering practice is often a collection of common-sense, heuristic approaches based on experience and use of legacy designs. There have been advances in the domain of systems engineering science in academia, in some industrial segments such as automotive, and in some tool companies, but the overall knowledge of these advances and of their potential in the system industry is at best spotty.

This paper gives the motivation and goals for an industry-academic partnership called Industrial Cyber-Physical Systems (iCyPhy) that is addressing these challenges. ICyPhy was formed in December of 2012 with the industrial founding partners being United Technologies Corporation (UTC) and IBM, and the academic ones being the University of California at Berkeley, and the California Institute of Technology.

UTC is a conglomerate that deals with multi-physics systems in several vertical application domains, mainly in the aerospace and building domains. In its role in the consortium, it represents companies that host designers and builders of complex systems, which we refer to in this paper as "systems houses." IBM is a global computing infrastructure and service company that is increasingly looking at planetary scale problems. Its role in iCyPhy is that of a "technology provider," serving large systems manufacturers in automotive, aerospace, and electronics, as well as cities and nations in their attempt to optimize services such as water, energy, health, and traffic management. IBM develops systems engineering tools, with emphasis of being "the integrator" of multiple engineering disciplines, tools, and application providers. In addition, IBM is active in specific systems engineering verticals such as requirement management, architecture management, quality management and collaboration.

UC Berkeley, which leads the consortium, brings broad expertise in systems design, modeling, and analysis. Berkeley has a proven track record of changing industries through improvements in design methodologies and tools, as evidenced by its impact on electronic design automation. Caltech brings key expertise in rigorous approaches to complex, multi-physics, cyber-physical systems design and analysis, most particularly by combining the principles of control systems engineering with those of formal verification.

2   Gap Analysis

The research topics are based on an analysis of the gaps that System Houses and Technology Providers are experiencing.

2.1   System House Gap Analysis

It has always been a goal of diversified systems houses to find synergies among apparently different industrial domains. These synergies often exist at the business level, but they are more difficult to achieve in engineering. A foundational assumption of iCyPhy is that general system-level design approaches lead to substantial rationalization in design, yielding processes that are leaner and more effective while substantially reducing time-to-market by re-using components and employing correct-by-construction methods.

2.1.1   Requirements Capture, Analysis and Domain Specific Modeling

Requirements capture plays an important role in today's development processes. Requirements capture is largely natural language based and leads to many iterations due to requirements ambiguity and lack of standard requirements libraries. The specific needs include:

- We need semi-formal and formal languages that reduce ambiguity and enable analysis for integrated systems. The level of requirements often is non-homogeneous and includes system performance requirements, safety requirements, system constraints, and customer-specific preferred solutions. The new methods should support different types of requirements to create formal (executable and analyzable) models for multiple domains.
- Because requirements are evolving throughout programs, we need analysis techniques that determine the impact of requirements changes on large interconnected systems.
- We need synthesis of early views of a system to reason about requirements validity internally, and with customers and suppliers. This requires domain-specific views of the requirements (e.g. mechanical, electrical, software, embedded hardware) at different levels of abstraction and the ability to capture cross-domain relationships.
- We need requirements modeling methods to support refinement into detailed design phases such that the design artifacts can be reused and designers can quickly iterate across abstraction levels.

2.1.2   System Integration and Views

Today's methods for system modeling are insufficient to allow for a formal model-based design flow. Most methods and tools are limited to single domains, and interconnecting these methods and tools is difficult or in some cases impossible. Existing methods and tools for cross-domain modeling and analysis lack clear semantics. More specifically:

- System modeling needs to be able to capture relationships between different domains (mechanical, electrical, software), as well as to enable analysis that crosses these domains (e.g. system reliability, performance, robustness). In particular, methods and tools are needed:

  - To reason about fault tolerance of systems and the impact of system degradation. This includes physical systems and failure modes, control system functionality, and the allocation of functionality to embedded platforms.
  - To capture and explore designs that cross multiple domains. The current practice is to limit the design space early by fixing certain decisions based on legacy knowledge and architectures and solutions from prior programs. This impedes design-space exploration and is not sufficient for programs where new architectures are introduced.

- Methodologies are needed that support the parallel nature of development programs. This requires both bottom-up and top-down capture of interfaces and constraints (e.g. through contracts).
- In addition to the integration of different views of the models, integration of different analysis methods within these views is needed. For example, the correlation between simulations and formal timing analysis is done today using independent models that have no formal relationship. Changes in architectures or requirements often make it impossible to quickly reuse such analysis.

2.1.3   Risk Management

Risk management is the identification, assessment, and prioritization of risks (defined in ISO 31000 as the effect of uncertainty on objectives), followed by coordinated and economical application of resources to minimize, monitor, and control the probability and impact of unfortunate events or to maximize the realization of opportunities [9, p. 46]. Risks can come from uncertainty in project failures (at any phase in design, development, production, or sustainment life-cycles), legal liabilities, accidents, natural causes and disasters as well as deliberate attack from an adversary, or events of uncertain or unpredictable root-cause.

Risk mitigation is a complex task. Among the risks that have to be considered in design, the ones due to uncertainties in the design parameters and in the specifications are of particular interest. The robustness of the design with respect to uncertainties in these two spaces must be addressed. Maintaining the correct operation of the system requires design centering, i.e., the choice of the nominal design variables so that the constraints are satisfied even if these variables drift from their nominal value due to the manufacturing process, product aging and adversarial environments. Sometimes centering the design is not sufficient to achieve the desired yield. In such cases, the choice and the determination of the range of tuning parameters is critical in allowing a company to adjust the design after manufacturing so that the constraints can be satisfied even though they are not satisfied in the non-tuned configuration. We are interested in design methods that allow to improve the yield of the design with respect to parametric variations and product capability of coping with an unpredictable environment.

2.2   Technology Provider Gap Analysis

Today, systems houses require engineering processes and tools that go way beyond what is known and available today. This is both a methodology and technology-intensive endeavor.

2.2.1   Formality and Usability

Historically, tools and methodologies that have had the most impact are those with strong formal foundations. Consider for example the classical engineering discipline of feedback control, a key enabler for many engineered systems. The discipline depends on a long history of mathematical models and analytical techniques that make stability and robustness analysis possible. Consider also the field of electronics, where today's circuits have a level of complexity and reliability that was unimaginable a few short years ago. The enabling tools and methodologies in this area have solid foundations in algorithms, discrete-event systems theory, synchronous-reactive concurrency theory, and automata theory. These foundations led in the 1980s to automatic layout, logic synthesis, and RTL-based design methodologies, which were developed primarily at Berkeley and deployed in industry through close collaboration with a number of companies.

Fundamentally, formal foundations enable automated analysis and synthesis. A model that is formal is analyzable by machine, whereas an informal model is only analyzable by humans. For example, it is routine today in electronic design automation to apply formal analysis methods that effectively verify behaviors of systems over all possible input conditions (e.g. model checking).

Usable tools do exist today, but they are often based on *ad hoc* heuristics that do not offer the sorts of guarantees enabled by machine analysis. Specifically, the gaps are:

- Use of natural language in requirements capture causes inconsistency and ambiguity. The scale of contemporary projects is only aggravating the problem.
- Interfaces between subsystems lack of formal definitions. Static and dynamic properties of such interfaces are often poorly characterized. As a result, system providers cannot perform or cannot trust analysis or synthesis of systems.
- Applying formal verification in system design today usually requires construction from scratch of new models of the systems. The models used in practice by engineers are not amenable to such verification. But the construction of these new models requires deep expertise in both the application domain and in the formal modeling techniques. This combination of expertise is rare among engineers. The skill gap impedes the use of formal verification.

The need for formal techniques as a way to bridge these gaps is evident. In the context of cyber-physical systems, tools with formal foundations do exist, but, in their present form they are not usable by an average system designer. Engineers need user-friendly languages with solid formal foundations to define requirements, behaviors, constraints, and interfaces. Once they have this, with enough coverage and expressivity, they will be able to make significant advances into virtual integration and analysis as well as into generative design of complex systems.

Formal methods and tools do not take humans out of the design process. It is well known that many success stories in formal verification have arisen not from the machine analysis enabled by formal foundations, but rather from the process of constructing the new models that today's verification techniques require. The (human) process of model building exposes flaws in the design. Using languages with formal foundations favors designer intuition because models become more readily understood by the designer. The subsequent machine analysis of the models that is enabled by their formal foundations then builds confidence in the revised designs. To favor designer intuition, tools need to integrate natural and intuitive front end languages with a formal back-end.

Formal methods and tools also do not reduce the need for simulation. On the contrary, they increase the value of simulation. When simulators are based on languages with well-understood and rigorous semantics, engineers can have confidence in the results of simulation.

Better tools and methodologies will be shorten project times, reduce redesign cycles, reduce the cost of testing, and enable better (more capable, robust, safe, and inexpensive) systems. Formal foundations will let model-based systems engineering (MBSE) live up to its full potential, and will speed its adoption.

2.2.2   Multitude of Domains

Several domains are already model driven (e.g. mechanical, electrical, and aeronautical), in the sense that engineers build and use models as an integrated part of the design process. The challenge today is the integration of diverse domains. Integration of diverse domains is intrinsic in todays complex, cyber-physical systems. It cannot be avoided. Today, this is often done locally in an ad-hoc manner, often with homegrown tools. This reduces the value of such integration, because it is hard to build confidence in the results. Or worse, integration is not done at all until late in the design process, when manufactured prototypes are put together for the first time.

In cyber-physical systems, one of the key challenges is that models must include both models of physical systems, which are rooted in continuous space and time, with models of computational systems, which are rooted is discrete, algorithmic (step-by-step) models [14]. The underlying principles of these two classes of models are well established, but many of the formal properties of the individual model evaporate when they are combined. For example, feedback control analysis in a time continuum may prove a system stable, while a software realization of the controller may yield instability. A pressing need today is to connect continuous-time physical device models with discrete models of computing systems. This integration must take as input the current industrial state-of-the-art tools and standards (e.g. SysML, Modelica, and Simulink) as entry points for such integration.

2.2.3   Complexity

The use of embedded computers and networks in complex systems today (such as cars and aircraft) has dramatically improved both the functionality and quality of these systems. These systems are more capable than before, and indeed, market pressures for ever more elaborate capability are enormous. But increased capability usually also comes with increased complexity. The key questions in product design becomes not those about affordability and return on investment, but rather about feasibility. The question is whether it is even *possible* to design the next generation products.

3   The Principles and Long-Term Focus of the Consortium

We believe the most promising means to address the challenges in systems engineering of cyber-physical systems is to employ structured and formal design methodologies that seamlessly and coherently combine the various dimensions of the multi-scale design space (be it behavior, space, or time), that provide the appropriate abstractions to manage the inherent complexity and heterogeneity, and that can provide correct-by-construction implementations.

The following issues are being addressed by the consortium:

- *Design Methodologies*
- *Heterogeneous Modeling and Tool Integration*
- *Formal Control Synthesis*
- *Design-space Exploration*
- *Design Drivers*

We believe that it is essential to address the entire process and not to consider only point solutions of methodology, tools, and models that ease part of the design. The consortium is addressing both foundations of modeling and algorithms and solutions involving tools, flows, and methodologies. The two parts are strongly interdependent, thus forming a unified body of work that is intended to transform radically the way we do design today.

This research requires novel principles. Academic research in methodology and design frameworks has been successful in the electronics domain because of the early involvement of industry in driving the needs and testing the results. In the case of systems engineering, academic work has not yet had the industrial impact that many hoped for. Given the size and complexity of the problems, a significant number of researchers must be involved. For this reason, the consortium is structured as a close collaboration between industry and academia. The research is driven by industrial needs, but cannot ignore foundational issues.

3.1   Research Summary

The taxonomy shown in Figure 1 gives a perspective on high-level research topics, how these research topics link together, and how they address specific industrial concerns (the light bubbles). The adjacencies can be exploited to derive solutions that are more performing than they would if developed in isolation. The scope of research is large. Foundational work is pervasive and relies upon an overall emphasis on formal methods and a rigorous approach to design.

The topics are briefly described below and are grouped according to the diagram shown in Figure 1. In each topic we will underline the foundation and the solution components.

3.2   Design Methodologies

This section addresses model-based systems engineering for cyber-physical systems.

3.2.1   Platform-Based Design for Multi-physics Systems

Driven by the industrial needs to integrate different physical domains, and by the diversity of modeling and simulation tools and methods for these physical domains, our focus is on principles of composability, abstraction, and refinement. Recent

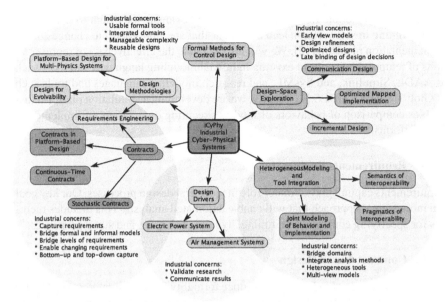

Fig. 1 Taxonomy of research needs

progress in object-oriented modeling languages like Modelica [6] suggests directions for such principles that can embrace mechanical, electrical, and thermal domains, for example. Key challenges include how to define precisely the notion of composability of components, and how to characterize the interfaces of components so that they can be composed in a natural and correct-by-construction fashion. Further, the notions of abstraction and refinement need to be developed, possibly relating these concepts to the mathematical approach to reduced-order modeling. The analysis and verification of components and of integrated systems need to leverage best-of-class tools that deal with different physical phenomena. For example, to analyze structural properties together with thermal behavior of a system may require composing significantly different tools and methodologies. How to relate the various domains so that they maintain consistency across layers of abstraction is an open problem, especially when the hierarchy may be different across physical domains. The composition rules and the relations among different viewpoints will be captured using contracts (see below).

3.2.2 Design for Evolvability

We are working to endow modeling languages with semantics that adequately express required behavior in a technology-independent way, and to provide synthesis tools that yield a multiplicity of implementations that, by construction, have the behavior specified in the models. Such semantics must include an ability to express temporal behavior and temporal requirements independently from the underlying system implementation. In the near term, we will evaluate the extent to which controlling timing in software can lead to designs that are more robust to changes in the implementation platform [15]. We will also evaluate the effectiveness of representations of temporal behavior in existing industrial modeling languages and tools such as AADL, Simulink, and SysML, and research modeling languages and tools such as Giotto [8] and PTIDES [4]. Finally, we are developing a simulation platform that enables comparison of behaviors of models executing on a variety of implementation platforms.

3.2.3 Requirements Engineering

Requirements capture plays a major role in industrial design processes. Our key goal is a more formal approach that will enable automated analysis. This includes methods for specifying constraints on timing, dynamic behavior, and static properties.

3.2.4 Contract-Based Design

Driven by the needs of industry to introduce formality in the design process, iCyPhy is focusing on contracts as a formalization of the rules for composition, abstraction, and refinement. Making contract-based design a technique of choice for system engineers, the team is developing mathematical foundations for contract representation and requirement engineering that enable the design of frameworks and tools. For example, a type-theoretic system for building domain-specific ontologies and annotating components with constraints on such ontologies enables better compatibility checking between separately developed components [18].

Contracts in Platform-Based Design. To integrate methodologies and tools, iCyPhy is merging contract-based design with platform-based design [24] to formulate the design process as a meet-in-the-middle approach, where design requirements are implemented in a subsequent refinement process using as much as possible elements from a library of available components. Contracts are formalizations of the conditions for correctness of element integration (horizontal contracts), for a lower level of abstraction to be consistent with the higher ones, and for abstractions of available components to be faithful representations of the actual parts (vertical contracts).

A typical use of contracts in cyber-physical system design would be to govern the horizontal composition of the cyber and the physical components and to establish the conditions for correctness of their composition.

Continuous-Time Contracts. To use contracts for heterogeneous domains, we need a theory for continuous-time contracts. Effective continuous-time contracts require (i) models of time that can semantically distinguish discrete events and continuous change, (ii) expressing bounds on continuous behaviors in terms of discrete constraints, and (iii) provide for interaction between simulation models and models for semantic analysis and design. In fact, a full-fledged continuous-time contract theory should support dynamical models as well as structural and performance models so as to be practical for CPS design. The long-term goal is to provide the conceptual framework for such contracts and software prototypes that demonstrate its efficacy.

In the near term, the plan is to extend techniques that have shown promise in the context of mixed-signal (analog/digital) integrated circuit design [22] to encapsulate the physical portion of the CPS and provide a "generalized" interface to its "cyber" counterpart. Such a generalized interface should offer:

- Composition constraints and rules for the interaction of a subsystem with its environment; such constraints are formalized with horizontal contracts;
- Simplified discrete-time and amplitude-quantized behavioral models for efficient design exploration and co-simulation of a subsystem with its environment; the range of validity of behavioral models are defined by bottom-up vertical contracts;
- Information about the capabilities of the subsystems in terms of timing, power consumption, size, weight and other physical aspects (performance models) that need to be transmitted to the system assemblers to allow for early detection of design errors; the range of usage of such performance models is defined by top-down vertical contracts.

Stochastic Contracts. To address the industrial needs of fault tolerance and robustness of design, we need to consider stochastic contracts. Complex systems are stochastic in nature. In fact, several parameters impacting both the behavior and the performance of these systems are subject to variability due to manufacturing tolerances, usage, and faults. Moreover, models and abstractions that are normally used to design multi-physics systems inevitably introduce inaccuracies, since either the dynamics of the several components are not perfectly known, or approximations are needed to guarantee efficient explorations. As a consequence, robust system design can often imply costly characterization based on Monte Carlo simulations or expensive overdesign to guarantee large safety margins.

The long-term objective is to provide a conceptual framework for such contracts to support analysis and design stochastic heterogeneous systems, and the software prototypes that demonstrate its efficacy.

In the near term, iCyPhy is focusing on the broad set of existing stochastic models that can capture both the continuous and discrete dynamics of cyber-physical systems, such as Markov jumps linear systems, piecewise deterministic Markov processes, stochastic hybrid systems, and switching diffusion processes, on which contracts need to be formulated. Because of the heterogeneity of stochastic hybrid systems, several models can indeed be adopted, depending on which dynamics are affected by uncertainties.

3.3 Heterogeneous Modeling and Tool Integration

The challenge is to define models of computation (MoCs) that are sufficiently expressive and have strong formal properties that enable systematic validation of designs and correct-by-construction synthesis of implementations. A second challenge is to identify which of the many MoCs and variants are actually needed, to figure out how to educate the community to use them, and to articulate the choices into industrial standards such as SysML. The major innovation being pursued in iCyPhy concerns the interoperability of MoCs, thereby enabling heterogenous design with rigorous foundations.

Semantics of Interoperability. MoCs are built by combining three largely orthogonal aspects: sequential behavior, concurrency, and communication. Similar to the way that an MoC abstracts a class of behavior, "abstract semantics" abstract the semantics of the MoC itself [16]. The concept is called a "semantics meta-model" in [25], but since the term "meta-model" is more widely used in software engineering to refer instead to models of the structure of models (see [21] and http://www.omg.org/mof/), we prefer to use the term "abstract semantics" here. The concept of abstract semantics is leveraged in Ptolemy II [5] and Metropolis [1] to achieve heterogeneous mixtures of MoCs with well-defined interactions.

The key challenge is providing actor-oriented MoCs [17] with well-defined semantics. All too often, the semantics emerge accidentally from the software implementation rather than being built in from the start. One of the key challenges is to integrate actor-oriented models with practical and realistic notions of time. To address, for example, modeling distributed behaviors, it is essential to provide multiform models of time. Modeling frameworks that include a semantic notion of time, such as Simulink and Modelica, assume that time is homogeneous in the sense that it advances uniformly across the entire system. In practical distributed systems, even those as small as systems-on-chip, however, no such homogeneous notion of time is measurable or observable. In a distributed system, even when using network time synchronization protocols (such as IEEE 1588 [10]), local notions of time will differ, and failing to model such differences could introduce artifacts in the design.

Pragmatics of Interoperability. Despite considerable progress in languages, notations, and tools, major problems persist. In practice, system integration, adaptation of existing designs, and interoperation of heterogeneous subsystems remain major stumbling blocks that cause project failures. We believe that model-based design, as widely practiced today, largely fails to benefit from the principles of platform-based design [24] as a consequence of its lack of attention to the semantics of heterogeneous subsystem composition.

Many previous efforts have focused on tool integration, where tools from multiple vendors are made to interoperate [19, 7, 11]. This approach is challenging, however, and yields fragile tool chains. Many tools do not have adequate published extension points, and maintaining such integration requires considerable effort.

iCyPhy believes a better approach is to focus on the semantics of interoperation, rather than the software problems of tool integration.

Nevertheless, a purely semantics-based approach will fail to have practical impact in industry because it will not embrace industry-standard tools. A promising recent development is the evolving Functional Mockup Interface (FMI) standard (see https://fmi-standard.org), which aims to enable model exchange and co-simulation between continuous-time models. The iCyPhy consortium is actively involved in the development of this standard with the goal of ensuring that it is capable of support a sound semantics of interoperation.

Joint Modeling of Behavior and Implementation. The Metropolis project [1, 3] has introduced the notion of a "quantity manager," a component of a model that functions as a gateway to another model. For example, a purely functional model that describes only idealized behavioral properties of a flight control system could be endowed with a quantity manager that binds that functional model to a model of a distributed hardware architecture using a particular network fabric. By binding these two models, designers can evaluate how properties of the hardware implementation affect the functional behavior of the system. The iCyPhy consortium is further developing this concept, generalizing it as a form of aspect-oriented modeling [12] and integrating it with the Ptolemy framework.

3.4 Formal Methods for Control Design

The co-design of controllers and certificates of correctness is emerging as a promising approach for "correct by construction" design that addresses issues in verification and integration [2, 13, 28]. There are a number of broad research directions available based on the initial work we have done in this area.

Performance Specifications. Current techniques for control protocol synthesis often provide correct behavior but with no regard for performance. It will be important to add in the ability to include cost (or reward) functions in temporal logic planning, allowing protocols that satisfy a set of (hard) constraints as well as minimizing a cost function associated with the continuous or discrete states [27]. A related issue is allowing optimization of the probability that certain specifications are met, to move away from pure worst-case performance [26]. Including the ability to specify real-time properties (such as the amount of time between an environmental event and the response of the control system) is also an area in where performance specifications must be generalized. This might build on work in the computer science literature on timed automata and real-time temporal logics, but also incorporate continuous dynamics and control actions [29].

Controller Architecture. Most existing techniques focus on the synthesis of a single, centralized controller. It will be important to develop methods for designing and validating formal interface specifications between subsystems (horizontal contracts) that allows verification and synthesis to be performed at the subsystem level, with guaranteed system level requirements [23]. We must also develop hierarchical control structures that make use of a demand-response architecture and formal interface specifications between layers (vertical contracts) to achieve a system-level goal. Preliminary work in control of autonomous vehicles provides a starting point for this work [28].

Over the long term, we seek to derive and implement algorithms for synthesis of control protocols that can be applied to cyber-physical systems. Key elements are increasing our ability to capture dynamics, uncertainty and feedback in our theory and integrating new algorithms for solving the types of problems identified above into the TuLiP (or other) open-source software packages.

3.5   Design Space Exploration

Communication Design. System designers today are leveraging as much as possible computing and networking technology to gain capability and performance and to reduce costs. One such approach is to select Ethernet as the physical layer for communication in embedded systems such as airplanes and cars. Since the protocol used with Ethernet is asynchronous, there are serious concerns about the safety implications of this choice. A solution to this problem is to use an additional protocol layer on top of Ethernet that would provide a synchronous platform for the applications. TTEthernet, Arinc 429/717, and Audio Video Bridging (AVB) Ethernet are all defining synchronous (time triggered) Ethernet-based standards. One approach being pursued in iCyPhy is PTIDES, which leverages these networking developments to provide a foundation for distributed software with controllable timing properties [4].

Optimized Mapped Implementation. If we want to perform automatic optimal mapping of behaviors onto architectural elements, we must embed behavior and architecture in the same semantic domain. For example, in automatic logic synthesis register-transfer level (RTL) descriptions and gate representations are mapped into a particular form of Boolean representation called the Boolean network. Doing this, we can optimize the mapping process by using a covering algorithm. We maintain that this process is indeed applicable to all layers of abstraction provided that a common semantic domain that makes the mapping algorithm effective is found.

Incremental Design. This research topic is related to the need to understand the impact of design changes on the performance, cost, and time to market. Albeit using formal languages and synthesis has been a key methodology approach in moving VLSI design to a level of productivity that was unimaginable before this approach was introduced, it did expose design implementations to instability with respect to changes in the sense that a small change in behavior can result in a large, unpredictable variation in the logic implementation. This yielded a large amount of redesign in subsequent implementation steps and in particular, in the layout of the integrated circuit under design that created havoc with schedule and chip size estimations.

This effect was the result of the optimization process that is notoriously unstable with respect to input changes if not constrained. A method that was developed in the VLSI domain was to limit the re-design due to the changes to a part of the previous implementation. Which part to choose was the research problem to be faced and that was only partially resolved in that application domain.

The consortium is studying the stability problem in design space exploration by limiting the degrees of freedom that are used in the optimization steps and in particular, the allocation of functionalities to architectural blocks.

3.6 Design Drivers

An effective industry-academic collaboration leverages the real-world systems experience of industry to test and refine academic models and tools. ICyPhy has focused on two richly heterogeneous system problems, namely the electric power and air management systems (EPS and AMS) of advanced aircraft.

Electric Power Systems. The EPS is a key subsystem of an aircraft vehicle management systems (VMS) [20]. Its function is to generate, regulate, and distribute electrical power throughout the aircraft.

EPS design poses several challenges, including controller architecture definition, contactor and sensor optimization, safety and fault tolerance, and efficient load management. Given an EPS topology, typically captured by a so-called single-line diagram (SLD), there is clearly no unique solution for the bus power control unit (BPCU), since both the controller inputs (sensor number and location) and outputs (contactor number and location), essential elements for controller design, are underspecified and left as design choices. The initial SLD structure itself may not guarantee the desired reliability level and may require modifications in terms of component and path redundancy.

An EPS system also has multi-physics aspects. The dynamics of generators and loads can get quite complex. The mechanical parts also affect system behavior, where latency and bounce in contactors can affect behavior. An it has cyber-physical aspects, since EPS systems today are implemented using networked microcontrollers.

Nearly every thrust within iCyPhy can be applied to an EPS design. Given a set of loads, the power system can be built out of a library including, among other components, generators, buses, power converters, sensors and contactors. System requirements are expressed in terms of safety, reliability, and availability constraints. EPS design is framed as an optimization problem where the selected candidate topologies and controller architectures satisfy all the requirements, while optimizing quality factors such as weight, efficiency, complexity, and cost. To achieve correct-by-construction design, we formulate contracts at different articulation points in the design flow. Controller synthesis techniques from Section 3.4 ensure realization of logical specifications. Horizontal contracts will formalize the conditions under which component integration is correct; vertical contracts will formalize the conditions under which an implementation is consistent with its abstraction, or an abstraction is a faithful representation of an implementation. Co-simulation of functionality and architecture enables studying how, for example, network architecture affects dynamics. Multi-physics simulation enables analysis of how network dynamics affects electrical dynamics. Finally, stochastic contracts can establish conditions under which safety and reliability requirements are guaranteed.

Air Management Systems. A second application driver is the air management system of an aircraft. This application has more diverse multi-physics aspects, since thermodynamics, fluid dynamics, and mechanical geometry all come into play. A significant challenge is to develop techniques that enable use of best-of-class industry-standard modeling and simulation tools together with the new tools and methodologies being developed.

4 Conclusion

Progress in systems engineering for cyber-physical systems requires a deep collaboration between industry and academia. The iCyPhy industry-academic partnership has been formed to develop a new generation of system modeling, design, and analysis tools and methodologies that will enable more effective design of more capable systems.

Acknowledgements. Thanks to John Arnold for helping to form the vision in this paper.

References

1. Balarin, F., Hsieh, H., Lavagno, L., Passerone, C., Sangiovanni-Vincentelli, A.L., Watanabe, Y.: Metropolis: an integrated electronic system design environment. Computer 36(4) (2003)
2. Belta, C., Bicchi, A., Egerstedt, M., Frazzoli, E., Klavins, E., Pappas, G.: Symbolic planning and control of robot motion (Grand Challenges of Robotics). IEEE Robotics & Automation Magazine 14(1), 61–70 (2007)
3. Davare, A., Densmore, D., Meyerowitz, T., Pinto, A., Sangiovanni-Vincentelli, A., Yang, G., Zhu, Q.: A next-generation design framework for platform-based design. In: Design Verification Conference (DVCon), San Jose, California (2007)
4. Eidson, J., Lee, E.A., Matic, S., Seshia, S.A., Zou, J.: Distributed real-time software for cyber-physical systems. Proceedings of the IEEE (Special Issue on CPS) 100(1), 45–59 (2012)
5. Eker, J., Janneck, J.W., Lee, E.A., Liu, J., Liu, X., Ludvig, J., Neuendorffer, S., Sachs, S., Xiong, Y.: Taming heterogeneity—the Ptolemy approach. Proceedings of the IEEE 91(2), 127–144 (2003)
6. Fritzson, P.: Principles of Object-Oriented Modeling and Simulation with Modelica 2.1. Wiley (2003)
7. Gu, Z., Wang, S., Kodase, S., Shin, K.G.: An end-to-end tool chain for multi-view modeling and analysis of avionics mission computing software. In: Real-Time Systems Symposium (RTSS), pp. 78–81 (2003)
8. Henzinger, T.A., Horowitz, B., Kirsch, C.M.: Giotto: A time-triggered language for embedded programming. Proceedings of IEEE 91(1), 84–99 (2003)
9. Hubbard, D.: The Failure of Risk Management: Why It's Broken and How to Fix It. John Wiley & Sons (2009)
10. IEEE Instrumentation and Measurement Society. 1588: IEEE standard for a precision clock synchronization protocol for networked measurement and control systems. Standard specification. IEEE (November 8, 2002)

11. Karsai, G., Lang, A., Neema, S.: Design patterns for open tool integration. Software and Systems Modeling 4(2), 157–170 (2005)
12. Kiczales, G., Lamping, J., Mendhekar, A., Maeda, C., Lopes, C.V., Loingtier, J.M., Irwin, J.: Aspect-oriented programming. In: Akşit, M., Matsuoka, S. (eds.) ECOOP 1997. LNCS, vol. 1241, pp. 220–242. Springer, Heidelberg (1997)
13. Kress-Gazit, H., Wongpiromsarn, T., Topcu, U.: Correct, Reactive, High-Level Robot Control. IEEE Robotics & Automation Magazine 18(3), 65–74 (2011)
14. Lee, E.A.: Cyber physical systems: Design challenges. In: International Symposium on Object/Component/Service-Oriented Real-Time Distributed Computing (ISORC), Orlando, Florida, pp. 363–369. IEEE (2008)
15. Lee, E.A.: Computing needs time. Communications of the ACM 52(5), 70–79 (2009)
16. Lee, E.A.: Disciplined heterogeneous modeling. In: Petriu, D.C., Rouquette, N., Haugen, O. (eds.) Model Driven Engineering, Languages, and Systems (MODELS), pp. 273–287. IEEE (2010)
17. Lee, E.A., Neuendorffer, S., Wirthlin, M.J.: Actor-oriented design of embedded hardware and software systems. Journal of Circuits, Systems, and Computers 12(3), 231–260 (2003)
18. Lickly, B., Shelton, C., Latronico, E., Lee, E.A.: A practical ontology framework for static model analysis. In: International Conference on Embedded Software (EMSOFT), pp. 23–32. ACM (2011)
19. Liu, J., Wu, B., Liu, X., Lee, E.A.: Interoperation of heterogeneous CAD tools in Ptolemy II. In: Symposium on Design, Test, and Microfabrication of MEMS/MOEMS, Paris, France (1999)
20. Moir, I., Seabridge, A.: Aircraft Systems: Mechanical, Electrical, and Avionics Subsystems Integration, 3rd edn. AIAA Education Series. Wiley (2008)
21. Nordstrom, G., Sztipanovits, J., Karsai, G., Ledeczi, A.: Metamodeling - rapid design and evolution of domain-specific modeling environments. In: Proc. of Conf. on Engineering of Computer Based Systems (ECBS), Nashville, Tennessee, pp. 68–74 (1999)
22. Nuzzo, P., Sangiovanni-Vincentelli, A.: Robustness in analog systems: Design techniques, methodologies and tools. In: Proc. IEEE Symp. Industrial Embedded Systems (June 2011)
23. Ozay, N., Topcu, U., Murray, R.M.: Distributed power allocation for vehicle management systems. In: Proc. IEEE Control and Decision Conference (2011)
24. Sangiovanni-Vincentelli, A.: Defining platform-based design. EEDesign of EETimes (2002)
25. Sangiovanni-Vincentelli, A., Yang, G., Shukla, S.K., Mathaikutty, D.A., Sztipanovits, J.: Metamodeling: An emerging representation paradigm for system-level design. IEEE Design and Test of Computers (2009)
26. Wolff, E.M., Topcu, U., Murray, R.M.: Robust control of uncertain markov decision processes with temporal logic specifications. In: Proc. IEEE Control and Decision Conference (2012)
27. Wolff, E.M., Topcu, U., Murray, R.M.: Optimal control of non-deterministic systems for a computationally efficient fragment of temporal logic. In: Proc. IEEE Control and Decision Conference (2013)
28. Wongpiromsarn, T., Topcu, U., Murray, R.M.: Synthesis of control protocols for autonomous systems. Unmanned Systems 1(1), 21–40 (2013)
29. Xu, H.: Design, specification, and synthesis of aircraft electric power systems control logic. PhD thesis, California Institute of Technology (2013)

SoS and Large-Scale Complex Systems Architecting

Dominique Luzeaux

Abstract. In the last decades, the systems designed in the following domains (banking, health, transportation, space, aeronautics, defense…) have been increasingly larger. With the growing maturity of information and communication technologies, systems have been interconnected within growing networks, yielding new services through combination of the system functionalities. This leads to an increasing complexity that has to be managed in order to take advantage of these system integrations. More broadly, the need to ensure the interoperability of systems with varied origins, not designed for interoperability, in order to valorize existing systems which are not best suited to current operational needs, calls for system-of-systems engineering practices.

In this paper we will focus on a specific part of the systems engineering process, dealing with "architecture". We will in particular discuss the specificities of architecting systems-of-systems and large-scale complex systems.

1 Adapting to a Changing Environment

1.1 Increasing Complexity, Globalization of Business

Current times are characterized by increasing complexity, scope and size of the systems of interest, their context, and the organizations creating the systems. There are increasing dynamics: shorter time to market, more interoperability, rapid changes and adaptations in the field. Furthermore, many of today's systems are developed by distributed teams at multiple locations, by multiple vendors, and across multiple organizations.

Dominique Luzeaux
Director for Land Systems Acquisition,
Direction Générale de l'Armement, French Ministry of Defense
e-mail: dominique.luzeaux@polytechnique.org

M. Aiguier et al. (eds.), *Complex Systems Design & Management 2013*,
DOI: 10.1007/978-3-319-02812-5_3, © Springer International Publishing Switzerland 2014

Going hand in hand with the evolution of techniques and economical models, society has changed, with the emergence of new working methods (telecommuting, virtual communities, etc.) and of social links alternatively favoring combination and complementarity depending on the opportunities: the former to increase one's flexibility and therefore one's value chain via the sharing of resources (each part being more efficient when grouped with the others), the latter to favor collective productivity and join a strongly integrated systemic set (the total achieving more than the sum of its parts).

Flexibility, exchanges, partnerships, opportunity alliances are becoming new models, constructed around the constant innovations in the new information and communication technologies. Thus, collaborative working tools, on the one hand, and digital transaction platforms, on the other, have been developed, thereby accelerating the spreading of these new models.

1.2   A Particular Field: Defense

Let us first underline the specificity of this field's economic environment: the acquisition cycles are very long compared to the technological maturity cycles of the elementary components. From technological study to production, a weapons system is frequently developed over a decade, and the weapons system itself will be used for three or four decades, depending on renovations. To highlight this dilemma, let us compare those delays to the ones of certain embedded technologies (embedded computing and electronics) whose life cycle is in the range of two or three years! Moreover, the field of defense is much less competitive than the civil market, and the investment and operation budgets do not follow economic principles. In France, for example, the ministerial budgets (including investments and operation costs) are allocated on an annual ratio annually, within a budgetary framework defined by law for six years and strongly influenced by the presidential decisions, which might therefore evolve depending on the terms of office. Besides, the long-term planning of new systems requirements or system renovation happens on a scale of fifteen or twenty years. From it spawns the inherent difficulty in making the various life cycles tally, against terms of use which are by necessity evolutionary, or even inventive.

Taking new threats (post-cold war, OOTW: operations other than war, 4th generation warfare or not war-related) into account calls for high flexibility and adaptability in the exploration of the concepts of defense systems. Indeed, the logics developed during the Cold War concerned symmetrical fights, in which bulks of metal faced off in wide battlefields or in the air; in terms of acquisition, this approach favored continuity of armament, with added technological sophistication in each new generation, but the number of weapons being more important than the individual differences between systems, the length of the acquisition cycles did not have a qualitative impact on a nation's capacity. On the other hand, the new threats highlight the possibility of asymmetrical conflicts: the potential adversary does not a priori possess the same off-the-shelf military technologies, but might put the common civilian technologies to inventive use, which would be a threat not on the level of the individual weapons system, but on the level of the opposing army's meta-system. The environment's evolutionary nature and the need for the system to adapt to those various evolutions contribute to its complexity.

All this is all the more noticeable since systems engineering (as a formalized set of best practices with reference processes and activities) takes its roots in the management of defense systems. Therefore there lies some logic in the fact that an evolution in the grounding field should trigger changes in the resulting processes and activities.

A changing international environment, with the need for more advanced weapon systems and limited resources, puts strong constraints on the acquisition of defense systems. The acquisition services' decision-makers must be able to consider future fighting situations when they design new weapons systems, while evaluating their performance and the production capacities so as to minimize the risks, delays and costs. Even if the new information and communication technologies spur technological advancements which make new methods of productions possible, more efficient and less costly, when faced with the financial stakes one has to "product the right product" on the first try.

On the individual system's level, it is therefore necessary to rapidly implement solutions that can be used in a few years instead of fifteen to twenty years from now. Incremental implementation steps must be designed so as to mirror the evolution of the various technologies' maturity and take the feedback into account. Only this can help people identify the material solutions which need to cooperate within a highly connected set of collaborative resources. The goal is to find the best global solution, rather than the optimization of a specific system to answer the need.

This leads to the consideration of systems as basic components which must be organized within what can therefore be called a *"system-of-systems"*, depending on the desired effect. On the level of the system-of-systems, one must take into account the asynchronous nature of the acquisition of individual systems (the life cycles might or might not partially overlap), and facilitate the co-evolution of said systems in terms of doctrine, organization, training.

Moreover, the current geopolitical context stresses the necessity of acquiring systems which can interoperate with the systems of other partners: interventions happen a priori within multinational operations, within coalitions whose composition changes both in time and depending on the missions. In addition, the systems purchased must be operational within minimum delays and therefore need a reduced staff, easily deployed and with minimum needs for logistical support.

1.3   Towards a Capability-Driven Approach

Military units are deployed, depending on circumstances and enemy movements, following plans or concepts of operations. The latter are verbal or graphic statements that clearly and concisely express what the joint force commander intends to accomplish and how it will be done using available resources (DOD 2010). They refer to scenarios that implement the strategic objectives and serve as a reference for units, material, logistics, by defining the unitary action sequences.

Each unit can thus fulfill its *"capability"* when it executes its part of a scenario. Recall that a "capability" (definition from the US Department of Defense) is the ability to achieve a specified wartime objective (to win a war or a battle, destroy a target). Transposed in business, a capability is the ability of an entity (department, organization, system) to achieve its objectives, specifically in relation to its overall mission by using the available resources – human, material, immaterial – in a coordinated manner.

This relies not only on communications but on a fluid interaction where every unit is compatible with the neighboring units and complete and reinforce each other. The key driver to success is therefore not only a network issue, but rather more the relationship between all components and their smart collaborative exploitation.

The evolution towards a capability rationale of the defense tool orients the acquisition process towards systems-of-systems. This rationale is built on the idea of operational capability, which consists in reaching a desired state via the effective combination of resources (staff, materials) and processes to create a set of tasks. Such an ability is expressed in terms of mission types and operation contexts, desired effects (which themselves can be described functionally or through levels of expected performances), doctrines, organization, training, equipment, formation, infrastructures, policies, etc. In the field of defense, an example of this is the "projection ability", which consists in being able to send a predefined potential of armed forces and material in a predefined time across a predefined distance to a foreign field of operation. Expressing the needs in such a way rather than through the numbers of planes with such or such technical capacity (action range, fret volume and weight, etc.) belongs within the rationale of the evolution of products towards services. We can therefore imagine – actually, more than imagine, since France has signed such a contract – that to fill the previous capability, a contractor could rent, on request, aircraft usually destined to carry other types of merchandise, even if this necessitates being able to rapidly modify them. This means that the contractor offers availability as a service rather than selling aircrafts which the army military would then have to operate during their entire life cycle, and therefore maintain, repair and upgrade.

The ability to map the links between any type of resource and their use with quantifiable effects heightens strategic flexibility, since said capability is a priori unvarying against a specific context's requirements. The difficulty is obviously to determine capabilities, for, on the one hand, each of them must allow for tight couplings between the implemented components, and on the other hand the coupling must be as weak as possible between different capabilities. Moreover, if the capability shows the final purpose, the trajectory from the current situation to that purpose must be defined, through the definition of capability increments, and more generally through a true process of capability management, going through these three stages:

– definition of capabilities: conceptualization of new ideas on capabilities, with ideas on planning, acquisition, development, management of the life cycle of the staff, materials and processes potentially implemented within said capability. This study happens on a long-term basis, with a horizon of a decade or longer;
– capability maintenance: guarantee the level of availability of the capability, which happens on a medium scale, with a five year horizon;
– capability usage: plan and pilot the operations which use said capability. This obviously happens on a short term basis.

2 Systems-of-Systems: Maier Is Obsolete!

A system-of-systems is defined by Mark Maier in 1996 (Maier 1996) as having most of the following characteristics:

- the constituent systems have an operational independence (they can be used separately);
- the constituent systems have a managerial independence (they are managed and acquired by independent teams with independent budget lines);
- the definition and the configuration of the global system are evolutionary in nature;
- there exists emergent behaviors of the global system;
- the constituent systems are geographically distributed.

Maier's definition is one of the most widespread definitions, even if a certain number of variations can be found in literature (Luzeaux et al. 2010). However, it features a double defect. First, it only grants a raison d'être to the system-of-systems' constituent components, without mentioning what the system-of-systems itself brings in relation to its constituent elements, apart from the possibility of emergent behaviors. Secondly, it is restricted to the technical criteria of geometrical geographical distribution and possible new behaviors: the capability aspect is not addressed, except very partially from the angle of sought effects measured in terms of technical performances. It is, moreover, very much oriented towards the systems' acquisition phase, and in particular neglects any notion of organization of the technical and human resources during the utilization phase.

This leads us to offer the following new definition, more general than Maier's, and which covers a priori all the variants found in literature, as well as the more recent ones which concern extended enterprises.

Definition: *a system of systems is an assembly of systems which can potentially be acquired and/or used independently, and whose global value chain the designer, buyer and/or user is looking to maximize, at a given time and for a set of conceivable assemblies.*

In this definition, we follow the general definition of the "value chain" made popular by Michael Porter, namely the set of interdependent activities whose pursuit creates identified and, if possible, measurable value (which means the customer is ready to pay to get the product or service). This includes every stage, from the purchase of raw materials to the actual use, or even the after-sale service if necessary. The value chain's efficiency essentially relies on the coordination of the various agents involved, and their ability to form a coherent, collaborative and interdependent network.

The introduction of the value chain in the definition, and the requirement for its optimization in a potentially changing context, is the key to defining what a system-of-systems is in relation to constituent systems. Some value chains are not necessarily predominantly technical, something which allows to insist on potential added values of systems-of-systems other than their simple technical performance. Following the usual breakdown of business management in three variables – people, processes, products –, value chains can be defined, which are predominantly organizational, functional, and technical. They include the following potential features that can have a major direct impact on the subsequent large-scale complex system engineering and project management:

- Political (e.g. nationalization, change of administrations);
- Economic (outsourcing, new business rules, but also crises...);
- Societal (user acceptance, way of living);
- Technological;
- Ecological (sustainability issues, recycling obligations);
- Legal (regulations constraining technical choices or contracting issues or management rules...).

Another issue that our definition deals with is the discrepancy in the lifecycles of the systems constituents (legacy, modified, new) and the fact that this discrepancy will evolve in time: a legacy system available today might be retired tomorrow and therefore be unavailable in the assemblies at hand tomorrow. The value chain captures these variations in time of the added value of such or such assembly. Stated differently, in relation to the architecting issue which will be our focus in the next paragraphs, the value chain helps evaluate all possible assemblies and choosing the right solution at the right time.

3   Architecture: A Key Concept

An architecture of a system is a set of fundamental concepts and properties in a specified environment, embodied in elements, relationships and principles to guide system design and evolution (as defined by the IEEE Std. 1471-2000, IEEE Recommended Practice for Architectural Description of Software-Intensive Systems). It provides the conceptual definition of the logical, physical structures, behavior, temporal relationships, and/or other aspects of a system and allocations among alternatives (e.g., physical elements, software, and operations; and/or functions in system-of-interest versus enabling system), cf. ISO/IEC/IEEE 42010.

A central idea of the IEEE 1471 Standard is that the description of an architecture should be expressed by describing multiple views each governed by a defined viewpoint to deal with various concerns of stakeholders. The standard does not provide the viewpoints; they should be selected based on the needs of the system.

An architecture is, on the one hand, the invariable representation of the object to design and develop, notwithstanding inevitable changes either in the context or made by those involved; on the other hand, it is the means of exchange and sharing between these parties throughout the life of the system. It is a representation, as it is a set of static and dynamical descriptions (it would be an error to exclude time factors or transformation functions which may operate on a given architecture to produce a different architecture), a set not strictly reduced to a juxtaposition, but more a composition or interweaving. It also constitutes a means for exchange and sharing as the formalized architecture transmits codified information, comprehensible to all who adhere to the standard of expression of information used. Thus we have a representation which overcomes spatial, temporal, cultural and generational limits and a priori has the same meaning for all. It may then be used by each individual to particular ends, but based on a general acceptation.

3.1 Reference Architectures and Architecture Frameworks

The activity of architecting aims to establish an agreement between all parties involved and to lay the foundations of the solution to the problem by a list of demands, covering technical, organizational and financial considerations.

An architecture is expressed by a set of viewpoints, the coherence and completeness of which must be mastered, over the whole lifespan of the solution under consideration. To do this in a way that is useful not only for the constituent system of interest at a given time, but also at any time of its lifecycle, as well as for other systems that could interface with it, we must have access to standards, or at least to methods, formalisms and notations which allow us to carry out comparative evaluation of various possible architectures. This is offered by architecture frameworks (Minoli 2008), which may be broken down into two broad families: methodological frameworks, on the one hand (TOGAF, FEAF, ISO-15704/GERAM, etc.), and formal and denotational frameworks on the other (DoDAF, MODAF, NAF, etc.). The first set allow us to develop concepts, principles and processes applicable to the products or services which, in the long run, will make up the planned system. The second group defines reference structures which may be instantiated to represent a system and thus obtain a set of views.

These views or perspectives allow us to consider the solution on several levels: technological, functional, system and strategic. The technological level describes the different choices of physical implementation for the solution. The functional level describes functions carried out and their possible hierarchy (functions and sub-functions). At system level, we find a description of the organization of the system and information, energy and matter flows between different blocks. THE Strategic level describes the business and organizational process which regulate non-technical aspects, and replace the system and its environment in the global context of use, taking account of objectives and policies for use. The search for the added value to be produced by the system during its use concentrates on this level.

Clearly, these different perspectives, whatever the architectural frameworks, are only one particular solution for representing a system. As when representing a three-dimensional object, there is no set group of privileged viewpoints, but simply a requirement that the representation be constituted in a sufficiently complete manner and contain potential overlaps. In the same way, we cannot define *the* architectural framework or *the* list of viewpoints to use. However, it is useful to have shared practices and a shared reference in terms of viewpoints, or the means of transforming certain viewpoints into others, if we wish to reuse a system architecture when considering a larger system which includes the first system in part or in its entirety. This becomes a key issue when engineering systems-of-systems, where the engineering teams are potentially different, distributed in several locations at various times.

Beyond architecture frameworks which provide only formal guidelines in order to design an architecture, reference architectures provide a template solution for a particular domain. A reference architecture is a resource containing a consistent set of architectural best practices for use by all teams within an extended organization. It captures the accumulated knowledge of many (Cloutier et al. 2010), ranging from "why" (market segmentation, value chain, customer key drivers, application), "what" (systems, key performance parameters, system interfaces, functionality, variability), to "how" (design views and diagrams, essential design patters, main concepts).

3.2   Architecting Systems-of-Systems

In order to manage architectural issues, there is an arising necessity of having guidelines when conceiving such architectures. These guidelines allow building a common understanding, sharing between different teams throughout the entire cycle of life of a system, even when teams have changed. This is enhanced by the multisite, multivendor feature of projects dealing with systems-of-systems, where collaborative environment with potentially multicultural engineering teams calls for standardization of the knowledge management process at each level of the systems engineering process.

This is illustrated by the current drafts of the revision of the ISO/IEC/IEEE 15288 "Systems and Software Systems Engineering – System life cycle processes", where the architecture is addressed by several technical and non-technical processes. Among the non-technical processes, more precisely within the "Organizational Project-Enabling Processes", the Portfolio management process identifies multi-project interfaces and dependencies that must be managed or supported by the project. In addition, a representation of domains for the reapplication of knowledge should be defined. This activity consists of the following tasks:

- Establish a representation for domain models and domain architectures: definition of the boundaries of domains and their relationships to others; definition of domain models capturing essential common and different features, capabilities, concepts, functions; definition of an architecture for a family of systems within the domain, including their common and different features.
- Define or acquire the knowledge assets applicable to the domain, including system and software elements: knowledge assets may include system elements or their representations (e.g. reusable code libraries, reference architectures), architecture or design elements (e.g. architecture or design patterns), processes, criteria, or other technical information (e.g. training materials) related to domain knowledge.

By minimizing the risks of ambiguity and by fostering knowledge management and transmission, architecture frameworks and reference architectures embody the building principles of the various architectures of the systems which will be assembled and integrated. As stated in (Cloutier et al. 2010), they effectively help create new product lines, facilitate multi-site, multi-organization, multi-vendor creation and life-cycle support, achieve interoperability between many different and evolving systems. This is done by providing architecture baselines and blueprints, by capturing and sharing patterns, by providing a common vision as well as a lexicon and taxonomy, by explicating modeling of functions above systems level, as well as decisions about computability, upgrade and interchangeability.

Architecture requirements refer to technical performance concerns but also to functional or business preoccupations that are addressed at enterprise level: issues dealing with processes, resources, assets, systems. They follow general principles, like transparency, openness, ethics, subsidiarity, modularity... Each such principle is to be implemented as a set of rules that formalize the various (functional, technical, applicative...) choices and the necessary tradeoffs in a non-ambiguous consistent (and hopefully complete) way.

The rules indicate the implementation choices and can eventually change, whereas the principles are stable. Principles are dictated by the enterprise strategy and deal predominantly with non-technical issues. They constitute what can be seen as the "enterprise architecture": an approach based on the business processes, on the strategic orientations, on the technical legacy. Actually they focus on all the dimensions that might appeal to any of the stakeholders and aim at providing sort of a lingua franca common to the whole enterprise while addressing its diversity and both its present and its future.

E.g. for a military system-of-systems, this would include the capability-driven vision (which lists strategic objectives that are an answer to the current geopolitical), the capability implementation plan (the list of the various capabilities and how they complement each other, and some information on existing capabilities or how they could be adapted to fulfill easily the desired role). This implementation plan is not a detailed technical architecture of the systems that will implement some of the capabilities; it focuses rather on the budgetary or the organizational issues that are a preliminary to any further detailed consideration. For instance, in the military case, some questions to be taken into account are the general organization of the involved Services (are they independent, are there any transverse units, how large are they in terms of number of units…), the key budgetary milestones (gates for decision-makers and parliamentary approbations, orders of magnitude), the political context (interoperability with extra-national organizations, bilateral or multilateral cooperation, sharing and pooling…).

Several architecture frameworks (e.g. DODAF, MODAF, NAF) propose views which are essential for managing systems-of-systems, such as:

- the *capability views*: strategic picture of how capability is evolving in order to support capability management and equipment planning,
- the *operational views*: the operational processes, relationships and context to support operational analyses and requirements development),

as well as other views which are key drivers for the project management of the system-of-systems, like for instance:

- the *acquisition views*: they support the acquisition program dependencies, timelines and lines of development status to inform program management,
- the *supplier view*, including the work breakdown structure,
- the *technology maturity view*: which attempts establishing a roadmap describing the impact of incremental maturing technologies on the systems architecture or the capabilities at the system-of-systems level.

4   Conclusion

Architecting a system-of-systems is a key step towards managing its complexity, as well as the first stage of the work breakdown structure and the allocation of various engineering and development tasks to the different teams. This is all the more important when potentially many teams and various organizations are in charge of the various tasks. However the success of such an approach relies on the reliability of the upstream steps performed before the architecting, chiefly the requirement process. The latter should be done thoroughly since it is the prerequisite to a successful engineering activity; the specificity of the system-of-systems has to be addressed, namely capability requirement engineering as well as extensively listing all stakeholders requirements. As we developed in (Luzeaux et al. 2010, 2011, 2013), this step is all the more essential, when a system gains in complexity.

Reference architectures and architecture frameworks facilitate reuse, harvest potential savings through reduced cycle-times, cost, risk and increased quality. As a knowledge repository, they contribute to the strategy of the enterprise, and as such they must be actively managed by the organization, with updates based on the evolution of the governing standards, emerging technology, changing stakeholder context, and maturing business. The architecture is a way to think our societal, technical, organizational system in a different way, by addressing them in their totality, integrating all strategic, economical, business, human and technological dimensions. This is all the more fundamental for large-scale complex systems, such as systems-of-systems.

Actually, the idea is to capture commonalities and essence, to provide constraints while opening opportunities, to remain at a sufficient abstract level while allowing further instantiation. Once these first principles are addressed, it is possible to delve into more technical details, as engineers are used to do, such as focusing on the means to implement the capabilities with existing systems or systems to be modified or to be developed. However at this point, the leitmotivs are: reuse and develop only as needed.

Architectures (if done in a proper way, i.e. avoiding any stove-pipe temptation and focusing on traceability of all intermediate decisions or trade-offs) are a great tool for innovative design. They allow the exploration of radical alternatives in a short amount of time, thus expanding the number of alternatives to be considered. Current research develops architecture transformation methods, which could be a powerful help to designing large-scale complex systems while taking advantage of reuse: this is greatly facilitated by defining layered architectures that address separately but jointly all platform independent functional and platform dependent technical aspects. An efficient architecture-based system engineering process should enable to: integrate legacy systems with new systems; exploit technology advances to provide desired capabilities to the users.

As a final conclusion, it is worthwhile mentioning that all the previous "theoretical" principles and ideas are currently applied (successfully, at least up to now!) to perform the systems engineering activities of the multi-billion euro military program Scorpion for the Army (Luzeaux 2012, Luzeaux 2013).

References

(Capirossi 2011) Capirossi, J.: Architecture d'entreprise. Eyrolles (2011)

(Cloutier et al. 2010) Cloutier, R., Muller, G., Verma, D., Nilchiani, R., Hole, E., Bone, M.: The concept of reference architectures. Systems Engineering 13(1) (2010)

(DoD 2010) Department of Defense Dictionary of Military and Associated Terms (2010), http://www.dtic.mil/doctrine/news_pubs/jp1_02.pdf

(Luzeaux et al. 2010) Luzeaux, D., Ruault, J.-R.: Systems of systems: concepts, illustrations, standards and methods. Wiley (2010)

(Luzeaux et al. 2011) Luzeaux, D., Cantot, P.: Simulation and modeling of systems of system. Wiley (2011)

(Luzeaux et al. 2013) Luzeaux, D., Ruault, J.-R., Wippler, J.-L.: Complex systems and system-of-systems engineering. Wiley (2013)

(Luzeaux 2012) Luzeaux, D.: Tutorial on SoS and application to the SCORPION Program. In: International Symposium of the INCOSE, Rome, Italy (July 2012)

(Luzeaux 2013) Luzeaux, D.: Tutorial on SoS and application to the SCORPION Program. In: Seminar of the Försvarets Materielverk and the Swedish chapter of INCOSE, Stockholm, Sweden (November 2013)

(Maier 1996) Maier, M.: Architecting Principles for Systems-of-Systems. In: 6th Annual International Symposium of the International Council of System Engineering, Boston (1996), http://www.infoed.com/Open/PAPERS/systems.htm

(Minoli 2008) Minoli, D.: Enterprise Architecture A to Z: frameworks, business process modeling, SOA, and infrastructure technology. CRC Press, Auerbach Publications, Boca Raton (2008)

Policy Design: A New Area of Design Research and Practice

Jeffrey Johnson and Matthew Cook

Abstract. Policy design is a new area of inquiry that takes the methods and traditions of design into the world of social, economic and environmental policy. Even though they may not know it, policy makers are designing future worlds and implementing these designs in the hope of realising their visions of the future. However, the methods of design are different to the methods generally used in the formation and execution of policy. In design requirements coevolve with the generation and evaluation of new systems. In policy some requirements may be ideologically fixed and pre-empt good overall solutions to. Assuming that policy design is indeed an important new area of design there are implications and opportunities for the design community. Since most policy makers have little formal knowledge of design, in the short term designers must engage in policy if policy-as-design is to be formulated in a designerly way. At the same time there is a need to educate policy makers in the theory and practice of design. The combination of research, applications, computer aided policy design, and design education in policy design creates great opportunities for the design community. When policy makers address their policy design task as designers, we can expect better policies with better outcomes.

1 Introduction

At its most extreme design is the creation of new things from nothing, from a blank sheet of paper to a blueprint; from a blueprint to a working system and its maintenance. Thus design has three phases: (i) establishing requirements, and the generation, evaluation and selection of hypothetical systems to satisfy those

Jeffrey Johnson · Matthew Cook
Design-Complexity Group
The Open University
Milton Keynes, UK
e-mail: jeff.johnson@open.ac.uk

M. Aiguier et al. (eds.), *Complex Systems Design & Management 2013*,
DOI: 10.1007/978-3-319-02812-5_4, © Springer International Publishing Switzerland 2014

requirements; (ii) implementation of a selected design as a real system; and (iii) maintenance of designed systems into the future. Each of these phases is characterised by interacting cycles of activity that can make them complicated and unpredictable, requiring human judgment and decisions.

Design has many dimensions that can make it complex (Alexiou *et al*, 2010). Many designed systems are themselves complex, as exemplified by cities, the internet, and the financial instruments that recently destabilised the world economy. The environment of design is complex, where this includes markets, fashion, regulation, standards, and dealing with clients who do not know what they want or what is possible. Design processes are complex where these include manufacturing or construction, supply chains, and managing the transition from blueprint to working system. Finally, design itself is a complex cognitive social process.

Designers manage this complexity extraordinarily well. Although design is often taught within particular application domains such as textiles, architecture, urban planning, electronic engineering, mechanical and electrical engineering the process of design is the same for them all.

We live in a world replete with designed objects and systems. Most of the time we are surrounded by many thousands of objects, each of which has been designed. These include buildings and the rooms we occupy, the furniture in them, the clothes we wear, the documents we read, the many small personal objects such as watches and mobile phones, and so on. The *clipboard challenge* of design involves asking someone to write down as many different designed objects that they can see. Try it yourself now. You will stop long before you have written down all of them because there are too many. Almost all these objects will have been created by professional designers, and their existence and success is due to the application of design methods.

Our thesis is that, although it not usually considered to be so, *policy* is a design domain and those making policy would benefit from education and training on the nature of design and the practicalities of implementation. In other words there is need to recognise a new discipline of *policy design*. This is argued by considering models of design in Section 2. Although there are many variants, for all but the simplest systems it can be argued that design *always* involves identifying requirements for new systems, *always* involves a cycle of generation and evaluation of possible new systems, and *always* involves revisiting the requirements as the system being designed becomes better understood during the process. As explained in Section 3, changing the requirements underlies a coevolution between what designers think is wanted and what designers think is possible. In this context Section 4 introduces policy and gives the example of designing policy to care for an aging population. Section 5 considers the relationship between design, policy and politics where there are clear differences and similarities between the problem-solution coevolution process of design and of policy formulation, *e.g.* a difference being possible resistance to reformulating policy requirements when this clashes with ideology. Section 6 suggests that that there will be a new Computer Aided Policy Design in the context of *policy* informatics. Section 7 takes a didactic approach to our proposition that policy is a new area of design research and practice, and presents a formal argument in favour. Section 8 gives our conclusions, which include the need for designers to engage in policy, and that great opportunities will be created by the new field of policy design.

2   Models of Design

Although there are many variants (see for example Cross, 1985), the simplest model of design involves the identification of *needs* or *requirements*, and the *generation* and *evaluation* of alternative ways of satisfying those requirements. Usually there are many dimensions for judging designs, with no overall optimum for them all. This requires the design problem to be *satisficed* by suboptimum trade-offs between the judgment criteria (Simon, 1969). These could include, for example, the processes available to make the object, costs, physical feasibility, social dimensions, and so on.

Many designed systems are hierarchical, conceived as collections of subsystems that work together to make the whole. For example, a jacket has a front, back, sleeves, buttons, while an aeroplane has mechanical, electrical, computer, seating, and many other subsystems.

In this context the design process involves an abstract concept of 'the new object or system' at the highest level of representation, and more tangible component objects existing at lower levels of representation. In the simplest case the designer devises new ways of combining pre-existing components to make new artefacts. More generally, all the parts required to design a system do not exist *a priori* and some have to be designed as subsystems. The specification for these subsystems comes from abstractions at higher levels and implicit or explicit hypotheses that "if a new component existed with a given specification, then when assembled with existing components in the way specified by the designer, the whole will have the desired emergent behaviours". Such hypotheses are effectively *predictions* or *forecasts*. In engineering such predictions are based on scientific principles and numerical calculation. In systems such as fashion predictions of fabric movement dynamics or market success of are made on the basis of more qualitative principles and calculations.

When the design process begins for a completely new object or system there are many uncertainties and unknowns. Design involves making explicit what was previously implicit as the system being designed becomes better formulated and understood. Sometimes this involves accumulating existing knowledge and sometimes it involves creating new knowledge relevant to the project in hand. For example, a candidate design may involve using materials in a way never tried before, and this may involve laboratory tests. Similarly, a candidate design may assume unknown user preferences that require empirical user research.

Seen this way, the generate-evaluate-generate cycle can be viewed as a helix through time, with each generate-evaluate iteration contributing new knowledge on which to base subsequent iterations.

The design of multilevel systems is characterised by the top-down questions of "what might be the conceptual system components, how might they fit together and what might be the emergent behaviour of the system" and the bottom-up question "if these tangible components are assembled in a given way will the new whole have the behaviour hypothesised top-down?" At some stage in the design process the abstract components hypothesised top down meet the tangible subsystems formed bottom up, and the higher level abstractions are instantiated. The result is a fully instantiated description of the new system, or blueprint.

During this top-down bottom-up design process assumptions are often made that turn out to be incorrect as the designer learns more about what they are designing, and the evaluation stage may reject an evolving design. This has costs for the unproductive work done, and to avoid them designers try to identify flawed assumptions as early as possible.

3   The Coevolution between What Designers Think Is Wanted and What Designers Think Is Possible

Design involves a form of problem-solving that is different from problem-solving in other areas. Very often the stated requirements for a new artefact or system are over-constrained with no solution or under-constrained with too many solutions. For example, the requirements for a new town house of having four bedrooms and costing less than €100,000 cannot be satisfied in most cities, while the requirements of having two bedrooms and costing less than €1,000,000 has too many options.

In most design projects the requirements are periodically revisited. Some requirements may be found to impose such severe constraints that a design cannot be found that satisfices them in an acceptable way, and one or more requirement must be relaxed or abandoned. But this changes the design problem, which means that design is not just the search for a solution to a given problem, but is also the search for a problem that has an acceptable solution. In other words, design is a process in which the requirements *coevolve* with the generation of possible ways of satisficing those requirements. Design is the coevolution between what the designer thinks is wanted and what the designer thinks is possible.

4   Policy Design

Policy involves creating a vision of the future and taking actions to make it into a reality. In this sense, policy *is designing the future*. More precisely, policy involves imagining new social, economic and environmental structures to make the world as it *ought* to be (Simon, 1969).

In democracies the *requirements* of the population are decided by political processes that give elected politicians the mandate and the money to make changes. Typically the requirements include social provision such as housing, employment, health and education and the way these *ought* to be depends on the values of the ruling politicians and their electorates.

City planning gives an example of policy-as-design for the built environment (Cook *et al*, 2013). Cities are systems that are constantly being designed but are never finished. The same applies to the social and economic systems that must function within this infrastructure.

As an example, consider social policies addressing the problem of caring for an ageing population in England: "For the first time, there are more people aged over 60 than children under 16 in the UK. ... The shift in proportion, composition and attitudes of the older age group has profound implications for public services. ... Those whose health has begun to fail also deserve to enjoy life as fully as possible and we need to find new ways to support them. ... but the response of public services is often limited. ... focused on a narrow range of intensive services that support the most vulnerable in times of crisis. ... We need a fundamental shift in the way we think about older people, from dependency and deficit towards independence and well-being. ... Interdependence is a central component of older people's well-being; to contribute to the life of the community and for that contribution to be valued and recognised. ... The challenge to respond to the needs and aspirations of a large and growing section of our community is not a marginal one. Much is straightforward and expectations are unexceptional. It is therefore all the more surprising that comprehensive, systematic approaches to older people are still relatively rare. In future, local councils and their partners should expect to be judged on their ability to build communities that support older people to live active, fulfilling lives." (UK Audit Commission, 2004).

The British concept of 'Care in the community' has a long history: "The 1989 community care White Paper marked a watershed in social work for adults in the UK. Its full title—Caring for People: Community Care in the Next Decade and Beyond (Department of Health, 1989)—signified the intention to set the direction of policy for many years. That this was ideologically driven is undisputed: the then Conservative government was determined to introduce the market into public services and the expanding world of social care seemed ripe for marketization. ... Nevertheless, there was broad agreement that significant change was needed. A series of policy reports throughout the 1970s and 1980s had pointed to failures in key aspects of the delivery of health and social care, and the escalating costs of residential and nursing home care were blamed for a soaring Social Security budget. ... Further change, ushered in by the 'new' Labour government from 1997 onwards, did not reverse the processes of the market economy of welfare but rather changed the message about what represented quality in service provision and the best ways to achieve this. The argument was that 'modernization'—in the shape of user-centred, 'joined-up' services—was needed if the system was to be 'fit for purpose' to meet the health and social care needs of the twenty-first century (Department of Health, 1998)." (Holloway and Lymbery, 2007).

Today the problem remains that old people are admitted into hospital due to illness or injury, and continue to occupy those hospital beds while they are recovering or after they have recovered. No system has yet been designed and implemented delivering 'joined up' care from a combination of providers including the National Health Services, local welfare services, and members of the community including family and friends, and volunteers. This last group fall under the Conservative Party's *Big Society* initiative: "We are helping people to come together to improve their own lives. *The Big Society is about putting more power in people's hands* - a massive transfer of power from Whitehall [UK Central Government] to local communities. We want to see people encouraged and enabled to play a more active role in society." (Conservative Party, 2013).

How might a design perspective deliver an affordable and effective system of care for elderly people? First we note the conflicting requirements of providing high quality personalised care at a bearable cost. In this paper the focus will be on the design of new systems. In this a designer would take a *user-centred* approach in which there are no 'average' users, and the users of the system include all those involved including professional staff and unpaid carers and helpers. The term *client* will be used to distinguish those who receive the care from those who provide it. A major classification can be made between those clients who are mobile and those who are not, and clients who have clinical health issues that require medical treatment. The unpaid carers and helpers can also be classified as, for example, spouses and partners, adult children, relations, healthy or unhealthy, own transportation, and so on.

The professionals in the system being designed will have their own chains of command, with some reporting to clinical departments, some reporting to welfare departments, and some reporting to other departments or agencies. These professionals will work together in formal and informal teams and the designer must think through their dynamics. Formal teams may be easier to define but the dynamics of emergent self-organising structures can be very important in systems with unpredictable behaviours. These team structures need to be designed in ways that do not disrupt the *a priori* internal structure of departments and agencies. E.g. ratification of a decision made by a three-person multi-agency team may require three phone calls to the respective superiors, with a high probability of delay due to one or more superiors not being immediately available.

Apart from people, the system being designed will involve locations and equipment. Locations typically include hospitals, nursing homes, and the client's own home, and equipment can range from something simple such as a handrail to complicated things such as a stair lift or a device for getting a person out of bed.

Even for the individual there are many parts to a support system to keep a client at home. Let the collection of relevant parts be written as lists enclosed by angular brackets, for example, <bedroom, bathroom, kitchen>. The system is multilevel, *e.g.* with the bathroom designed as a configuration of <bath/shower, hoist, chair, WC, sink, etc.>. Clearly it is important that the bathroom is well designed for the individual client from an architectural perspective, and it is important that the rooms form a well-designed unit to facilitate safe and comfortable movement. However, robust social structures also have to be designed.

Generally social structures are combinations of people, spaces and equipment, *e.g.* <bed, client, nurse, water, towels, etc.> for a 'blanket bath' in bed. Importantly, if any part of the structure is missing the system breaks down. In social systems the most crucial parts of the structure involve people. For example, if the nurse does not arrive at the house and the helpers present do not have the necessary nursing skills, the client cannot have the blanket bath. System breakdown can be more severe when more people are involved. For example, consider a case review involving the structure <client, spouse, nurse, doctor, equipment specialist> where the equipment specialist fails to arrive. Then valuable resource is lost in the time of the nurse and doctor, possibly causing knock-on failures for other clients. Even though this appears relatively simple at the level of one client, there is the possibility of cascades of failure causing stress and frustration for the professionals, the clients, and their carers.

Good design involves recognising constraints and producing solutions that are robust to component failure. As just discussed, the care system is dynamic involving many combinations of people and things through time. Inevitably parts of the system will fail, for example the nurse may get stuck in a traffic jam, a carer may be taken ill, or a piece of equipment may be faulty. A well designed system will anticipate these failures and have remedial actions to minimise the overall system disruption and maintain delivery of services.

For example suppose the programme of care for an individual involves the combination <client, nurse, physiotherapist, carer>, where this is planned to be instantiated as <Ann, Tom, Gill, Bill; team> but Tom is detained at an emergency with his previous client. Then if information is communicated efficiently and another nurse, Maria, is available, the team can be reconfigured as <Ann, Maria, Gill, Bill; team>. Scheduling the allocation of resources is a well known problem in the design of complex systems, and is increasingly approached through agent-based modelling when the resources are heterogeneous and numerous. This is discussed further in Section 6.

A common problem in social systems is that they are not 'joined up' so that responsibility and authority can be ambiguous and the necessary combinations are not formed. At a higher level of aggregation care at home has to fit into a well-designed administrative structure that makes the connections at all levels. In particular the system has to have sensors to detect component failure, and it has to be designed to respond to component failure. Where responsibilities are combined between units, the way the system functions and copes with failure has to be codesigned between those units.

5   Design, Policy and Politics

This paper argues that policy design is a new domain of application for design. In a sense this is obvious. Policy involves the creation of *artificial systems* in the sense of Herbert Simon (1969). By definition, artificial systems are designed. The problem with policy as formulated and implemented today is that its practitioners mostly have no education in the theory and practice of design and do not reflect on the systems they create from a design perspective (Schön, 1983).

Of course some policy makers understand very well the part that design can play in policy. For example, the British Member of Parliament, Barry Sheerman, co-chair of the Associate Parliamentary Design & Innovation Group (APDIG) and a member of the Design Council, writes that "Too many people still think that good design means a beautiful table or chair or a new piece of architecture, such as the Shard. There is a whole body of expert design capacity in this country that could help design services, particularly public services... good design, as shown in a new publication from the Design Commission, could help recovery in this country" (Sheerman, 2013). Sheerman is talking about design for policy, which is different to policy for design as illustrated by the European Commission's report on *Design for Growth & Prosperity* (EC, 2012).

Policy design goes beyond policy for design. Policy can be about *anything* and policy design can be about the design of any system. A more subtle distinction can be made about the design of the policy, as opposed to the design of the system which is the subject of the policy.

For politicians ideology and policy may be the same thing, for example "the rich ought to be taxed higher/lower, and this is the policy". Such an approach is not holistic and may overlook the way different policies interact to give the emergent behaviour of the whole, and miss creative design solutions that benefit all stakeholder users. In contrast designers know that requirements may change. Furthermore they know that the process of creating new systems to fulfil evolving needs takes time.

Not all policy is ideological, and often policy makers are looking for the best way to design systems to give the outcomes they and their electorate want. Then policy is usually conducted as *narrative*, or stories about the way individuals and societies work. These narratives form the theoretical basis on which to design social systems and to *predict, forecast*, or *anticipate* the outcome of policy interventions.

A practical understanding that policy makers can take from design is that, when faced with a design problem, it is very rare that the eventual design solution is found early in the process. Designers expect to generate many possible solutions and to evaluate those solutions critically, rejecting many or most of them. Furthermore when designs are implemented it may be discovered that some of the underlying assumptions were incorrect. In this case the design may be modified to accommodate the new knowledge. When policy is seen as design, it is more natural to change the underlying assumptions, even when this goes against ideology.

Experience shows that during implementation some of the assumptions underlying a design were incorrect. Macho politics may inhibit policy makers from admitting such errors, and thereby deny them the possibility of correcting them and designing better policies.

6   Computer Aided Policy Design and Policy Informatics

Computer Aided Design (CAD) has had a major impact across the design domains over the last four decades. In particular CAD allows the dynamics of new systems to be analysed in detail before they are fabricated, supports costs analyses, and facilitates communication by allowing specialists and non-specialists to view graphical representations of systems.

CAD is today essential in the design of mechanical and electronic systems, in architecture and the design of environmental systems, in textile design and manufacture, and many other areas. As Policy Design becomes better understood it too will benefit from the creation of bespoke CAD support. This already happens on a day to day basis in land use and transportation planning through the use of Geographic Information Systems (GIS) and computer simulation. It is certain that new computer-based support, or *policy informatics*, will emerge for policy design.

To illustrate this consider again support for the elderly. It was required to be robust in the face of disruption through the failure of components and subsystems at various levels. Modern telephony makes detecting such failures much easier. For example, mobile phones can act as sensors with apps that report their location automatically. An information system can know that Tom is not where he needs to be in order to join the team <Ann, Tom, Gill, Bill; team> planned to provide a service at a given time, even if Tom is too busy coping with an emergency to phone in and report it. With this information a system could be designed to locate another nurse, Maria, and reconfigure an alternative structure <Ann, Maria, Gill, Bill; team>. This is just one aspect of the Big Data revolution that will enable the design of new kinds of organisation of socio-technical systems.

The design of many systems involves predictions of their behaviour. Prediction in social systems is different to prediction of physical systems. For example, Finite Element Analysis allows precise predictions of the dynamical behaviour of physical systems. Point predictions are usually not possible in social systems, *i.e.* it is rarely possible to say with certainty that a social system will be in a particular state at a particular time.

The models that underlie prediction or forecasting in social systems are often expressed as narratives rather than mathematical formulae. This is analogous to areas of design such as fashion, interior design, graphic design and even golf course design, where there are rigorous principles underlying the narrative. An outstanding challenge in the science of social systems is the formulation of narrative models that can be implemented within computers, and this is an important area of research for computer aided policy design.

To some extent agent based modelling and computer simulation implement narratives of social interaction and investigate the emergent behaviour of many heterogeneous interacting agents. Such simulations often give unexpected outcomes for given inputs, and in this respect computer simulation is one of the only ways that may be able to forecast the unknown unknowns. In this respect agent based simulation can be seen as a policy analogue to finite element modelling of physical systems.

7 Proposition: Policy Is Design, and Policy Is a New Area of Design Research and Practice

This paper asserts that policy design is a new area of design. To make this argument explicit we reason as follows:

> Thesis: it is true that policy involves design
> Anti-thesis: it is false that policy involves design
> Synthesis: reject anti-thesis: is it is true that policy involves design
> Corollary: policy design is a (new) area of design research and practice

For the thesis we argue that policy involves (i) the identification of requirements, (ii) the generation of new systems to satisfice those requirements, (iii) the evaluation of the new systems, (iv) when designs are rejected cycling back to generate new systems, and (v) compromise between interest groups and stakeholders that involves changing the requirements when acceptable satisficing solutions cannot be found. These are all the characteristics of design.

Against the thesis it can be argued that policy (i) does not involve identifying requirements, which is clearly false – there is no need for policy if there are no unsatisficed requirements; (ii) does not involve the generation and (iii) evaluation of alternative systems, which is again clearly false; (iv) does not involve the generating new policies when others are rejected, which again is clearly false , and (v) does not involve compromise and changing the requirements, but this is central to the art of politics.

The synthesis rejects the anti-thesis leading to an emphatic conclusion that policy has all the elements of design, and is an area that involves design. Whether or not this conclusion is new is a matter of opinion. For more than four decades the journal *Planning and Design* has shown the natural relationship between architecture – an undisputed design discipline – and urban and regional planning where the physical environment is designed in the context of policy. Here at least, *policy design is an area of design research and practice.*

In contrast to saying that policy is the context of design, we say that *policy itself is the outcome of a design process*, and this is true for areas of policy not conventionally considered to be design. For example, financial instruments are designed, medical treatments are designed, housing allocation systems are designed, care in the community involves design, and so on. The design of these systems is currently not informed by design theory and practice, and we believe that the outcomes would be much better if they were. For example, in retrospect it can be seen the design of financial instruments was intended to benefit the banking system designers rather than the public who would normally be considered to be the users of these systems. As another example, in this paper we have sketched the possibility of care for old people being treated as a design problem.

That it can be contentious to suggest that policy at large is design suggests that this is a new direction for design research and the application of design thinking.

This supports our corollary that policy design is indeed a new area of design research and practice.

This conclusion presents an exciting challenge to the design community. Policy design is a new area that currently engages few design theorists or practitioners outside the area of environmental planning. Most policy makers come from intellectual traditions that do not embrace design and do not make policy using the methods of design. This suggests the possibility of a proactive programme to take the theory and knowledge of design into the policy making realm. In one or two decades it may be common to hear policy makers in town halls and ministries discussing the formation of policy in design terms. We believe that this will result in policies that are better designed and more fit for purpose than the failed policies we see in many areas of social, economic and environmental policy today. This is an important opportunity for the discipline of design play a leading role in the design of the future.

8   Conclusions

In this paper we have argued that policy design is a new area of inquiry that takes the methods and traditions of design into the world of social, economic and environmental policy.

Although there are many variants we have given a characterisation of design that always involves the identification of requirements and cycles of generation and evaluation of ways of satisficing those requirements. Furthermore it is common to revisit and change the requirements during the design process. This means that design is a coevolutionary process between what the designer thinks is wanted and what the designer thinks is possible.

Even though most of those involved do not know it, policy makers are designing future worlds and implementing their policies in the hope of realising their visions of the future. The methods of design are different to the methods generally used in the formation and execution of policy. In design requirements coevolve with the generation and evaluation of new systems to satisfy the requirements. In policy some requirements may be ideologically fixed and pre-empt good overall solutions to societal problems.

Designers have led other disciplines in the application of computers to solve real problems and a new era of Computer-Aided Policy Design is already emerging under the heading of policy informatics with important opportunities for design.

Assuming that our thesis is correct, and that policy design is indeed an important new area of design there are implications and opportunities for the design community. Since most policy makers have little formal knowledge of design, in the short term designers must engage in policy if policy-as-design is to be formulated in a designerly way. At the same time there is need to educate policy makers in the theory and practice of design. The combination of research, applications and education in policy design presents great opportunities for the international design community.

When policy makers address their policy design task as designers, we can expect that better policies will be created with better outcomes.

References

Alexiou, A., Johnson, J., Zamenopoulos, T.: Embracing complexity in design. Routledge, Oxford (2010)

Conservative Party, 'Big Society',
http://www.conservatives.com/Policy/Where_we_stand/
Big_Society.aspx (viewed June 30, 2013),
http://archive.audit-commission.gov.uk/auditcommission/
sitecollectiondocuments/

Cook, M., Johnson, J., Rauws, W.S.: Towards an evolutionary approach to understanding the role of design in collaborative planning. In: European Conference on Complex Systems, Brussels (September 2012)

Cross, N.: Engineering design methods: strategies for product design. Wiley and Sons, Chichester (1998)

European Commission, Design for Growth & Prosperity Report and Recommendations of the European Design Leadership Board, DG Enterprise and Industry of the European Commission. Edited by Michael Thomson, Design Connect and Tapio Koskinen, Aalto University Unigrafia, Helsinki (2012)

Holloway, M., Lymbery, M.: Editorial - Caring for People: Social Work with Adults in the Next Decade and Beyond. British Journal of Social Work 37, 375–386 (2007)

Johnson, J.H.: Science and policy in designing complex futures. Futures 40, 520–536 (2008)

Schön, D.: The Reflective Practitioner: How professionals think in action. Temple Smith, London (1983)

Sheerman, B., http://www.policyconnect.org.uk/apdig/people/
barry-sheerman-mp

Simon, H.: The sciences of the artificial. MIT Press, Cambridge (1969)

UK Audit Commission, Older people – independence and well-being: The challenge for public services, Audit Commission, 1 Vincent Square, London SW1P 2PN (2004),
http://archive.audit-commission.gov.uk/
auditcommission/sitecollectiondocuments/
AuditCommissionReports/NationalStudies/
OlderPeople_overarch.pdf

UK Department of Health, Modernising Social Services, Cm 4149, London, HMSO (1998)

UK Department of Health, Caring for People: Community Care in the Next decade and Beyond, Cm 849, London, HMSO (1989)

A Virtual Web Net to Eco-manage Food Packaging Waste

Clara Ceppa and Gian Paolo Marino

Abstract. Latest data published by Eurostat (the statistical office of the EU) showed that, in Europe, 80 million tons of packaging waste were produced in 2008. We can also notice that Italy alone produces about 8 tons of packaging waste per year arising from the food industry. This paper deals with the problem of waste deriving from packaging produced in pre- and post-production phases of food sector. Through the creation of a "web-based software" which is specifically designed on the basis of the methodological concepts of Systemic Design, whose fundamental principle is that waste (output) of a process becomes resource (input) to another one, the waste of pre- and post-production phases can be managed and reused by other industries. Through the capabilities of the portal, users can thus minimize costs of waste disposal by selling rejects on the Net, use more resources already existing on the spot, improve logistics, expand their market, strengthen ties with other companies and boost the whole local economy.

Keywords: web net, food-pack waste, company virtual network, software.

1 Introduction

Human behaviour is currently not guided by a sense of belonging but by an attempt to dominate Nature for the sake of economic growth and development. As pointed out by Janet Abramovitz of the Worldwatch Institute: "Many ecosystems have been worn out to the point of losing their capacity to recover and are no longer able to withstand natural perturbations. The key to reach an operative definition of ecological sustainability are found in understanding the fact that sustainability does not refer to a state of immobility but a dynamic process of co-evolution.

Clara Ceppa · Gian Paolo Marino
Politecnico di Torino, DAD - Department of Architecture and Design, Italy
e-mail: {clara.ceppa,gianpaolo.marino}@polito.it

M. Aiguier et al. (eds.), *Complex Systems Design & Management 2013*,   63
DOI: 10.1007/978-3-319-02812-5_5, © Springer International Publishing Switzerland 2014

We need to make a relationship between humans and Nature that is profitable; it is necessary to create an eco-compatible society based on a lifecycle of products (from farming to package, from logistic to sales, from consumption to disposal) that is consistent with the environmental needs and equipped with a socioeconomic apparatus capable of responding to human needs while consuming few resources" [1].

If we stop and think that over 90% of the water used in a brewery does not end up in the bottle, and over 20% of the grain after threshing is buried [2], we can understand how dramatically urgent it is to do more with the resources that we have in hand.

If we look the pre-production phases, we could realize what impact company's habits of consumption have on the environment.

How much is the amount of waste arising from packaging? From a probe, made by Coldiretti (Italian association of farmers), is emerged that 12 million tons of waste from packaging were produced in Italy per year.

The food industry, with over 2/3 of these 12 million tons, is the most responsible for the production of packaging waste.

These data are confirmed also at a European level. On the basis of latest data published by Eurostat (the statistical office of the EU), Europe, in 2008, produced 80 million tons of packaging waste.

Nowadays, for the health of the environment and human being, a total rethinking of manufacturing processes and waste management made by companies becomes fundamental.

This paper does not address the issue of proper disposal of waste by consumers; on the contrary, it studies the problem of the waste that occurs before the products distribution phase, since it has not been sufficiently addressed enough. The focus, therefore, pertains to pre-production phase in the food packaging industry. Currently, rejects generated by that sector represent a large amount of waste upstream of the primary production cycle.

2   How to Eco-manage Food-Pack Waste: The Systemic Design Methodology

Currently we are forced to spend huge amounts of money to treat and dispose waste. The Systemic Design methodology proposes that it is possible to add value to the discarded materials and consider them as new ones that could be used again. In this way we can eliminate waste disposal costs and create a network for selling the outputs to other companies as raw materials.

To do this, we need to examine the current production lines not by breaking them down into single independent sequential phases that are separate one from each other, but by studying the complexity of the whole of the parts according to the studies developed by the ecologist brothers Eugene P. and Howard T. Odum [3]. These men were among the first to conduct analyses on energy and the transfer of materials in complex natural ecosystems. These ecosystems were also observed by English botanist Arthur Tansley as if they were an open-loop system that exchanges material and energy with other ecosystems. To understand the complexity of a system made up of relations between various players and a heart of energy and material flows, we are helped by examining "The System" par excellence: the Nature. In fact, in nature there is no concept of waste because what one species eliminates is what another species uses for its own nourishment [4]. Even surpluses are metabolized by the system itself. If these conditions, essential to any living system, are transferred to the world of production by applying the first principle of Systemic Design according to which the waste (output) of one productive system can be used as a resource (input) for another one, we will be promoting the type of production that moves towards zero emissions [5]. In such a scenario all the actors involved in the production chain will start to reason by connections.

Facing the incapacity to introduce new techniques for managing the problem of waste, we are still using dumps. Nonetheless these should be considered as a transitory and temporary solution. Therefore systemic methodology proposes a new approach that stimulates people and companies to reduce all forms of waste and helps appraising the remaining outputs by giving them a new economic and legislative value. This way not only the so-called waste products are elevated to a status of materials worthy of proper, controlled and more sustainable management, but they can "move" within the production chain with new positivity and dignity.

The perspective of Systems Design [6] challenges current industrial setups, emancipates us from a consumerist vision based exclusively on the product and proposes a new production methodology.

This generates greater profits and creates jobs and wealth in the community by spawning new entrepreneurial initiatives, developing businesses and improving the already existing businesses.

This process can be applied to any production sector. It is deliberately being proposed locally not only to appreciate the territory potentialities and specificities and reinforce the connection with tradition, but also to avoid useless transportation costs and the air pollution created by it.

2.1 Case Study (First Step): Definition of the Methodological Approach and Operative Phases to Design and Develop a Website to Eco-manage Waste from Food Packaging

This paper presents the definition, design and realization of a complex website for processing information based on evolved data-base systems that can acquire, catalogue and organize information related to the productive activities in the area of study (Piedmont, Italy), the outputs produced and the inputs required as resources.

Main data were acquired and organized in terms of quantity/type/quality of waste and geographical location of companies on the territory. All the data were all correlated by the Systemic Design logic.

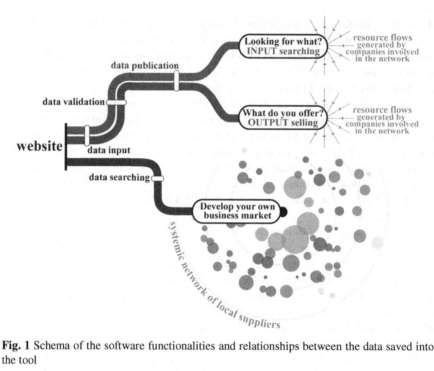

Fig. 1 Schema of the software functionalities and relationships between the data saved into the tool

The huge amount of data obtained by using this website will be then a precious asset and a vital platform for local administrators and, obviously, for entrepreneurs. The last mentioned actors will then be able to work in a more sustainable way.

The main functions of this system were defined as follows [7]:

- "What do you offer?";
- "Looking for what?";
- "Develop your own business market".

"What do you offer?": the website will tell users which companies produce outputs that they can use as resources;

"Looking for what?": with this option producers of waste will be able to determine which local companies could use their outputs as resources in their production process.

"Develop your own business market": it is a function that allows companies to visually capture their real opportunities for new connections with other users. The difficult understanding of a bulleted list is solved through the use of an infographics schema of immediate comprehension and a map, such as the one below, where:

- the circles (black and coloured) represent the companies involved in the network;

- each colour symbolizes the different categories of resources (goods, waste, energy surpluses, ...);
- the placement of circles, within different radius, depicts the geographic distance between the company that is using the website (black circle) and the other ones (coloured circles);
- the size of each circle represents the amount of resources that the company using the network (black circle) exchanges with companies with which it is related to (coloured circles).

This is a simple and intuitive way to understand if the company can become a supplier of another activity or if it can reduce its supply-chain radius.

By using the geo-localization function, the system can ascertain which local activities are situated within the range of action defined according to the search criteria in the website.

Then it positions them exactly on a geographical map to show the users which companies can be part of the network of enterprises which could enter into a reciprocal relationship so to increase their own business and maximize their earnings through the sale of their outputs and the acquisition of raw materials from other local production activities.

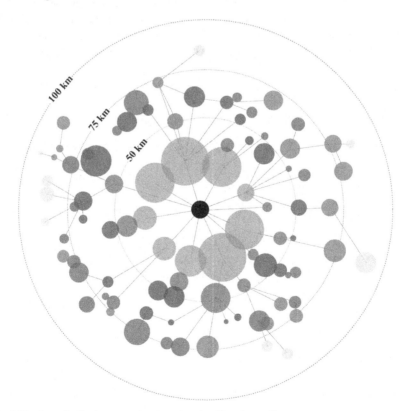

Fig. 2 The hypothetical new systemic network of local suppliers

The consultation of the system has been expressly designed by following the systemic approach and made usable by means of Web 2.0 technologies; this approach has made possible the future publication of the interactive web portal as a facility that can be used by operators who want to consult it and interact with it. We chose to use Web 2.0 technologies because they enhance the role of each single user and, consequently, the social dimension of the network: Tim O'Reilly argues that the value of the network does not lie in technology but in content and services, and that the strength of the network is represented mainly by its users. The enhancement of the social dimension of the network, through instruments capable of facilitating interaction between individuals and the user transformation into active creators of these services, allows that the quality of the offered services improves as the number of users involved in their use [8].

The website system has been designed to provide an informative service dedicated to two types of operators for whom different functions and typologies of access have been developed. These operators are represented by companies seeking to optimize their production processes through the exchange of resources with other entrepreneurs, and by both researchers and public bodies who can analyze the material flow in the area and the distribution of productions so to facilitate the development of the territory and of the local economy as well. This website presents two distinct interfaces connected to a single database capable, through a set of features of low level and processing algorithms, to contextualize the information about a specific territory with the requests of the operators.

Fig. 3 Cloud of research key-words

2.2   Case Study (Second Step): How to Solve the Waste Problem of an Italian Artisanal Chocolate Workshop by Using the Systemic Website

The case study will consider the problem of waste management that occurs earlier in the distribution of finished products, that is to a level BtoB (Business to Business) of edible goods and their associated packaging.

Raw materials (which are necessary to obtain finished products) and packaging (useful for their retail) reach, in turn, industrial plants and workshops wrapped in additional packaging.

Currently, these latter represent a large amount of waste, upstream of the primary production cycle. Through a systemic analysis, however, waste may instead represent an economic resource to be used in both the industry and energy sectors, thereby triggering new business with low environmental impact.

The specific case study refers to an Italian artisanal chocolate workshop located in North-West of Italy (company size: 5 employees; turnover: $\leq \text{€}$ 1 million) and shows how it can be possible to successfully use (under an economic, environmental and, last but not least, social point of view) all those packages that would not be usable otherwise thanks to the potentialities outlined by the systemic website which has been developed on the basis of what emerged during the "first step" of this analysis.

In specific, currently, food raw materials (e.g. cocoa beans, spices, etc.) come to the workshop packed in sacks made of different materials and sizes; once emptied, they are collected separately.

The workshop produces the following amount of waste per year:

- 66 jute sacks = 43 kg;
- 160 Nylon sacks (polyamide) = 10 kg;
- 350 LDPE sacks (low-density polyethylene) = 10.5 kg;
- 350 PAP sacks (cardboard) = 93 kg.

Those packaging which are intended to protect finished products (e.g. tapes, boxes, etc.), also come, in turn, wrapped in other packages.

Thus, every year the workshop produces the following amount of waste:

- PAP secondary packaging cartons (cardboard) = 1,400 kg;
- PAP reels (cardboard) = 8 kg;
- PP reels (polypropylene) = 4 kg;
- HDPE supports for sticky labels (high-density polyethylene) = 18 kg;
- CA supports for sticky labels (paper based polylaminate) = 9 kg.

However, using the proposed website, problems described above can be faced and improved thanks to the basic functions previously detected at the very beginning of the research: "Looking for what?", "What do you offer?" and "Develop your own business market". For this purpose, new flows of matter were outlined in order to allocate the waste produced by the workshop at local companies, which can then reuse them as raw materials in their production processes. Thus, the 1,500 kg of paper and cardboard derived from boxes, supports of sticky labels, sacks and cores of reels can be valued in a company, located 59 km away, so to produce sound deadening boards and thermal insulation flakes cellulose for walls, roofs and attics to be used to improve the indoor comfort of buildings. With both recovered paper and cardboard, then, it could also be possible to produce insulators for electrical sector as well as cones and diaphragms for loudspeakers: that would either imply the transportation of thrown away paper material to companies currently located at a distance less than 250 km away from the workshop, or boosting new companies that are sited within shorter distances in order to contain costs and emissions.

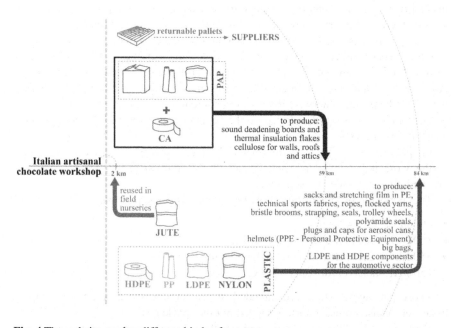

Fig. 4 The website results: different kinds of output reuse

3   Conclusions

Through the creation of this "web-based software"[1] (therefore usable without installation and available in any location connected to the Internet), designed on the basis of the methodological concepts of Systemic Design, the pre-production waste can be reused by multiple industries.

---

[1] The website has been developed thanks to the work done during the master thesis of Fabio Conte; it is currently being tested.

Thanks to this virtual platform companies will have the opportunity to interact and discuss with each other so to create new networks in which local producers and neighbouring suppliers will work together to achieve immediate and tangible economic and environmental benefits. Thanks to the features of the platform, registered users can:

- minimize costs of waste disposal, selling them on the Net;
- use more local resources;
- improve their logistics;
- expand their market;
- strengthen ties with other companies registered to site;
- revitalize the local economy.

The actual benefits, that could be immediately obtained, are:

- less use of raw materials;
- increased use of resources already on site;
- a greater traceability of the resources used in industrial processes so as to ensure, by implication, a high quality finished product.

The design of a virtual network like the one proposed entails other benefits:

- viability of companies that can freely interface and discuss among themselves by creating a virtuous circle, economically and environmentally profitable at a local level;
- reduction of CO_2. Actually its production is caused by a supply of raw materials and semi-finished products for packaging manufacturing which happens long away from the points of final packing;
- reduction in energy consumption due to unnecessary transport;
- encouraging the production of packaging made by natural materials and systems in relation to the real needs of production and storage of food and to existing local resources;
- drastic reduction of secondary and tertiary packaging or constant re-use of them in according to the regulations for the protection of human health and organoleptic qualities of food;
- encouraging the creation of new small businesses in relation to input-output managing of the packing material, so to favour the creation of new jobs;
- optimized use of available resources (outputs) and reduction of packaging production and unnecessary surplus or under-used and poorly waste.

References

[1] Brown, L., et al.: State of the World. Worldwatch Institute Book, Washington, D.C. (2001)
[2] Pauli, G.: Breakthroughs-What business can offer society. Epsilon Press, Surrey (1996)

[3] Chelazzi, G., Provini, A., Santini, G.: Ecologia dagli organismi agli ecosistemi. Casa Editrice Ambrosiana, Milano (2004)

[4] Capra, F.: The Science of Leonardo. Doubleday, New York (2007)

[5] Bistagnino, L.: Ispirati alla natura. In: Slowfood n. 34, Slow Food Editore: Bra (Cuneo) (June 2008)

[6] Bistagnino, L.: Design Sistemico/Systemic Design, 2nd edn. Edizioni Ambiente, Milano (2011)

[7] Conte, F.: Nuova applicazione web per lo sviluppo di un sistema di relazioni commerciali a livello territoriale. Il caso studio dell'Atelier del Cioccolato di Guido Castagna. Master thesis, Politecnico di Torino, DAD-Department of Architecture and Design, Torino (2011)

[8] Rasetti, A.: Il filmato interattivo. Sperimentazioni. Time & Mind Press, Torino (2008)

Open Architecture for Naval Combat Direction System

Denis Janer and Chauk-Mean Proum

Abstract. France has adopted an "Open Architecture" approach for its Naval Combat Direction Systems (NCDS) in order to reduce the system's total cost of ownership, to improve the system flexibility and to ensure system interoperability with existing or future systems. DCNS has been contracted by DGA, French Ministry of Defence, to develop this approach by defining a global process to specify and qualify the openness level of its Naval CDS. This process is based on the modelling of a reference architecture using NAF views and the creation of a requirement repository for system and technical openness specification and qualification. This study aims to increase the standardization of Naval CDS, above all from system view point.

1 Introduction

France has adopted an "Open Architecture" approach for its Naval Combat Direction Systems (NCDS), the real-time part of the Combat Management System (CMS). Comforted by international state of the art and similar studies in other countries, French vision relies on an open architecture built on a modular architecture and the use of open standards.

Like for other nations, Open Architecture is deemed:

- to reduce the system's total cost of ownership
- to improve the system flexibility and evolutivity in order to face operational need evolutions and technology changes

Denis Janer · Chauk-Mean Proum
DCNS
BP 403 (Le Mourillon)
83055 Toulon cedex, France
e-mail: {denis.janer,chauk-mean.proum}@dcnsgroup.com

M. Aiguier et al. (eds.), *Complex Systems Design & Management 2013*,
DOI: 10.1007/978-3-319-02812-5_6, © Springer International Publishing Switzerland 2014

- to ensure the system maintainability
- to ensure system interoperability with existing or future system

DCNS has been contracted by DGA, French Ministry of Defence, to develop this approach based on the following tasks:

- Standards Registry definition. This first task aims at:
  - identifying relevant state of the art principles and patterns for architecting and designing a CDS
  - identifying relevant standards
  - defining openness categories or levels and the appropriate selection of principles / patterns and standards for each openness category
- Requirements Registry definition. This task covers:
  - the functional and technical decomposition of a NCDS
  - the requirements definition of interfaces with external systems: sensors, effectors, TDL (Tactical Data Link) and CSS (Command Support System)
  - the requirements definition of interfaces related with Force Level Capabilities
- Openness Qualification definition. This task includes:
  - an Assessment Process definition
  - a Qualification Process definition
  - a Qualification Tooling specification

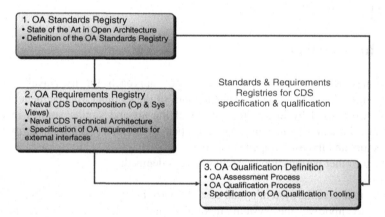

Fig. 1 OA Study for naval CDS

This paper will present the objectives of OA Standard registry and openness category definition approach and will focus on the work done on NCDS decomposition, external interface and qualification process. As a conclusion, it will expose the perspectives and continuation of this program.

2   Standard Registry and Openness Categories

Based on international OA state of the art and previous studies conducted by DGA and DCNS, the first step of the study was to identify:

- Architecture and design principles or patterns that should be recommended to contribute to architecture modularity aspects
- Standards and technologies that should be recommended for openness aspects

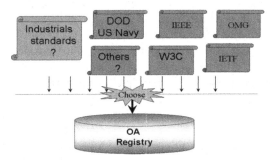

Fig. 2 OA Standard registry

These recommendations are not limited to a technological viewpoint. They address operational, system and technical viewpoints.

Indeed, the state of the art in terms of architecting and engineering reveals the need to consider the different viewpoints of a System of Interest as depicted in the following schema.

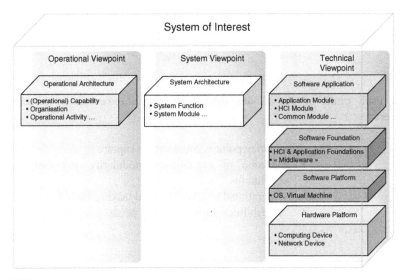

Fig. 3 Viewpoints

Yet, current practices in Open Architecture do not address those different viewpoints:

- Some approaches only address a single viewpoint. The US Navy OACE (Open Architecture Computing Environment) for example focuses only on the technical viewpoint (the HCI – Human Computer Interface part is not addressed either).
- Other approaches define qualities (e.g. openness, interoperability, modularity, extensibility, reusability, maintainability, composability but without their characterisation in those viewpoints.

The innovative approach taken by this study is to consider for each of the previous viewpoints, different openness categories or levels defined by a set of required architectural or engineering principles and patterns and open standards.

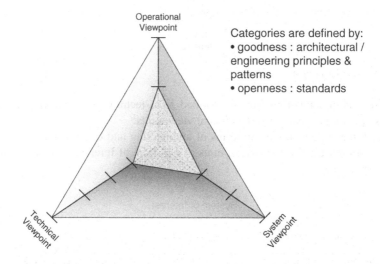

For each point of view, both modularity and standardization have been considered:

- Operational viewpoint is necessary to capture
  - Mission or capabilities modularity and their future evolutions
  - Operational organisation and tasking variants
  - Capabilities interoperability needs

- System View point enables the translation from operational architecture and its modularity into system functional decomposition and deployment
- Technical viewpoint enables system to be built according to a layered technical architecture where application layer relies on software foundation, ensuring the independence of application modules from technologies. This viewpoint is similar as OACE but adds a category related with HCI and plug and play.

Standards and technologies are considered not only for their contribution to system openness but also for their maturity and their diffusion. This analysis and these set of recommendations make possible to define Openness categories for each viewpoint depending on modularity and openness/standard use objectives.

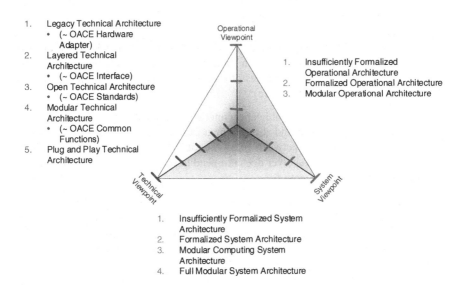

1. Legacy Technical Architecture
   - (~ OACE Hardware Adapter)
2. Layered Technical Architecture
   - (~ OACE Interface)
3. Open Technical Architecture
   - (~ OACE Standards)
4. Modular Technical Architecture
   - (~ OACE Common Functions)
5. Plug and Play Technical Architecture

Operational Viewpoint

1. Insufficiently Formalized Operational Architecture
2. Formalized Operational Architecture
3. Modular Operational Architecture

1. Insufficiently Formalized System Architecture
2. Formalized System Architecture
3. Modular Computing System Architecture
4. Full Modular System Architecture

Fig. 4 Openness categories definition

The following examples will illustrate some of the architecture / engineering patterns that have been selected for each viewpoint's category.

The current doctrine for military operations is expressed as "Command by Veto". This doctrine, also known as "Centralized planning and direction, decentralized execution" still applies for network centric warfare operations.

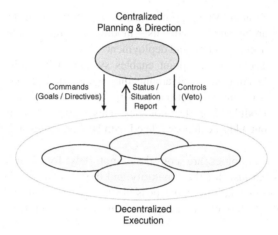

The doctrine highlights two different interaction patterns:

- the request / reply interaction for commands and controls between the commander and his subordinates. In such interaction, an acknowledgement, the reply, is required for each request to inform the requester of the result of the request (e.g. success or failure).

- the event notification / data publication interaction for status and situation reports between the subordinates and the commander. In such interaction, the event / data are transmitted between the sender and the receiver without acknowledgement. The receiver does not need to inform the sender of its activity upon the event / data reception.

For the highest openness category in the Operational Viewpoint (Category 3 – Modular Operational Architecture), the distinction and the formalisation of these two different patterns are required between (operational) activities.

These interaction patterns relate with the following architecture patterns in the System Viewpoint:

- SOA (Service Oriented Architecture) pattern where the interaction between participants (i.e. between service consumer and service provider) are done through request / reply. This pattern is particularly suitable for controlling a given participant (e.g. for resolving conflict).

- DOA - Data Oriented Architecture pattern where the interaction between participants is done through data publish / subscribe. This pattern is particularly suitable for exchanging / sharing sensor data between functions.

The distinction and the formalisation of these two architecture patterns are required at Category 3 – Modular Computing System Architecture in the System Viewpoint, and enables a seamless mapping from the Operational Viewpoint's Category 3.

Regarding open standard examples, DoDAF and NAF architecture frameworks have been selected for the description of both Operational and System viewpoints, middleware and real-time standards from the OMG (Object Management Group) like the DDS (Data Distribution Service) have been selected for the Technical viewpoint.

It's worth noting that this first part of the study is not specific to naval domain and could be reused in other domains for real-time systems similar as NCDS.

3   OA Requirements Registry

The second part of the study has been conducted to identify how and which elements of the standard registry have to be applied to a naval CDS. To do so, operational, system and technical architectures of a naval CDS have been considered. As a consequence, this part is more specific to the naval domain.

3.1   NCDS Modelling

A NCDS decomposition and modelling have been provided for these 3 viewpoints. Modelling has been made based on NAF formalism and a subset of NAF-views.

- Operational point of view : as previously seen for operational categories, the aim was to initiate a first operational modelling defining typical capabilities and operational activities (see figure 5). The perimeter of the study covers Anti Air Warfare, Anti Surface Warfare and Force Level capabilities (cooperative situation awareness, engagement on remote data).
- System point of view : functional decomposition is aiming especially at identifying openness supports related to external interfaces based on the system data flows and external interface definition (see figure 6).
- Technical point of view : the aims is to identify openness supports across the 4 layers model distinguishing different kind of application module (HCI, repository, gateway) and taking into account for each of them a typical technical decomposition capturing dependencies between technical modules across the 3 lowest layers (see figure 7).

The results of NCDS modelling is an identification of potential openness supports.

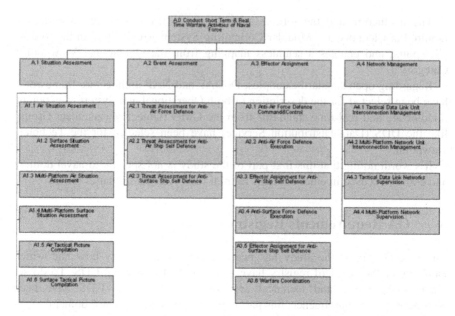

Fig. 5 Operational view

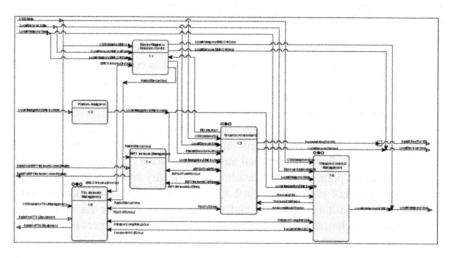

Fig. 6 System view

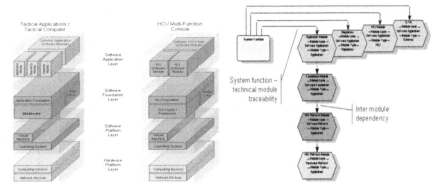

Fig. 7 Technical views

3.2   External Interfaces

The interfaces of a system form natural openness supports. This part of the study consisted in closely examining the flows exchanged between the functions of the NCDS reference architecture and the combat system equipment items in order to retain system exchanges presenting the most common features with respect to the diversity and variability of the equipment items.

Openness supports are then specified according to OA principles and forms an initial definition of unitary interface with a set of gateways that are designed to adapt these external interfaces to the specific nature of the equipment items. Gateway approach is fully compliant with gateway stereotype approach recommended for physical and software architecture, even despite interface normalization issues.

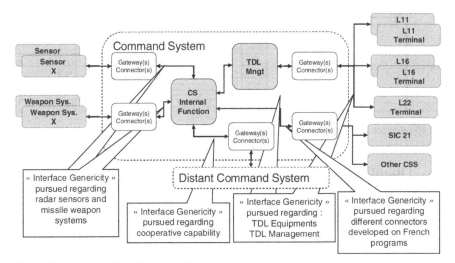

Fig. 8 CDS External Interfaces candidate for genericity

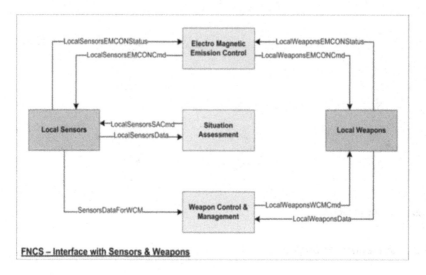

Fig. 9 Sensors and Weapons Interfaces Modelling

Each openness support is characterized by requirements specifying applicable principles and patterns and associated unitary interfaces list. Each unitary interface is characterised by the SOA or DOA pattern, static and dynamic definition and applicable principles and patterns.

Openness supports			C/S S		Interfaces (called OA)	DOA	SOA	Standards
S_501 Sensors and Weapons				I_0010	Stop/Start own ship emissions		x	
				I_0030	Manage emission and firing areas		x	
				I_0060	Receive digital video	x		
			x	I_0130	Perform kill assessment request		x	
	S_502 Sensors			I_0040	Receive sensor plots	x		
				I_0050	Receive sensor tracks	x		
				I_0080	Manage extraction zones		x	
			x	I_0100	Control sensor states and modes		x	
				I_0020	Manage radar frequency usage		x	
		S_503 Radars		I_0070	Receive jamming interference reports from radar	x		
			x	I_0090	Perform radar measurement request		x	
		S_504 Fire Control Syst.	x	I_0120	Command target acquisition		x	

Fig. 10 Interfaces Openness support example

4   Qualification Process

The purpose of the qualification process is to check and qualify the degree of openness of a CDS in relation to the requirements baseline previously defined. This qualification process can form part of a contractual context in which the customer defines, in its requirement specification, a CDS openness objective in line with one of the categories defined in the normative baseline. The customer then validates, through the CDS definition, the applicable openness supports and their specifications based on the requirements baseline.

The result of this process must enable the effective attainment of an openness category according to each operational, system and technical viewpoint to be verified and provide information concerning the work that would need to be implemented to attain the higher category if this has not been reached.

The figure 11 presents the sequence of the steps to estimate and check the requirements.

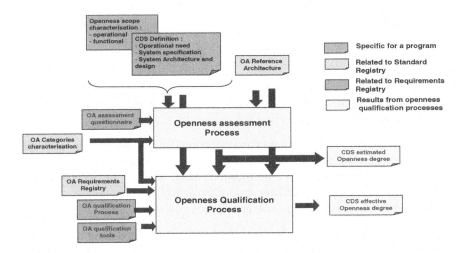

Fig. 11 Openness Qualification process

The prerequisite to run the qualification process is to estimate the degree of openness attainable through an estimation process questionnaire. The estimation step represents an initial qualification step as it is used to:

- Identify the category estimated to be attainable according to each viewpoint
- Check the evidence provided in the responses to the questionnaire

The qualification process itself is executed by implementing the steps to check the coverage of the applicable requirements.

The figure 12 presents an overview of the tool chain upon which the process is based.

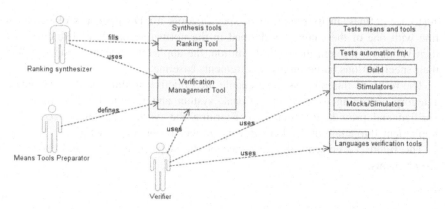

Fig. 12 Tools chain

5   Conclusion

DCNS and DGA have developed an innovative open architecture approach for Naval CDS development addressing a full scope:

- An openness characterisation and categorisation not only for the technical viewpoint but also for the operational and system viewpoints
- An Openness support specification based on set of recommended principles, patterns, standards and technologies
- An Openness assessment and qualification process

It provides a first reference Naval CDS architecture modelling addressing operational, system and technical point of view. The Openness categories definition could be applied in other domains and is not Naval specific.

This study has shown an insufficient standardization initiative from system view point. Except for Tactical Data Link, standardization remains mostly confined to transverse aspects like Alert Management. OA standard registry, CDS modelling and external interfaces requirements form a first analysis that should support and contribute to NCDS interface standardization.

Work is still going on to update standard registry with some emerging new principles or standards and to extend OA standard registry recommendations and related requirements to a wider operational and functional scope.

References

[1] IBM, Open Architecture Technical Principles and Guidelines (2008),
     https://acc.dau.mil/CommunityBrowser.aspx?id=398489
[2] Open Architecture Computing Environment (OACE) Design Guidance V1.0 (August 23,
     2004), http://www.everyspec.com/USN/NSWC/
     OACE_DSN_GUIDANCE_VER-1_11546/
[3] Open Architecture Computing Environment Technologies and Standards V1.0

Application of an MBSE Approach for the Integration of Multidisciplinary Design Processes

Nicolas Albarello and Hongman Kim

Abstract. In the current practices of system developments, there is a large gap between systems engineering activities and engineering analyses. In particular, there exists no practical link between architecture-level models, developed using languages such as SysML, and disciplinary models described in specialized tools. In order to close the gap, a capability was developed that integrates SysML models with analysis models. This work discusses application of the capability to the design of an aircraft propulsion system. System requirements and architecture were defined in a SysML model and analysis models were imported to the SysML model to set up parametric diagrams. Engineering analyses were performed from the SysML model to check requirements compliance of the system. When introduced to industry design processes, the technology can streamline requests for engineering analyses from system architecture models.

The paper will first introduce the integrated modeling and analysis capability and will then show an application of it on an industrial test case.

1   Introduction

In the development of large and complex systems, systems engineering approaches are used to manage system complexity and to ensure that the delivered system will meet all requirements. Model-based systems engineering (MBSE) is an

Nicolas Albarello
EADS Innovation Works, France
e-mail: nicolas.albarello@eads.net

Hongman Kim
Phoenix Integration, United States
e-mail: hkim@phoenix-int.com

M. Aiguier et al. (eds.), *Complex Systems Design & Management 2013*,
DOI: 10.1007/978-3-319-02812-5_7, © Springer International Publishing Switzerland 2014

approach to improve the traditional document-based systems engineering approach through the use of a system model [1]. System Modeling Language (SysML) [2], [3] is a graphical modeling language that was created as an international standard to support MBSE. To achieve the full benefit of MBSE, there is a critical need to create links between systems engineering and disciplinary/domain engineering.

Currently, system level analysis from within the SysML modeling tools is generally limited to the evaluation of simple parametric equations. This means that while SysML models are capable of describing a given system configuration with a high degree of detail, it is difficult to properly evaluate how well the design meets the requirements or to perform important trade-offs between performance, cost, and risk. The lack of an easily accessible analytical capability makes it difficult for systems engineers to quickly understand consequences of inevitable changes in requirements and system configuration and take necessary actions.

On the other hand, domain/disciplinary engineers (structural, thermal, electrical, software, cost, etc.) routinely use a wide variety of sophisticated analysis tools to analyze and design the system. Because these tools are not connected to the system model, it is difficult to use the system model to setup an analysis problem or to update the system model using analysis results. If this gap can be bridged, domain/disciplinary engineers would be able to use the MBSE data repository to obtain the design information needed to create their analytical models, and conduct analyses in the support of system development [4].

An integrated modeling and analysis capability was developed that bridges the gap [5], [6], [7]. The technical approach was based on integrating SysML modeling tools with process integration and design optimization (PIDO) framework such as ModelCenter® [8]. This approach has the advantage of using the common interface provided by the PIDO framework to connect SysML with various engineering analysis tools such as CAD/CAE, legacy codes, mathematical solvers, and spreadsheets. The integrated toolset has been developed under close collaborations with engineers in industry, and it supports both of the distinct perspectives of systems engineers and domain/disciplinary analysts.

From an industrial perspective, the adoption of the MBSE approach has improved understanding of the system and communications between the system team and subsystem/component suppliers. Also, it ensures that the architecture definitions are consistent and coherent. The MBSE approach, combined with analytical capabilities such as the one presented here, is a key enabler for the early validation of systems and for trade-studies at architecture level.

2   Integrated Modeling and Analysis

The integrated modeling and analysis capability is based on a few general principles. First, the capability needs to support models of different levels of abstraction. SysML provides a number of modeling constructs to support model abstraction, but model abstraction needs to encompass analysis models, as well.

Second, a right balance needs to be sought between generic systems engineering models and domain specific models, because they have their own strengths and weaknesses. While SysML is very useful in defining system architecture and relationships, there are many specialized modeling and analysis tools for specific domains. Compared to generic SysML models, domain specific models can be more effective to accurately describe particular aspects of a system. Third, engineering analyses need to be performed in the context of overall system development. It is easy to lose track of the big picture when performing analyses in the development of a complex system. Fourth, both top-down and bottom-up approaches are needed in the creation of models. SysML models are typically created using a top-down approach, evolving from high level abstractions to more details. But SysML models may need to be updated based on inputs from domain/disciplinary engineers who are responsible for creating analysis models.

2.1   Defining Engineering Analyses in SysML

To accurately model analysis components that will be executed in the PIDO framework, a number of SysML stereotypes were defined, which were packaged into a profile. *MC_Component* stereotype was defined to specify the location of an executable analysis model for a constraint block (e.g., analysis block). *MC_Variable* stereotype was defined to create mapping between SysML ports (e.g., parameters) and variables in analysis models. An enumeration type, *Direction*, is used to specify causality (input or output) of parameters that are stereotyped by *MC_Variable*. To automatically check requirements compliance, *RequirementVerification* stereotype can be applied to a requirement block. The stereotypes and their usage will be discussed in more detail in the following sections.

SysML parametric diagrams are used to specify parametric relationships among system properties. A parametric diagram uses analysis blocks (called constraint blocks) that represent physical or logical relationships between system properties in the model. However, the out-of-the-box capability of constraint block is limited to algebraic equations. To overcome this limitation, SysML constraint blocks are extended so that they point to black box analyses that can be scripts, spreadsheets, CAD/CAE tools, or legacy analysis codes. In this work, analysis models are hosted by Analysis Server®, which is capable of sharing engineering analyses and running them remotely. Each black box analysis hosted by Analysis Server is identified by a URL (Uniform Resource Locator). The *MC_Component* stereotype was used to associate the URL with constraint blocks.

2.2   Automatic Generation of Analysis Models

When a parametric diagram is set up using the aforementioned stereotypes, it has necessary information to create an analysis model that can be executed through the PIDO framework. Figure 1 shows an example of parametric diagram using the stereotypes, created for a brake design problem. The diagram uses four analysis blocks including a cost analysis based on an Excel spreadsheet, and performance analyses for caliper, pad, and stop distance that are based on C++ programs.

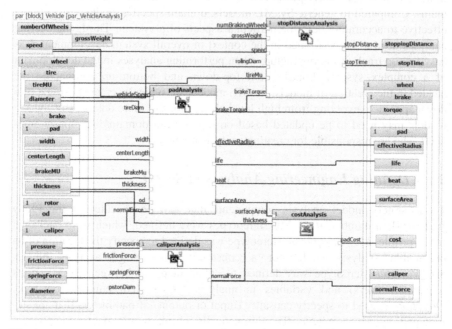

Fig. 1 Parametric diagram for a brake pad analysis

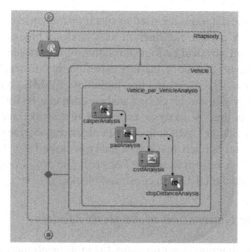

Fig. 2 Analysis model automatically created from a parametric diagram

A capability was developed that automatically creates an analysis model from a parametric diagram. Figure 2 shows the analysis model generated from the parametric diagram in Figure 1. For the analysis model generated, the user can change values of input parameters and execute the analysis model. It should be noted that the automatically generated model is connected to the SysML model. This approach allows using the SysML model as an authoritative source of design information for engineering analyses. Input parameter values are initialized from the values defined in the SysML model. It is also possible to update the SysML model using new values computed from the analysis model.

The bridge between SysML and executable models can be used either within SysML tool environment or within ModelCenter environment. This was done to support distinctive perspectives of engineering analysts and systems engineers. Engineering analysts, who prefer working within PIDO environment, may use the bridge available as the MBSE Plug-In of ModelCenter. Systems engineers, who prefer working within SysML tool environment, may use the bridge available as an extension of the SysML tool, called MBSE Analyzer. This allows engineers to work within tool environment they are familiar with.

2.3 Requirements Modeling

One objective of this work was to develop techniques to use engineering analyses to check requirements compliance. A challenge is that requirements are typically defined in textual formats that require human interpretation before they can be verified in an automated fashion. This work took an approach of manually interpreting requirements and attaching lower/upper bounds to each requirement. Consider a textual requirement: "The brake pad shall have a projected life of at least 72,000 miles under normal driving conditions." When it is interpreted, a lower bound of 72,000 miles can be attached to the requirement. Then, the requirement can be verified by comparing the lower bound with the pad life computed by an engineering analysis.

SysML tags were used to attach lower/upper bounds to requirements. Figure 3 shows the example of the brake pad life requirement, for which a *lowerBound* tag was defined. The requirement is related to the *life* property of the pad through a *satisfy* relationship. Since the value of the *life* property is smaller than the lower bound, the requirement is not satisfied. A capability was created that automatically checks compliance status of requirements. If a requirement is not satisfied, it is automatically flagged.

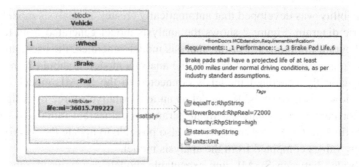

Fig. 3 Lower/upper bound tags are used to automatically verify requirements

2.4 Importing Analysis Models into SysML Models

The discussion so far was focused on converting SysML models to engineering analysis models. But conversion in the reverse direction is often desired. For instance, if engineering analysis models already exist, it is beneficial to reuse them. To this end, a capability was created to import analysis models into a SysML model as constraint blocks. This capability allows browsing available analyses on the network, and creating constraint blocks in a selected package in the SysML model. All necessary stereotypes are automatically set up when an analysis is imported. A ModelCenter model that may consist of many analysis components can be imported as a constraint block. In this case, the ModelCenter model is treated as a black box analysis. If there many variables in the analysis model, the user can select only important variables, and expose them as SysML constraint parameters.

The import capability also allows updating existing constraint blocks. This is useful when analysis models are still in flux and the SysML model needs to be synchronized often as the analysis models are being refined. In practice, systems engineers may create initial constraint blocks in SysML as specifications for engineering analyses. Then, engineering analysts will use the specification and create engineering analyses. If the original specification turns out to be inaccurate, engineering analysts may recommend changes or they can import analytical models to update the existing constraint block. This provides a way for engineers to work together to refine SysML models and engineering analyses models and connect them.

3 Application

In order to evaluate the benefit of using the capability described above in an MBSE approach, EADS Innovation Works applied the proposed technique to one of its case studies. The main expectation was to be able to link SysML models representing the requirements and the system architecture with engineering models in order to compute the key performance indicators (KPI) of the given architecture and to perform trade studies.

3.1   Overview of the Case Study

The selected case study was the design of an aircraft with a novel propulsion system. The case study was inspired by the PhD thesis of Cyril de Tenorio performed at the Georgia Institute of Technology [9]. The work, funded by the EADS Foundation, aimed at using an MBSE approach in order to couple multidisciplinary sizing models and run an optimization process. ModelCenter was used for the integration of sizing models but the link with the architecture model was realized through an ad-hoc interface with the IBM Rational Rhapsody SysML modeling software.

3.2   Integration of Multidisciplinary Design Processes

In order to test the capabilities of the integrated modeling and analysis, a real architecture design process was reproduced for the aircraft example. The SysML model was built using Rhapsody 7.6.1. The following details how the different activities of the process were implemented.

3.2.1   Define/Import Requirements

One of the main motivations of adopting an MBSE approach is the capability to keep traceability between requirements and design. In SysML, requirements are modeled as model elements with an identifier and a textual description. Requirements can be linked to design elements by the *satisfy* relationship.

In order to be able to check compliance status of a requirement through numerical analysis, the *RequirementVerification* stereotype was used that defines the allowed range of a given value property. Consider an example of a weight requirement stating that *"The mass of each fan shall be less than 8000 kg"*. This requirement is satisfied by the *Mass* attribute of the *Fan* block. By using this extension approach, the requirement will be verified for each instance of the *Fan* block present in the architecture model.

3.2.2   Define Architecture and Variants

SysML block definition diagrams (BDD) and internal block diagrams (IBD) were used to define system architectures. Figure 4 represents an architecture with two electric fans, a DC bus, a DC generator and a power plant. Design alternatives at architectural level (i.e., different organizations of the system structure) may be defined as different architectures and may be evaluated using different analyses.

Mission data (altitude, velocity, required power, etc.) required definitions of multiple values that represent different stages of the mission. The mission data were used as array inputs to engineering analyses. To model the mission data, SysML properties with multiplicity were used. Then, instance specifications were used to define multiple values with indices for each property.

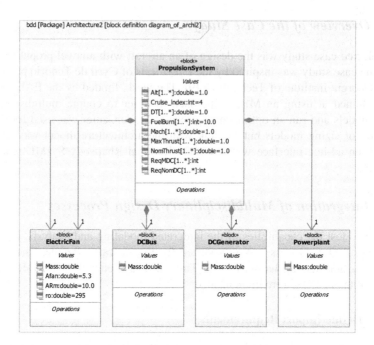

Fig. 4 Definition of the propulsion system architecture in SysML

3.2.3   Define MOEs and Request Analysis Models

In order to evaluate the quality of candidate designs, measures of effectiveness (MOE) must be defined. These MOEs are criteria that can be used to compare different design alternatives. They can be modeled as SysML value properties. For instance, the mass and the fuel consumption of the propulsion system are the main MOEs considered in this case study.

In SysML, the computation of MOEs (and of any properties) can be defined using the *parametric diagram* and *constraint properties*. Parametric relations between properties are typically defined using mathematical equations in SysML, and the mathematical equations are solved by specific solvers supported by the SysML tool (e.g., Maxima or Matlab for Rhapsody). In industrial projects, this is a huge restriction since:

- Existing models in other languages cannot be reused
- More complex analyses may be needed.

The main goal of the integrated modeling and analysis is being able to compute these MOEs and verify requirements by using analysis models defined in domain specific languages and tools. The architect may create SysML constraint blocks as specifications for engineering analyses. Using the specifications, domain experts may create detailed analysis models that will support evaluation of the system architecture.

3.2.4   Design Analysis Models

Domain experts are responsible for creating analysis models. In our example, the analysis models were all built using the open-source mathematical software Scilab. For each component, the model executes a sizing process in order to compute the mass of the component and its outputs such as requested power and fuel burn.

The analysis models were wrapped using ModelCenter. A wrapper is an interface around a program that permits ModelCenter to run the program. The wrapped analysis models were published through Analysis Server, a companion product of ModelCenter, so that they can be used locally or remotely by system architects. Scilab models were wrapped using the QuickWrap feature of ModelCenter which permits building wrappers around programs that take files as input and produce files as output.

3.2.5   Integrate Analysis Models with Architecture Model

The architect can import analysis models into an architecture model using the MBSE Analyzer plug-in of Rhapsody. These analyses models are imported as constraint blocks with the *MC_Component* stereotype. In these blocks, only the reference to the analysis model and its variables (inputs/outputs) are imported.
The architect can use the imported constraint blocks to specify the type of the constraint properties already existing in the architecture model. Figure 5 shows a parametric diagram created using this approach for the computation of MOEs of the propulsion system.

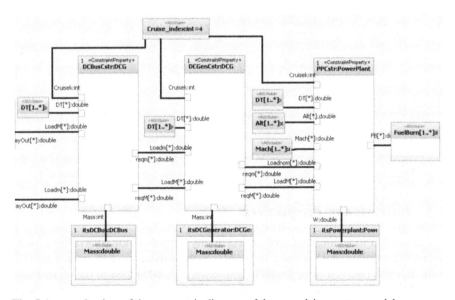

Fig. 5 A zoom-in view of the parametric diagram of the propulsion system model

3.2.6 Perform Trade Study

Once all analysis models are linked to design attributes, the architect can test different configurations of the architecture by running corresponding analyses. This trade study can be performed directly within Rhapsody using the MBSE Analyzer plug-in. The architect can modify parameters of the design and save configurations in the SysML model as new instance specifications. This approach allows storing considered designs and associated analysis results in the SysML model and capitalizing them for design reviews and trade-offs. The status of requirements (satisfied/violated/unknown) can be directly visualized in the graphical user interface (see Figure 6).

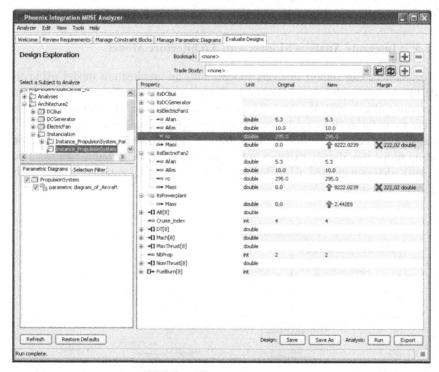

Fig. 6 Trade-study in the graphical user interface with visualization of requirement status

3.3 Interests and Perspectives

The integrated modeling and analysis capability permits linking a systemic view (including requirements) defined in SysML with analysis models created by domain experts. This is a large technological advantage compared to the current use of SysML tools for analysis, since analysis models can be created in domain specific languages and tools and can be distributed over a network.

The MBSE Analyzer plug-in in Rhapsody permits the architect to run trade studies from within the SysML tool. Currently, parametric study and design of experiments can be performed within SysML tool. In the future, this capability will be extended to support design optimization and probabilistic analysis.

However, the architecting of the holistic model is a key task that is still not supported by any method and that can become very difficult task. Indeed, the simulation architect must organize its simulation models so that they can be plugged together and that their assembly provides the answer to the question with the expected accuracy. A key step of this method would be the model requests for engineering analyses. Indeed, the quality of the model requests issued by system architects to domain experts is a key factor of the success of the design activities for complex systems. Following the process described previously, the architects only describe the requested inputs and outputs of the analysis models. Obviously, this is not sufficient to get the right model and a model request shall contain more information about the purpose of the model, the requested granularity, the domain of application, etc. The specification of model requests has a potential to greatly improve system development capabilities by facilitating information exchange between system architects and domain experts.

4   Conclusion

Tools and techniques were developed that bridge the gap between systems engineering models and engineering analysis. The capability supports distinct perspectives of systems engineers who need to evaluate system architecture and domain/disciplinary engineers who are performing detailed engineering analyses in the support of system development. The integrated capability was used to automatically check requirements compliance status using realistic analysis models.

The capability was applied to the design of a propulsion system of an aircraft. It was demonstrated easily integrating multi-disciplinary analysis models with a system model defined in SysML. This allows the architect to perform trade-studies on the system and to check requirements using analysis models created in domain specific tools. The proposed technique supports streamlined model requests for engineering analyses, which ensures that the input data used in detailed engineering analyses is consistent with the architecture level information.

References

[1] Fisher, J.: Model-Based Systems Engineering: A New Paradigm. INCOSE Insight 1(3), 3–4 (1998)
[2] OMG System Modeling Language, http://www.omgsysml.org
[3] Friedenthal, S., Moore, A., Steiner, R.: A Practical Guide to SysML – The System Modeling Language. Elsevier Inc. (2009) ISBN 978-0-12-378607-4
[4] Mosier, G.: Parametric Design and Analysis to Support Model-Based Systems Engineering Using SysML. NASA Tech Briefs, p. 80 (October 2010)

[5] Kim, H., Fried, D., Menegay, P., Soremekun, G.: Integrated Modeling and Analysis to Support Model-Based Systems Engineering. In: Proceedings of the ASME 2012 11th Biennial Conference on Engineering Systems Design and Analysis, ESDA 2012-83017, Nantes, France, July 2-4 (2012)

[6] Kim, H., Fried, D., Menegay, P.: Connecting SysML Models with Engineering Analyses to Support Multidisciplinary System Development. In: 14th AIAA/ISSMO Multidisciplinary Analysis and Optimization Conference, AIAA 2012-5632, Indianapolis, IN, September 17-19 (2012)

[7] Kim, H., Fried, D., Menegay, P., Soremekun, G., Oster, C.: Application of Integrated Modeling and Analysis to Development of Complex Systems. In: Conference on Systems Engineering Research (CSER 2013), Atlanta, GA, March 19-22 (2013)

[8] Phoenix Integration, Inc., "Improving the Engineering Process with Software Integration", a white paper (2002), http://www.phoenix-int.com

[9] De Tenorio, C.: Methods for collaborative conceptual design of aircraft power architectures (2010)

A Hybrid Event-B Study of Lane Centering

Richard Banach and Michael Butler

Abstract. A case study on automotive lane centering control is examined in Hybrid Event-B (an extension of Event-B that includes provision for continuously varying behaviour as well as the usual discrete changes of state). This allows aspects beyond the reach of a discrete Event-B treatment to be more deeply investigated. Lane centering offers particular challenges concerning how the monitoring of continuously varying continuous functions is handled and how this interacts with discrete mode-level decision making.

1 Introduction

These days, an ever increasing proportion of the equipment we interact with all the time involves digital control of analogue phenomena. The repercussions of this trend include an increasing number of devices that could do real harm if they malfunctioned. Besides this, the capabilities of digital control bring with it ever increasing complexity, with the risks that this inevitably brings.

It is by now well accepted that formal techniques, appropriately deployed, can help with both of these issues. However, in the main, formal techniques are strongly focused on purely discrete reasoning, dealing poorly with continuous behaviours. The *hybrid* and *cyberphysical* systems we speak of (see, e.g. [30, 33, 2, 32, 12]) are rather poorly served by conventional formal techniques. See, for example, the extensive review in [14], which covers a large number of approaches and their

Richard Banach
School of Computer Science, University of Manchester,
Oxford Road, Manchester, M13 9PL, U.K.
e-mail: banach@cs.man.ac.uk

Michael Butler
School of Electronics and Computer Science, University of Southampton,
Highfield, Southampton, SO17 1BJ, U.K.
e-mail: mjb@ecs.soton.ac.uk

M. Aiguier et al. (eds.), *Complex Systems Design & Management 2013,*
DOI: 10.1007/978-3-319-02812-5_8, © Springer International Publishing Switzerland 2014

accompanying tool systems, including: Charon [5, 6], CheckMate [29], HSolver [25], HyTech [7, 17], Modelica [23] and Simulink [22], among others. Although these techniques have enjoyed success over the years, most of them are either limited in their expressivity (typically driven by a desire to achieve decidability for a given language fragment), or they lack rigour by comparison with most discrete techniques (for example by employing simulation as a strategy for verification).

An exception to this general trend is KeYmaera (see [1, 24]). This is a system that combines formal proof of a quality commensurate with contemporary formal techniques, with the kind of continuous behaviour that is needed in the description of genuine physical systems. KeYmaera concentrates on the verification of properties of a defined system model. In this sense its focus is different from the refinement-based approach of the B-Method —our concern in this paper— since, although verification of some properties can often be seen as a kind of refinement, KeYmaera does not emphasise refinement as a *development approach* in the way that the B-Method emphatically does.

The increasing interest in hybrid and cyberphysical systems just noted, and the desire to achieve high dependability in their development, has led to attempts to use Event-B [3] for their development, for example during the DEPLOY project [16]. However it is fair to say that the absence of the ability to deal with continuous phenomena directly within the discrete Event-B framework during such work is keenly felt.

This need to deal with continuous phenomena within Event-B has prompted the development of an extension, Hybrid Event-B [11], treating discrete and continuous behaviour on the same footing. In this paper, we apply Hybrid Event-B to a case study previously done in discrete Event-B: the modelling of an automotive lane centering system first investigated during a collaboration between the second author and GM Research. Among other things, such revisiting of the case study can confirm the suitabilty of the Hybrid Event-B formalism to adequately deal with facets of the original discrete Event-B treatment that were less than ideal.

The lane centering case study explored in this paper forms a natural accompaniment to another automotive-based case study, on cruise control, done by the present authors in [10]. The contrast between the two is instructive. In cruise control, the problem can be seen as completely self-contained. The desired speed is set, and the car can easily monitor progress towards it, and how well it is being maintained. This gives rise to a situation in which convincing invariants can be established. In lane centering though, the road ahead is unpredictable in principle. So the problem is not self-contained any more. Progress towards the desired goal is constantly dependent on external information, which impacts the kind of invariants one can establish. These aspects are discussed in detail in Section 4.

In contrast to KeYmaera, there is at present no dedicated tool support for Hybrid Event-B. So a further benefit of case studies like this one is to confirm that Hybrid Event-B contains the right collection of ingredients for industrial scale modelling, before the serious investment in extending the Rodin Tool [4, 26, 27] is made.

The rest of this paper is as follows. Section 2 overviews the lane centering system, including how our case study differs in detail from the requirements tackled in the previous version. Section 3 then gives a pure mode based model for the lane centering control system, while Section 4 refines it to a model where both modes and continuous behaviour are fully defined. These models are related to one another using a suitable refinement. Section 5 discusses various issues raised by the preceding material, and concludes.

2   Lane Centering Controller Overview

A lane centering controller (LCC) is a software system which automatically keeps a car correctly aligned with respect to the centre of the lane in which it is travelling. It does this based primarily on information received from a path generator unit, which calculates the target and predicted paths of the car for a short period into the future, based in turn on information from an image processing unit (which looks for lane markings in the road) and other data available from the engine management system. The car's driver can engage the LCC, which will then attempt to discharge its lane centering obligations, but the LCC is an assistance system rather than a safety system, so when it is unable to perform its function, it issues a warning, at which point the driver must resume responsibility for steering the car. In addition, it must disengage upon request of the driver (or in case of a system fault). All of these are safety properties.

The left hand side of Fig. 1 gives a diagrammatic view of the LCC architecture, taken from [34]. We see the image processing unit, which feeds information about the car's lateral position in the lane, and information about the road curvature, to the path generator unit. This then receives further information regarding the yaw angle and rate, the steering angle, the lateral and longitudinal speed, and the driver-selected offset. From all this information, the target path, the predicted actual path and a safety margin, are calculated and fed to the LCC itself, which generates the actuated steering angle for keeping the car on target.

The right hand side of Fig. 1 illustrates how some of these parameters relate to the movement of the car. In reality, the LCC only works when an adaptive cruise control system is actively controlling the speed of the car, leading to a coupling between the two systems. As noted in the Introduction, we have considered cruise control earlier in [10], but for simplicity, in this paper we neglect the coupling between it and the LCC. In the same vein, we neglect the lateral offset parameter, and a number of more subtle considerations concerning circumstances under which the LCC has to curtail its activity because the information it has is insufficient, or is of insufficient quality. We cater for all such cases by switching the LCC off or by taking an error transition, offering some further discussion in the Conclusions.

With these caveats in place, in Fig. 2 we see the state transition diagram for the LCC at an intermediate level of description, comparable to, but simplified from, the LCC description in [34].

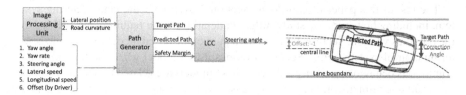

Fig. 1 Architecture of a lane centering controller (LCC) on the left. On the right, a schematic illustration of the geometric elements that figure in LCC computations.

The LCC starts in the *OFF* state, from where it can be made to *SwitchOn*, putting it into the *STANDBY* state. In this state the LCC can be made to *SwitchOff*. Alternatively, the driver may try to engage the LCC. If the motion of the car is too *UnAl*igned with the lane markings for the LCC to take over, the state remains *STANDBY*, but if the motion is adequately *Aligned*, the the LCC goes into the *ACTIVE* state, in which it actively steers the car.

Normal LCC working can be overridden by putting the *IndicatorOn*, which signifies that the driver will shortly turn out of the lane. Alternatively, the driver may try to steer the car manually by forcing the steering wheel out of the orientation determined by the LCC. In either of these cases the state becomes *OVERRIDE*.

Putting the *IndicatorOff*, or ceasing to try to steer the car manually causes the LCC to try to *Resume* normal working in the *ACTIVE* state — provided the car is adequately aligned; if it is not then the *OutOfAl*ignment transition switches the LCC off.

In addition to these ways of working, the driver can *SwitchOff* the LCC in the *ACTIVE* and *OVERRIDE* states (as well as *STANDBY*), and when in any of these three states, the LCC can undergo an *Error* transition into the *ERROR* state, for example if the path generator unit or image processing unit undergo some failure, or their information is too low quality to be relied on safely.

We take it for granted that the state transition diagram in Fig. 2 does an effective job of modelling the top level requirements of the LCC system. For example, it addresses the requirement that when it is switched on, the LCC does not start working

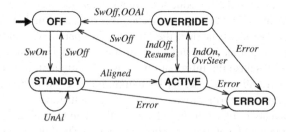

Fig. 2 The state transition diagram for a simplified LCC

immediately, but waits in the *STANDBY* state until instructed to take control of the steering. Likewise, the transitions between the *ACTIVE* and *OVERRIDE* states correspond to requirements embodying assumptions about the appropriateness of the use of the accelerator during LCC working. And so on.

In the following sections, we will develop a series of Hybrid Event-B machines to capture this design. We explain the technicalities of Hybrid Event-B as we go.

3   Lane Centering — Top Level Mode Oriented Control

In this section we start the development of the LCC system. Fig. 3 contains a straightforward translation of the state transition diagram in Fig. 2 to the discrete, mode oriented part of Hybrid Event-B. The nature of the translation is rather self-evident, so we restrict ourselves to a few essential comments.

The state of the system is recorded in the *mode* variable, which is restricted to the values we mentioned above (this observation constituting the sole invariant at this level of abstraction), and is initialised to *OFF*. Beyond this, the events of the model simply record the state changes permitted by Fig. 2 in a relatively obvious way.

The events merit three further technical observations at this point. Firstly, the design is *aggressive*, in the sense that although each event is guarded by a condition that permits its sensible execution, the states are not protected against inappropriate requests from the environment. In each state, only those events can be executed that make sense in our model, whereas in the real world, the driver, could, if a little unwisely, engage controls that one would not expect to see engaged there. At a higher level of abstraction than we model in this paper, the system would have to be protected against such stimuli.

Secondly, although we only appear to have discrete transitions, these *mode* events each have an input, whose value is essentially just the event's name (except for *UnAl* and *Aligned* where there is a genuinely nondeterministic choice), which is furthermore never used. The explanation for this is that Hybrid Event-B models real time behaviour, which therefore means that the occurrence times of mode events (which execute instantaneously) must be defined. For convenient modelling, Hybrid Event-B stipulates that unscheduled stimuli from the environment —such as the otherwise unspecified timings of occurrences of our mode events— are modelled by the arrival of inputs to the relevant mode events.

Thirdly, and following on from the previous point, there is also another, *pliant* event, *PliTrue*. Its nature is signalled by the 'STATUS pliant' tag. Unlike mode events, pliant events describe periods of continuous behaviour, taking place over nonempty intervals of time, in between mode events (occurrences of which are thus isolated in time). The semantics of Hybrid Event-B stipulates that a mode event can prempt a running pliant event as soon as the mode event becomes enabled, and upon the completion of the mode event, some pliant event should have become enabled and should be selected for execution to take forward the continuous evolution of the system.[1]

In our case, *PliTrue* stipulates in the COMPLY *INVARIANTS* clause merely that the continuous behaviour in between mode event occurrences should obey the invariants. With its help, the behaviour of the model is defined for all times after the initialisation point, since, having a trivial guard, whenever a mode event runs, *PliTrue* is re-enabled immediately afterwards.

4 Lane Centering — From Mode Control to Continuous Control

In this section we go beyond the pure mode oriented model of Section 3, to include a description of the continuous behaviour in the periods between occurrences of mode events, that a more complete model demands.

Our enhanced model appears in Fig. 4, completed in Fig. 5. We go through this element by element, giving, as we go, appropriate commentary, not only on the technical background of the case study itself, but also on relevant aspects of Hybrid Event-B and its semantics, especially when this is at odds with corresponding aspects of discrete Event-B, or raises interesting issues in its own right.

Fig. 4 starts with the INTERFACE *LCC_PG_IF* block. The INTERFACE construct is the Hybrid Event-B syntactic construct that enables different machines to be coupled to each other to form a larger system, working together in an integrated way, under the control of the required invariants. In our case study, we have taken the view that the LCC will 'own' those variables whose values it can control, whereas, variables which are set externally, or inferred indirectly from the car's environment via the image processing unit, will not be owned by the LCC.

The *LCC_PG_IF* interface contains these 'externally owned' variables, that need to be shared with the LCC. The first is a measure of the torque applied by the driver to the steering wheel, *trq*. The other two concern the target path and deviation from the lane centre, both of which need to be calculated from the visual inputs to the overall system.

For lack of space in this paper, we simplify matters considerably compared with the system architecture of Fig. 1. Thus the target path variable (for which one can imagine a large number of quite detailed formulations), is assumed to have already been converted to a target steering angle, θ_T. What we intend by this is the following: *if it were the case that the car was already exactly in the middle of the lane, then steering at θ_T would keep it exactly where the overall lane control system would*

---

[1] This *eager* semantics for preemption of pliant events by mode events gives another reason why an *asynchronous arrival* semantics for inputs to mode events is needed. In discrete Event-B as soon as an event has completed, its successor is enabled, since the state does not change in beteen event occurrences (which are assumed separated in time as a matter of *interpretation*). In Hybrid Event-B, real time figures explicitly, and a second mode event is not permitted to execute immediately after a first mode event, despite being enabled then. Giving the second mode event an asynchronous input gets round the problem.

want the car to go. We accept that this amounts to a rather specific treatment of the requirements of an LCC system, since in reality, such requirements would speak in more detail about what constitutes the kind of road that the system is expected to be able to cope with, and what kind of behaviour would be expected under those conditions. A further simplification in our model is the assumption that if the path generator is sending the value of θ_T to the LCC and has not called an *Error* transition, then it is safe for the car to proceed down the path defined by θ_T at its current velocity (e.g. the car has not driven into a muddy field that the image processing unit and path generator could not cope with).

The last variable in the *LCC_PG_IF* interface is d, which represents the measured deviation of the car from the lane centre. All of *trq*, θ_T and d are declared PLIANT, which means that they are subject to continuous change during the pliant transitions that interleave occurrences of the mode events at runtime.

The *LCC_PG_IF* interface also contains the invariants that these variables must obey. These say that *trq*, θ_T and d are all real-valued, and that they remain within statically determined bounds. Finally, the *LCC_PG_IF* interface also contains these variables' initialisations.

We come to the *LCC_1* machine itself, which is a refinement of *LCC_0* and CONNECTS to the *LCC_PG_IF* interface, making the latter's variables accessible. As well as the earlier *mode* variable, we have a pliant variable, θ, which holds the current steering angle (this being a representation, on the lines noted above, of the predicted path in Fig. 1). It is real-valued and bounded within static contraints.

Proceeding to the events, θ is initialised to an arbitrary value at system switch on time. After that, *PliDefault* is the refinement of the *PliTrue* event of *LCC_0* that demands no more than invariant preservation during pliant transitions when it is enabled. We observe that *PliDefault* is enabled during all modes other than *ACTIVE*, so in effect, all the additional design that is embodied in the refined system model is targetted at just the *ACTIVE* state.

Given the last remark, some of the earlier mode events remain unchanged. This applies for instance to the *SwOn* and *SwOff* events, that cater for the driver deliberately turning the LCC on or off.

The next two events deal with the driver trying to actively engage the LCC system. To facilitate the discussion, we introduce the *ACTIVE state tolerance condition*: $ASTC \equiv (|d| < \Delta_d \wedge |\theta - \theta_T| < \Delta_\theta)$. We make the assumption that the LCC is not to be used when the current conditions are such that the target path (as determined by the path generator) is too different from the current path. We model this in a simplified way by demanding that the target and current steering angle do not differ by too much $|\theta - \theta_T| < \Delta_\theta$, and that the lane centre deviation d is not excessive $|d| < \Delta_d$. Hence the *ASTC*.

The *ASTC* enters our model in a number of places. Firstly, when the driver attempts to engage the LCC, he cannot know in advance whether the *ASTC* will be satisfied or not — aside from anything else, it depends on internal constants, Δ_d and Δ_θ, that he does not know and would not be able to make use of during driving even if he did. So in the *LCC_0* machine earlier, this ignorance was catered for in a genuinely nondeterministic choice between the *UnAligned* and *Aligned* events,

MACHINE *LCC_0*
VARIABLES *mode*
INVARIANTS
   mode ∈ {*OFF*, *STANDBY*, *ACTIVE*,
       OVERRIDE, *ERROR*}
EVENTS
  INITIALISATION
   STATUS ordinary
   BEGIN
    mode := *OFF*
   END
  SwOn
   STATUS ordinary
   ANY *in*?
   WHERE *in*? = *swOn* ∧ *mode* =
OFF
   THEN *mode* := *STANDBY*
   END
  SwOff
   STATUS ordinary
   ANY *in*?
   WHERE *in*? = *swOff* ∧ *mode* ∈
    {*STANDBY*, *ACTIVE*, *OVERRIDE*}
   THEN *mode* := *OFF*
   END
  UnAl
   STATUS ordinary
   ANY *in*?
   WHERE *in*? = *tryAct* ∧
    mode = *STANDBY*
   THEN skip
   END
  Aligned
   STATUS ordinary
   ANY *in*?
   WHERE *in*? = *tryAct* ∧
    mode = *STANDBY*
   THEN *mode* := *ACTIVE*
   END
  IndOn
   STATUS ordinary
   ANY *in*?
   WHERE *in*? = *indOn* ∧
    mode = *ACTIVE*
   THEN *mode* := *OVERRIDE*
   END

  IndOff
   STATUS ordinary
   ANY *in*?
   WHERE *in*? = *indOff* ∧
    mode = *OVERRIDE*
   THEN *mode* := *ACTIVE*
   END
  OvrSteer
   STATUS ordinary
   ANY *in*?
   WHERE *in*? = *ovrSteer* ∧
    mode = *ACTIVE*
   THEN *mode* := *OVERRIDE*
   END
  Resume
   STATUS ordinary
   ANY *in*?
   WHERE *in*? = *resume* ∧
    mode = *OVERRIDE*
   THEN *mode* := *ACTIVE*
   END
  OOAl
   STATUS ordinary
   ANY *in*?
   WHERE *in*? = *oOAl* ∧
    mode = *OVERRIDE*
   THEN *mode* := *OFF*
   END
  Error
   STATUS ordinary
   ANY *in*?
   WHERE *in*? = *error* ∧ *mode* ∈
    {*STANDBY*, *ACTIVE*, *OVERRIDE*}
   THEN *mode* := *ERROR*
   END
  PliTrue
   STATUS pliant
   COMPLY *INVARIANTS*
   END
END

Fig. 3 Mode level description of lane centering control

INTERFACE LCC_PG_IF
PLIANT trq, θ_T, d
INVARIANTS
$trq \in \mathbb{R} \wedge |trq| \leq MAX_{trq}$
$\theta_T \in \mathbb{R} \wedge |\theta_T| \leq MAX_\theta$
$d \in \mathbb{R} \wedge |d| \leq MAX_d$
INITIALISATION
$trq \in [-MAX_{trq} \ldots MAX_{trq}]$
$\theta_T := 0$
$d := 0$
END

MACHINE LCC_1
REFINES LCC_0
CONNECTS LCC_PG_IF
VARIABLES $mode$
PLIANT θ
INVARIANTS
$mode \in \{OFF, STANDBY, ACTIVE,$
$\qquad OVERRIDE, ERROR\}$
$\theta \in \mathbb{R} \wedge |\theta| \leq MAX_\theta$
EVENTS
  INITIALISATION
    STATUS ordinary
    REFINES *INITIALISATION*
    BEGIN
      $mode := OFF$
      $\theta \in [-MAX_\theta \ldots MAX_\theta]$
    END
  PliDefault
    STATUS pliant
    REFINES *PliTrue*
    WHEN $mode \neq ACTIVE$
    COMPLY *INVARIANTS*
    END
  SwOn
    STATUS ordinary
    ANY $in?$
    WHERE $in? = swOn \wedge mode = OFF$
    THEN $mode := STANDBY$
    END
  SwOff
    STATUS ordinary
    REFINES *SwOff*
    ANY $in?$
    WHERE $in? = swOff \wedge mode \in$
      $\{STANDBY, ACTIVE, OVERRIDE\}$
    THEN $mode := OFF$
    END

UnAl
  STATUS ordinary
  REFINES *UnAl*
  ANY $in?, out!$
  WHERE $in? = tryAct \wedge$
    $mode = STANDBY \wedge$
    $\neg(|d| < \Delta_d \wedge |\theta - \theta_T| < \Delta_\theta)$
  THEN $out! := BEEP$
  END
Aligned
  STATUS ordinary
  ANY $in?$
  WHERE $in? = tryAct \wedge$
    $mode = STANDBY \wedge$
    $(|d| < \Delta_d \wedge |\theta - \theta_T| < \Delta_\theta)$
  THEN $mode := ACTIVE$
  END
LCC_Active
  STATUS pliant
  REFINES *PliTrue*
  WHEN $mode = ACTIVE$
  SOLVE $\mathcal{D}\theta = -C(\theta - \theta_T) - Kd$
  END
SwOff_Emrg
  STATUS ordinary
  REFINES *SwOff*
  ANY $out!$
  WHEN $mode = ACTIVE \wedge$
    $\neg(|d| < \Delta_d \wedge |\theta - \theta_T| < \Delta_\theta)$
  THEN $mode := OFF$
    $out! := BEEP$
  END
IndOn
  STATUS ordinary
  REFINES *IndOn*
  ANY $in?$
  WHERE $in? = indOn \wedge$
    $mode = ACTIVE$
  THEN $mode := OVERRIDE$
  END
IndOff
  STATUS ordinary
  REFINES *IndOff*
  ANY $in?$
  WHERE $in? = indOff \wedge$
    $mode = OVERRIDE \wedge$
    $(|d| < \Delta_d \wedge |\theta - \theta_T| < \Delta_\theta)$
  THEN $mode := ACTIVE$
  END

Fig. 4 Enhancing the mode level lane centering control to fully continuous control. First part.

OvrSteer
  STATUS ordinary
  REFINES OvrSteer
  WHEN mode = ACTIVE ∧
    |trq| ≥ THR_{ACTIVE}
  THEN mode := OVERRIDE
  END
Resume
  STATUS ordinary
  REFINES Resume
  WHEN mode = OVERRIDE ∧
    |trq| ≤ thr_{ACTIVE} ∧
    (|d| < Δ_d ∧ |θ − θ_T| < Δ_θ)
  THEN mode := ACTIVE
  END
OOAl_Ind
  STATUS ordinary
  REFINES OOAl
  ANY in?, out!
  WHERE in? = oOAl ∧
    mode = OVERRIDE ∧
    ¬(|d| < Δ_d ∧ |θ − θ_T| < Δ_θ)
  THEN mode := OFF
    out! := BEEP
  END

OOAl_OvrSteer
  STATUS ordinary
  REFINES OOAl
  ANY out!
  WHERE mode = OVERRIDE ∧
    |trq| ≤ thr_{ACTIVE} ∧
    ¬(|d| < Δ_d ∧ |θ − θ_T| < Δ_θ)
  THEN mode := OFF
    out! := BEEP
  END
Error
  STATUS ordinary
  ANY in?, out!
  WHERE in? = error ∧ mode ∈
    {STANDBY, ACTIVE, OVERRIDE}
  THEN mode := ERROR
    out! := BEEP
  END
END

Fig. 5 Enhancing the mode level lane centering control to fully continuous control. Second part.

indicated by having them both demand the same input value *tryAct*. In the *LCC_1* machine, the choice between *UnAl* and *Aligned* is further refined by the truth of *ASTC*. If the *ASTC* is true, then *Aligned* is selected, whereas if the *ASTC* is false, *UnAl* is selected. On a technical note, we remark that the disjunction of the *LCC_1* machine's *UnAl* and *Aligned* guards is equivalent to the disjunction of the *LCC_0* machine's *UnAl* and *Aligned* guards, addressing both guard strengthening and relative deadlock freedom refinement requirements. Besides this, we note that *UnAl*, in refusing to perform a function that the driver is expecting (i.e. to engage the LCC), alerts him to this fact via an audible alarm, represented in our model by sending a *BEEP* on the output variable *out!*.

Assuming successful engagement of the LCC, the *LCC_Active* pliant event runs. It has no nontrivial guards (aside from the *mode*), since the relevant initial conditions are already confirmed by the *Aligned* event that enables it. The job of *LCC_Active* is to align the car's path to the target path. For simplicity, we model this using a straightforward negative feedback control law applied to the current steering angle, indicated in the SOLVE clause of the *LCC_Active* event: $\mathcal{D}\theta = -C(\theta − \theta_T) − Kd$. Assuming that steering to the right, and deviation to the right from the lane centre, are both measured positively, the time derivative ($\mathcal{D}$) of the current steering angle θ is set to a negative linear combination of steering angle excess (of current over target) and deviation, tending both to bring the car closer to the lane centre, and to align the steering angle with the target steering angle.

We regard the θ_T and d parameters in the above ordinary differential equation (ODE) as external signals. Doing this reduces it to a linear ODE with inhomogeneous term; see [8, 31]. Ordinary differential equations of this form have a standard solution. In this case it is: $\theta(t) = \theta(t_L)e^{-C(t-t_L)} + \int_{t_L}^{t} e^{-C(t-s)}[C\theta_T(s) - Kd(s)]ds$, where t_L is the symbol used in Hybrid Event-B to refer generically to the start time of any time interval during which a pliant event runs.

If, in the feedback system described, θ_T and d were both constant, then steady convergence of the car to the lane centre and of the current steering angle to the target steering angle would be guaranteed. A consequence of this would be that the *ASTC*, true at the start of an *LCC_Active* pliant transition, would be an invariant during any such transition. However, the fact that θ_T and d are both time dependent (preventing these terms from being extracted from under the integral in the $\theta(t)$ solution above) means that we have no such guarantee. If θ_T or d were to vary wildly enough during an *LCC_Active* transition, then the bounds in *ASTC* might be breached. Since we cannot prove that the bounds won't be breached, we have to make separate provision for the case where they are.

This is the purpose of the *SwOff_Emrg* mode event. It is enabled in the *ACTIVE* state when the *ASTC* fails. Since it has no input, like all mode events without input, it becomes eligible for scheduling as soon as its guard becomes true, and (if selected from among all the mode events whose guards become true at that moment, if there is more than one such event), it *preempts* the currently running pliant transition, *LCC_Active* in our case.

The effect of *SwOff_Emrg* is like that of *SwOff*, except that, being an event that is scheduled spontaneously rather than at the behest of the driver, there is an additional audible *BEEP* to alert the driver.

The next few events handle the driver's temporarily countermanding the *ACTIVE* state. Use of the indicator is modelled by *IndOn* and *IndOff*, mode events caused by the discrete actions of flicking the indicator on or off to enter or exit the *OVERRIDE* state. Of course, when the *ACTIVE* state is re-entered via *IndOff*, the *ASTC* must be checked.

The driver can also countermand the *ACTIVE* state by wilful use of the steering wheel. The *trq* variable tracks the torque applied to the steering wheel during a system run. If, in the *ACTIVE* state, this exceeds a threshold value THR_{ACTIVE}, the *OvrSteer* event changes the state to *OVERRIDE* while the driver takes control of the steering. Once the torque drops below the threshold value thr_{ACTIVE} again, the *Resume* event re-enters the *ACTIVE* state, having confirmed that *ASTC* holds. Normally, we would have that $thr_{ACTIVE} < THR_{ACTIVE}$ to prevent Zeno-like thrashing as the applied torque hovered around the threshold value.

Of course, since the *ASTC* has to hold if the *ACTIVE* state is to be re-entered, we have to contend with the possibility that it might not. Two new events cater for this. When using the indicator, event *OOAl_Ind*, scheduled when the indicator is flicked off by the driver (denoted using the $in? = oOAl$ input) but $ASTC$ does not hold, switches the LCC off, alerting the driver with a *BEEP*. When using the steering wheel, event $OOAl\_OvrSteer$, scheduled when the torque drops below thr_{ACTIVE} but $ASTC$ does not hold, switches the LCC off, alerting the driver with a *BEEP*. We need two $OOAl$ events at the LCC_1 level, since one has an input and the other does not, even if their actions are the same. Both events refine the LCC_0 level $OOAl$ event, the different I/O signatures in the $OOAl\_OvrSteer$ case being handled by a suitable witness relation. Finally, we have the LCC_1 level *Error* event, *BEEP*-enhanced compared with its LCC_0 counterpart. It completes our survey of the LCC_1 model.

5 Discussion and Conclusions

In the preceding sections, we overviewed the lane centering controller case study, previously examined using discrete Event-B in [34], with a view to creating an enhanced development using the richer facilities of Hybrid Event-B. We then presented such a development, based first on a mode level description in Section 3, subsequently refined to a description incorporating a definition of the required continuous behaviour in Section 4. The relatively simple modelling in these sections, partly a consequence of lack of space in this paper, raises two particular issues that deserve further discussion.

The first issue is that the simple modelling approach reduced path descriptions to a single real quantity, θ. This simplicity meant that controlling the path could be reduced to a simple linear feedback control law, with external input depending on the steering angle difference and lane centre deviation. Obviously, a path is actually a function from some parameter to position, i.e. a higher order concept, so more sophisticated representations can certainly be contemplated.

For example, the image processing unit may generate a moving representation of the road in front of the car —provided it is discernable (c.f. earlier remarks about muddy fields)— as a time dependent strip of varying width, length and direction. This could be communicated to the path generation unit via a few time dependent geometric parameters. The path generation unit could combine this with car velocity and direction information to derive the desired and predicted paths, as functions from time to position in the moving strip. According to the architectural diagram in Fig. 1, it would then be the responsibility of the LCC to synthesise the required steering angle from this information. In principle this is a problem in adaptive optimal control, and how it would be approached would depend crucially on the notion of optimality adopted (see, e.g. [15, 19, 20, 9, 28]).

We evaded the repercussions of all this potential complexity by assuming that the path generation unit already emitted a desired and safe steering angle and deviation from the middle of the road, and that it was sufficient for us to approach the required path via a relatively simple feedback law. However, if we were to take on board the more sophisticated modelling indicated in a more serious way, it would be appropriate to consider how the concepts involved would be handled in a system like Hybrid Event-B.

On the theoretical side there would be no problem, since higher order entities can be handled just as conveniently as basic variables can, within conventional continuous mathematics. On the practical side though, the relative dearth of analytic results for higher order entities, manifests itself in a smaller portfolio of cases that could be mechanised in an automated proving system. A further observation on the same point is that mechanical provers tend to perform much more poorly on higher order objects than they do on first order ones. So a completely abstract formulation of paths in the way we sketched it might need to be approached with caution in the context of mechanical proof.

The second issue concerns the nature of the *ASTC* that figured heavily around the *ACTIVE* state. We already argued that we could not guarantee that the *ASTC* would be maintained for arbitrarily long periods while the LCC wished to remain in the *ACTIVE* state, i.e. that the *ASTC* would be an invariant for such periods. However, our remedy, of preempting the *LCC_Active* pliant transition whenever the *ASTC* failed at runtime, does in fact guarantee that '*mode = ACTIVE ⇒ ASTC*' is indeed an invariant. Why then did we not include such an invariant in our model?

To answer this, we note that *ASTC* contains variables from both the interface *LCC_PG_IF* (i.e. θ_T and d) and the machine *LCC_1* (i.e. θ). In which of these then, should we put this candidate invariant? Note that neither of them declares *all* of the variables mentioned. Assuming that an invariant should only use variables that are declared in the syntactic unit it resides in, we would have to amalgamate *LCC_PG_IF* into *LCC_1* to declare the suggested invariant.[2] In our case study, we managed to sidestep this problem, since we were able to demand *ASTC* on entry to *LCC_Active*, and preempted *LCC_Active* whenever *ASTC* failed, which together are tantamount to the stated invariant.

Nevertheless, the wider problem, of desirable invariants straddling the boundaries of otherwise sensible partitionings of large systems, remains. Perhaps the most promising suggestion for improvement regarding this issue comes from the 'shared event' approach of Butler [13] in which bound variables carrying communicated values can enjoy nontrivial properties without breaking the syntactic structuring of separate components. A generalisation of this to shared variables would be widely applicable across the B-Method.

Thus, we can safely say that the case study undertaken here has amply demonstrated the suitability of Hybrid Event-B for formally describing the requirements, specification and behaviour, of the kind of hybrid system that Event-B is increasingly being applied to these days, and moreover, it has raised a number of issues for future consideration. This is valuable experience which acts as a further spur to the development of full mechanical support for Hybrid Event-B within the framework of the Rodin Tool [4, 26, 27].

---

[2] It is sometimes suggested that existentially quantifying the 'other' variable(s) in an invariant of this kind can solve the problem. Unfortunately it does not, since asserting that some values *merely exist* (that satisfy some property) is quite different from asserting that the *actual current values* satisfy it. This is particularly dangerous when the invariant is a nontrivial safety property.

From the present vantage point, we can envisage without too much danger of error, on how such mechanical support could be organised. The KeYmaera tool [1, 24] provides sound inspiration. The KeYmaera tool started life by integrating an adapted version of the KeY proof tool [18] with Mathematica [21], thus allowing all the continuous reasoning to be delegated to the Mathematica tool. Since then, a number of other provers have been integrated into KeYmaera. A similar strategy could be followed for Rodin. The fact that Hybrid Event-B distinguishes cleanly between mode events and pliant events, makes it particularly evident that the POs of Hybrid Event-B [11] separate cleanly those in which discrete reasoning can be used from those in which continuous mathematics is needed. Just as for KeYmaera, the latter can be delegated to Mathematica in the first instance.

However, the fact that Mathematica is proprietary, and its reasoning is thus not open to scrutiny by users, means that complete reliance on its conclusions might not be warranted in a context where the highest levels of dependability were demanded of a Hybrid Event-B development. In such cases, the requisite fragments of continuous mathematics could be developed in Rodin-specific rulesets, and the prover could be organised to use these in preference to Mathematica in situations where they were available. By this means, more and more of the problems tackled via Hybrid Event-B and Rodin could be covered by proofs that were fully open to inspection. The present authors intend to pursue the strategy just described for the mechanisation of Hybrid Event-B.

Acknowledgement. The authors would like to thank Sanaz Yeganefard for discussions, and for help with some of the figures. Michael Butler is partly funded by the FP7 ADVANCE Project (http://www.advance-ict.eu).

References

1. KeYmaera, http://symbolaris.com
2. Report: Cyber-Physical Systems (2008),
   http://iccps2012.cse.wustl.edu/_doc/CPS_Summit_Report.pdf
3. Abrial, J.R.: Modeling in Event-B: System and Software Engineering. Cambridge University Press (2010)
4. Abrial, J.R., Butler, M., Hallerstede, S., Hoang, T.S., Mehta, F., Voisin, L.: Rodin: An Open Toolset for Modelling and Reasoning in Event-B. STTT 12, 447 (2010)
5. Alur, R., et al.: Hierarchical Hybrid Modeling of Embedded Systems. In: Henzinger, T.A., Kirsch, C.M. (eds.) EMSOFT 2001. LNCS, vol. 2211, pp. 14–31. Springer, Heidelberg (2001)
6. Alur, R., Grosu, R., Hur, Y., Kumar, V., Lee, I.: Modular Specification of Hybrid Systems in CHARON. In: Lynch, N.A., Krogh, B.H. (eds.) HSCC 2000. LNCS, vol. 1790, pp. 6–19. Springer, Heidelberg (2000)
7. Alur, R., Henzinger, T., Ho, P.: Automatic Symbolic Verification of Embedded Systems. IEEE TSE 22, 181–201 (1996)
8. Antsaklis, P., Michel, A.: Linear Systems. Birkhauser (2006)
9. Astrom, K., Wittenmark, B.: Adaptive Control. Dover (2008)
10. Banach, R., Butler, M.: Cruise Control in Hybrid Event-B. In: Liu, Z., Woodcock, J., Zhu, H. (eds.) ICTAC 2013. LNCS, vol. 8049, pp. 76–93. Springer, Heidelberg (2013)

11. Banach, R., Butler, M., Qin, S., Verma, N., Zhu, H.: Core Hybrid Event-B: Adding Continuous Behaviour to Event-B (2012) (submitted)
12. Barolli, L., Takizawa, M., Hussain, F.: Special Issue on Emerging Trends in Cyber-Physical Systems. J. Amb. Intel. Hum. Comp. 2, 249–250 (2011)
13. Butler, M.: Decomposition Strategies for Event-B. In: Leuschel, M., Wehrheim, H. (eds.) IFM 2009. LNCS, vol. 5423, pp. 20–38. Springer, Heidelberg (2009)
14. Carloni, L., Passerone, R., Pinto, A., Sangiovanni-Vincentelli, A.: Languages and Tools for Hybrid Systems Design. Foundations and Trends in Electronic Design Automation 1, 1–193 (2006)
15. Clarke, F., Ledyaev, Y., Stern, R., Wolenski, P.: Nonsmooth Analysis and Control Theory. Springer (1997)
16. DEPLOY: European Project DEPLOY IST-511599,
   http://www.deploy-project.eu/
17. Henzinger, T., Ho, P., Wong-Toi, H.: HyTech: A Model Checker for Hybrid Systems. IJSTTT 1, 110–122 (1997)
18. KeY, http://www.key-project.org
19. Kirk, D.: Optimal Control Theory: An Introduction. Dover (2004)
20. Lewis, F., Vrabie, D., Syrmos, V.: Optimal Control. Wiley (2012)
21. Mathematica, http://www.wolfram.com
22. MATLAB and SIMULINK, http://www.mathworks.com
23. MODELICA, https://www.modelica.org/
24. Platzer, A.: Logical Analysis of Hybrid Systems: Proving Theorems for Complex Dynamics. Springer (2010)
25. Ratschan, S., She, Z.: Safety Verification of Hybrid Systems by Constraint Propagation Based Abstraction Refinement. In: Morari, M., Thiele, L. (eds.) HSCC 2005. LNCS, vol. 3414, pp. 573–589. Springer, Heidelberg (2005)
26. RODIN: European Project RODIN (Rigorous Open Development for Complex Systems) IST-511599, http://rodin.cs.ncl.ac.uk/
27. RODIN Tool, http://www.event-b.org/,
   http://www.rodintools.org/,
   http://sourceforge.net/projects/rodin-b-sharp/
28. Sastry, S., Wittenmark, B.: Adaptive Control: Stability, Convergence and Robustness. Dover (2011)
29. Silva, B., Richeson, K., Krogh, B., Chutinan, A.: Modeling and Verifying Hybrid Dynamic Systems using CheckMate. In: Proc. 4th International Conference on Automation of Mixed Processes: Hybrid Dynamic Systems (ADPM 2000), pp. 323–328 (2000)
30. Sztipanovits, J.: Model Integration and Cyber Physical Systems: A Semantics Perspective. In: Butler, M., Schulte, W. (eds.) FM 2011. LNCS, vol. 6664, p. 1. Springer, Heidelberg (2011), http://sites.lero.ie/
   download.aspx?f=Sztipanovits-Keynote.pdf
31. Walter, W.: Ordinary Differential Equations. Springer (1998)
32. White, J., Clarke, S., Groba, C., Dougherty, B., Thompson, C., Schmidt, D.: R&D Challenges and Solutions for Mobile Cyber-Physical Applications and Supporting Internet Services. J. Internet Serv. Appl. 1, 45–56 (2010)
33. Willems, J.: Open Dynamical Systems: Their Aims and their Origins. Ruberti Lecture, Rome (2007), http://homes.esat.kuleuven.be/~jwillems/
   Lectures/2007/Rubertilecture.pdf
34. Yeganefard, S., Butler, M.: Control Systems: Phenomena and Structuring Functional Requirement Documents. In: Proc. ICECCS 2012, pp. 39–48. IEEE (2012)

A Method for Managing Uncertainty Levels in Design Variables during Complex Product Development

João Fernandes, Elsa Henriques, and Arlindo Silva

Abstract. Uncertainty is a constant concern in the design and development of complex systems. Managing it is important for organizations since it fosters design process improvement and optimization. This paper presents a method aiming to support large organizations understanding, quantifying and communicating uncertainty levels that normally arise during the early stage of the design process. This stage is particularly affected by decision-making uncertainty because designers and engineers have not yet acquired sufficient knowledge or possess sufficient confidence to decide what is the precise value that should be assigned to design variables in order to satisfy the customer's needs in an optimal manner. This type of uncertainty has been coined in literature as imprecision. The method proposed in this paper relies on the statistical characterization of the typical level of imprecision that should be expected by designers and engineers based on the collection of historical records of change from past product development processes. We quantify the level of imprecision based on two proxy variables: the time between changes and the magnitude of change that can be expected in new projects. In addition, our method proposes a Matrix of Imprecision that is capable of communicating uncertainty levels to the participants involved in new product development projects.

Keywords: uncertainty managament, imprecision levels, process improvement, complex product development.

João Fernandes · Arlindo Silva
ICEMS, Instituto Superior Técnico,
TULisbon, Portugal
e-mail: joao.ventura.fernandes@ist.utl.pt,
      arlindo.silva@ist.utl.pt

Elsa Henriques
IDMEC, Instituto Superior Técnico,
TULisbon, Portugal
e-mail: elsa.h@ist.utl.pt

M. Aiguier et al. (eds.), *Complex Systems Design & Management 2013*,
DOI: 10.1007/978-3-319-02812-5_9, © Springer International Publishing Switzerland 2014

1 Introduction

Large technical systems, such as a jet engine or an aircraft, are inherently complex to design and develop. Because of that, they are normally designed through decomposition into many sub-systems and components. Decomposition and subsequent integration are cornerstone principles of the systems engineering approach to complex product development [1, 2].

Although necessary in order to manage complexity, decomposition creates many interdependencies linking design activities performed by different teams which are responsible by the synthesis of solutions for different parts of the system. This occurs since there are many design parameters shared between different levels of the system, at the same system level and at system interfaces.

These numerous interdependencies also imply that the level of uncertainty of one design parameter depends upon or affects the level of uncertainty of the parameters that it is connected to. Teams working concurrently to deliver sub-system design solutions are impacted by uncertainty in variables flown down by upstream levels, for instance. Another example of interdependencies is the existence of uncertainty in the definition of interface design parameters which may potentially affect several sub-system and component design teams. Uncertainty generates change and, due to the dependencies created from system decomposition, change can propagate. Recognizing that change propagation is an important issue for complex product development management, the topic has deserved considerable academic attention during the last decade [3, 4, 5].

The authors begun by conducting exploratory research about uncertainty in complex product development through interviews to engineers and managers at a major aerospace manufacturer. The development process realized by this firm has large teams of specialized engineers working concurrently to deliver and integrate design solutions synthesized for the system, sub-systems and components and it is characterized by long design lead times. We found from the interviews conducted at the manufacturer that lack of knowledge about the level of uncertainty in design parameters communicated by one design team and used as inputs to design activities by another team generates many practical management issues during this long and concurrent development process. Questions such as "shall I wait for accurate input information?", "how much change should I expect in the value assigned to this design parameter?", "if I begin now with this value, what is the risk of having to do it all over again?" or "will I get any time saving later if I begin to work with uncertain inputs?" arise often in practice. These questions clearly reflect the engineers' need of handling the uncertainty in the design information exchanged during concurrent engineering. They also indicate the desire of avoiding unnecessary rework, thus reducing the overall develoment lead time.

Based on the needs observed from this initial qualitative research, this paper aims to support designers and engineers taking such process management decisions with more information about the typical level of uncertainty that should be expected during the different stages of product development. Among these stages, we are particularly interested in the preliminary design stage, which is when uncertainty levels are normally higher. Section 2 follows with review and discussion of prior literature on this topic. Section 3 covers the novel method proposed for understanding, quantifying and communicating uncertainty levels and contains illustrative examples. Section 4 concludes and discusses further research.

2 Literature Review

Uncertainty quantification and management is a foundational topic that has concerned researchers across a wide range of sciences and applications. However, the topic is far from consensual. An in-depth survey performed across the social sciences, physical sciences, engineering and management sciences [6] has shown that definitions and taxonomies vary considerably across these fields.

Similarly to some of the authors [7], we view uncertainty as a *situational* property that depends upon the human's subjective interpretation of the quantity and quality of the information available describing a particular phenomenon or system of interest. Being a situational property, the perception of the "types and "causes" of uncertainty thus varies with the surrounding context. Being essentially context-dependent, this interpretation of uncertainty explains why a wide range of taxonomies has been found across the different sciences. This view thus also suggests that a universal definition is unlikely to be achieved and that uncertainty quantification methods need to be adjusted to the particular situation. Consequently, our view also supports that the choice of the quantification technique should take into account the type, quantity and quality of the information available and the language required by the user who will interpret the meaning of uncertainty [7].

2.1 An Uncertainty Definition

In this paper, we address the topic of uncertainty management in complex product development. Our research specifically approaches situations in development processes where design parameters of uncertain value are inputs to activities performed by multiple interdependent design teams. In this context, we define uncertainty as absence of sufficient *knowledge, definition* or *confidence* in design variables that prevent design process participants to specify or predict a priori the system being designed - its behavior and properties - in a deterministic and quantitative way.

There are two key ideas in this definition. The first is that uncertainty arises from lack of knowledge, definition and confidence about the value assigned to design variables. The second is that this uncertainty inhibits a deterministic approach to the design process. Nevertheless, our definition is sufficiently general to encompass the different types of uncertainty affecting design variables that have been previously discussed in design literature. The subsequent sections discuss them and provide our own view of the topic.

2.2 Uncertainty in Design Variables

2.2.1 Variability, Aleatory and Stochastic Uncertainty

Variability, aleatory uncertainty or stochastic uncertainty are alternative terms which have been used in prior literature to designate the amount by which a design variable is expected to vary about its *nominal* value as a result of processes that engineers do not fully control or comprehend. A common example is changes in part dimensions as a result of variability in manufacturing processes. Other examples include, for instance, changes in expected component life due to variability in raw material properties or in-service loads. It ultimately relates to lack of knowledge about factors that affect the true value of design variables. Variability affects the performance of technical systems and its study led to the development of robust design methods [8, 9] and various probability-based methods in reliability and safety engineering [10, 11] aiming to explicitly accommodate noise and uncontrolled variability in the design process of these systems.

2.2.2 Model Uncertainty

Model uncertainty refers to the amount by which the true value of a design variable differs from the value *computed* for it. Analysis and decision-making in engineering design is normally supported by some type of mathematical model [12]. These models contain uncertainty, since all embed mathematical and physical simplifications of reality and numerical and programming approximations.

There are many examples in the product development context. Mathematical and numerical simplifications in engine design are required, for instance, due to incomplete knowledge about complex physical processes such as combustion and turbulence. All researchers and practitioners involved in modeling and simulation have traditionally addressed the issues of model calibration, validity and accuracy [13] and thus the issue is transversal to all disciplines used in integrated product development environments, such as structural, aerodynamic, thermal and control systems modeling.

Another issue that arises is to understand and match different levels of model uncertainty resulting from multi-fidelity predictions [14]. The design process normally evolves from early estimates of design variables based on rough models to more accurate computations of their values based on complex models during later stages. However, situations occur in design processes when higher fidelity models are used earlier in conjunction with lower fidelity models. Since the former receives data inputs which are expected to contain higher model uncertainty, tuning the end result's fidelity is a considerable challenge even for model experts. In addition, experimental models used in physical validation of design features and behaviours and during prototype development also contain uncertainties of the same type.

2.2.3 Design Imprecision

Several authors [15, 16, 17] identified and defined another form of uncertainty in product design and development: design imprecision. This term has been coined to express variations in the values assigned to design variables due to updated *preferences* of the designer. The design process involves choosing values for such design variables within permissible ranges of the design space. To a reasonable extent, this process is similar to a decision-making process of choosing among alternatives. Imprecision thus expresses uncertainty in decision-making. The preferred value for a design variable is seen to *emerge* from the design process [16] and imprecision captures the decision-making uncertainty surrounding it.

This form of uncertainty is known to be highest at the preliminary design stage [15]. During this stage, requirements have not solidified yet and often change; engineers have more freedom to switch between alternative values that can be assigned to design variables; and design teams explore concurrently several concepts for the system, sub-systems and components. Introducing degrees of flexibility in the system's architecture or in some components has been a recent approach to mitigating the effects of this type of uncertainty in development projects [18]. Design iterations - which are normally faster during this stage due to the use of lower fidelity design tools - gradually reduce imprecision in design variables during the process and allow participants to gain knowledge and confidence about what is the best solution to choose among the alternatives.

This imprecision reduction process is illustrated in Figure 1. By the end of the preliminary design stage, engineers have typically committed to one concept (down

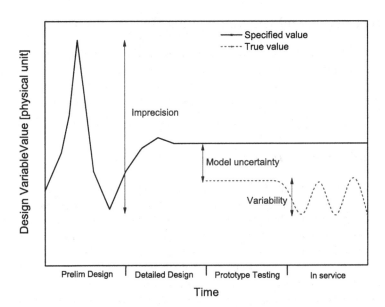

Fig. 1 Uncertainty surrounding the value of a design variable during complex product development

selected among alternatives) which is delivered to the detailed design stage. Model uncertainty and variability are also represented in Figure 1, showing how the true value may differ from the value specified to the variable during later stages of complex product development.

In addition, various advanced mathematical concepts have been used in prior literature to quantify uncertainty levels arising from imprecision. Fuzzy set theory [19] has been proposed to represent imprecision and compared against the frequentist probability approach [15, 17]. Interval analysis [20] has been explored to analyse, predict and optimize the performance of a brake system [21]. Designing for safety and reliability were examined by other authors [22, 23] through the use of possibility theory [24]. And subsequent research [25] argued that when uncertainty is large the use of imprecise probability theory [26] generates designs with higher expected utility than the traditional probability approach.

However, our experience with industrial firms shows that these mathematical concepts remain relatively unfamiliar to the engineers involved in practice in the design and development activities of complex systems. This is probably why, to the authors' knowledge, prior work has remained essentially confined to academic design examples. We claim that this constitutes a research *gap* and thus there is room for alternative approaches to uncertainty management that show its applicability in industrial environments. Consequently, our research aims to complement and extend prior literature on uncertainty in complex product development with the proposal of a novel method for managing uncertainty levels under the form of design imprecision.

3   Quantifying and Communicating Uncertainty Levels

This section presents the five main steps included in the proposed method for understanding, quantifying and communicating uncertainty levels under the form of imprecision during complex product development projects. We have coined the method as the Imprecision Management Method (IMM). Figure 2 illustrates the method and supports the discussion.

3.1   Collection of Historical Records of Change

The method starts with the collection of historical records of change characterizing the behavior of uncertain design variables - which are of interest to the organization - during past product development projects (Figure 2). We propose to characterize the level of imprecision in design variables using past records of changes. We argue that design variables change during the design process to reflect the evolution in design choices and preferences of participants. The information about changes is normally recorded in various project documents produced during product development, such as requirements documents, communication records, engineering memos and design reports. Each organization thus contains an historical body of knowledge about design changes which can be captured from such documents.

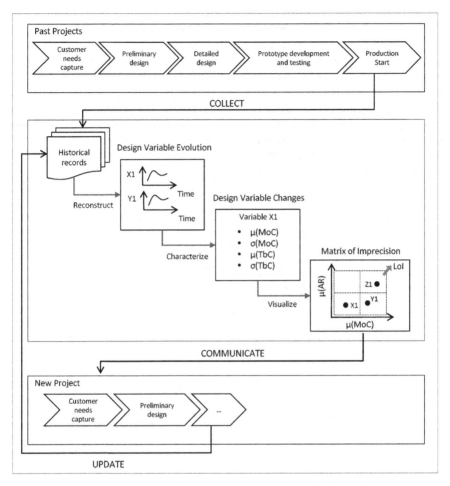

Fig. 2 The Imprecision Management Method. Legend: μ - arithmetic average; σ - standard deviation; MoC - magnitude of change; TbC - time between changes; AR - arrival rate of change; LoI - level of imprecision.

3.2 Reconstruction of the Variables' Time Evolution

The second step in the method uses historical records of change retrieved from past project documents to reconstruct the time evolution of the uncertain design variables that were selected by the organization. This time evolution allows us to identify and study the changes that occurred in each of the variables. Furthermore, the changes identified for each variable can be grouped into a sample that is eligible for statistical analysis. Because of that, it is important that changes are collected from *multiple* projects so that unbiased and statistically significant samples can be attained.

The benefit of the reconstruction of the variables' time evolution is illustrated in Figure 3, which represents a fictious time evolution for design variable X1 during a complex product development project. X1 is one of the variables presented in the overall representation of IMM (Figure 2). The general behaviour shown in Figure 3 is that X1 varies frequently and significantly during the early stage of the development process and stabilizes after a certain amount of time has passed. This step of the method thus allows us to divide the evolution of the design variable into different *time periods*, according to how the level of uncertainty is also seen to evolve during time. In this illustrative example, our analysis could lead us to study and charaterize the level of uncertainty of X1 during two periods, the one comprised between 0 and T1 and the one between T1 and T2.

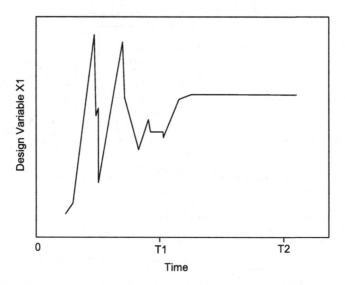

Fig. 3 An illustrative time evolution for design variable X1 during a complex product development project. The periods comprised between 0-T1 and T1-T2 can be identified as time periods of significantly different levels of uncertainty.

3.3   Statistical Characterization of Imprecision

The third step in the IMM is to perform a statistical analysis to the sample of changes found from several past projects. We view the level of change that occurred during a project as a *proxy* of the underlying level of design imprecision. The authors thus propose that the level of uncertainty under the form of imprecision in a design variable is inferred from:

(a) Statistical characterization of the **magnitude of change** (MoC), i.e., the amount which represents the shift in preference relatively to a value previously assigned. Figure 2 illustrates that this characterization can be performed using, for instance, the average amount of change ($\mu(MoC)$) and the standard deviation ($\sigma(MoC)$) found for a generic design variable X1.

(b) Statistical characterization of the **time between changes** (TbC), i.e., the span observed between two consecutive decisions of changing the value assigned to the variable. The average and standard deviation are proposed in Figure 2 to characterize the time between changes. In addition, the inverse of TbC - which represents an *arrival rate of change* (AR) - can also be used.

IMM characterizes these proxy variables using traditional statistical measures: average values, standard deviations, probability density functions, etc. However, it should be noted that both the MoC and the TbC are treated as random variables, which is an underlying assumption in the approach.

3.4 Communication of the Typical Levels of Imprecision to New Projects

The purpose of the approach is to communicate the meaning of design uncertainty under the form of imprecision in a language familiar to engineers. For instance, the typical level of imprecision in design variable X1 can then be characterized as the interval $X1 \pm \mu(MoC(X1))$. To promote a visual representation of the typical level of imprecision surrounding design variables - which facilitates communication in organizations - we propose in this paper a Matrix of Imprecision (MoI). This matrix plots the average amount of change and the average arrival rate of changes that should be expected during product development (Figure 2). The MoI is read similarly to a risk matrix: higher arrival rates and higher magnitudes of change relate to a higher level of imprecision (LoI). The MoI can thus inform designers and engineers engaged in new projects of the typical level of design uncertainty under the form of imprecision that is likely to be encountered during the project.

Figure 4 adds further details to the MoI represented in Figure 2. In order to enhance the matrix's communicational power, distinct imprecision *windows* can be added to position variables according to higher or lower typical levels of uncertainty. We present a proposal containing 4 windows: low AR/ low MoC; low AR/ high MoC; high AR/ low MoC; and high AR/ high MoC (Figure 4). The idea can be naturally extended to other topologies that may be defined by organizations.

Furthemore, Figure 4 depicts a set of illustrative data showing that different design variables - X1, Y1 and Z1, for instance - will be characterized by different levels of uncertainty. In our illustrative example, the matrix presented is able to show in a visual manner that the $LoI(Z1) > LoI(X1_{0-T1}) > LoI(Y1) > LoI(X1_{T1-T2})$. Engineers engaged in product development of components affected by X1, Y1 or Z1 could use the information contained in the MoI to delay the start of new design iterations involving any of the variables when a new change is expectable (according to the expected TbC or AR). Or they may decide to perform sensitivity studies and design for higher robustness to change in order to accommodate changes below a chosen threshold (the average MoC, for instance). Such strategies may be devised in practical terms using the outcome of this step of the IMM.

Figure 4 shows additionally that the MoI can present and communicate how the imprecision level of a particular design variable evolved during *different stages* of the product development process. The $LoI(X1_{0-T1})$ and $LoI(X1_{T1-T2})$ appear from the statistical characterization of the level of uncertainty in $X1$ according to the time periods identified in Figure 3 for this design variable (0 to T1 and T1 to T2). Similarly, we argue that the MoI can be used to compare imprecision levels between *different projects* performed across the organization.

The authors believe that this tool's main strength is joining *quantitative* and *qualitative* information into a single representation. We claim that it can support organizations managing complex design processes and be used simultaneously as a knowledge management tool.

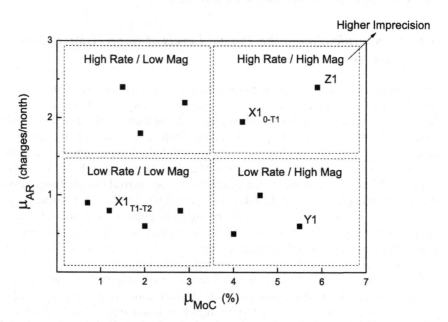

Fig. 4 A Matrix of Imprecision showing the typical level of uncertainty under the form of imprecision that can be expected in an illustrative set of design variables of interest to the organization, such as X1, Y1 and Z1

3.5   *Knowledge Update from New Projects*

The last step proposed in the method is a continuous update and refinement of the historical body of knowledge about change that supports it. We illustrate in Figure 2 that completion of a new product development project triggers a knowledge update step with new information about the changes observed in the design variables during that project. This ensures a continuous increase in change records and the size of the samples. Therefore, steps 1 to 4 shall be repeated in subsequent analysis with a larger dataset which leads to increasing accuracy in predictions with the progressive use of the method. This is an important step in the method, since previous steps in the IMM rely essentially on inference (future levels of imprecision are inferred from past events and from an available sample of observations of change).

4 Conclusions and Further Research

This paper contributes with a management method aiming to support industrial organizations engaging in complex product development projects understanding, quantifying and communicating uncertainty levels. Our motivation originates from the needs of designers and engineers. Design imprecision - a type of uncertainty related to decision-making during design - triggers many changes in the values assigned to design variables, particularly during the early stages of the process. Designers and engineers must decide how to organize design tasks and often do so without sufficient knowledge about the typical level of imprecision that they should expect in inputs to their activities.

The method proposed in this paper is able to quantify the level of uncertainty that should be expected in design variables and thus supports designers and engineers optimizing their processes. Our structured approach to imprecision management relies in five main steps: 1) collection of historical records of change in uncertain variables from past projects; 2) reconstruction of the design variables' time evolution using the records collected; 3) statistical characterization of design imprecision based on two proxy variables, the magnitude of change and the time between changes; 4) communication of imprecision levels to new projects based on a matrix of imprecision; and 5) continuous knowledge update from new projects.

The method's main strengths are twofold: its relative simplicity when compared to other methods based on advanced mathematical concepts which makes it easier to apply in industrial practice; and the caracterization and communication of uncertainty arising from decision-making during the early design phase based on metrics which are familiar to engineers and designers. The predictive power of future levels of uncertainty is its main limitation, since the method relies on the collection and analysis of past data about the behaviour of design variables. However, the knowledge update step included in the method - feeding back data arising from new projecs - ensures a continuous increase in change records and in the size of the samples and thus leads to increasing accuracy in predictions with the progressive use of the method.

Further research includes the application of the proposed method in industrial environments with the purpose of demonstrating its feasability and usefulness. An empirical case-study is ongoing at a major aerospace manufacturer in order to fullill these research goals.

Acknowledgments. Funding from the Portuguese Foundation of Science and Technology under the doctoral grant SFRH/BD/51107/2010 is gratefully acknowledged.

References

[1] Blanchard, B.S., Fabrycky, W.J.: Systems engineering and analysis, 4th edn. Prentice Hall (2006)
[2] NASA. Systems engineering handbook. National Aeronautics and Space Agency, 1st edn. (2007)
[3] Clarkson, P.J., Simons, C., Eckert, C.: Predicting change propagation in complex design. Journal of Mechanical Design 126(5), 788–797 (2004)

[4] Giffin, M., Weck, O.D., Bounova, G., Keller, R., Eckert, C., Clarkson, P.J.: Change
     propagation analysis in complex technical systems. Journal of Mechanical De-
     sign 131(8), 081001 (2009)
[5] Hamraz, B., Caldwell, N.H.M., Clarkson, P.J.: A multidomain engineering change prop-
     agation model to support uncertainty reduction and risk management in design. Journal
     of Mechanical Design 134(10), 100905 (2012)
[6] Thunnissen, D.P.: Propagating and mitigating uncertainty in the design of complex mul-
     tidisciplinary systems. PhD thesis, California Institute of Technology, USA (2005)
[7] Zimmermann, H.J.: An application-oriented view of modeling uncertainty. European
     Journal of Operational Research 122(2), 190–198 (2000)
[8] Tsui, K.L.: An overview of taguchi method and newly developed statistical methods for
     robust design. IIE Trans. 24(5), 44–57 (1992)
[9] Chen, W., Allen, J.K., Tsui, K.L., Mistree, F.: A procedure for robust design: mini-
     mizing variations caused by noise factors and control factors. Journal of Mechanical
     Design 118(4), 478–485 (1996)
[10] Kapur, K.C., Lamberson, L.R.: Reliability in engineering design, 1st edn. John Wiley
     and Sons, New York (1977)
[11] Cornell, C.A.: A probability-based structural code. Journal of American Concrete Insti-
     tute 66(12), 974–985 (1969)
[12] Goh, Y.M., McMahon, C.A., Booker, J.D.: Development and characterization of error
     functions in design. Research in Engineering Design 18(3), 129–148 (2007)
[13] Arendt, P.D., Appley, D.W., Chen, W.: Quantification of model uncertainty: calibration,
     model discrepancy, and identifiability. Journal of Mechanical Design 134(10), 100908
     (2012)
[14] Sinha, A., Bera, N., Allen, J.K., Panchal, J.H., Mistree, F.: Uncertainty management in
     the design of multiscale systems. Journal of Mechanical Design 135(1), 011008 (2013)
[15] Wood, K.L., Antonsson, E.K., Beck, J.L.: Representing imprecision in engineering de-
     sign: comparing fuzzy and probability calculus. Research in Engineering Design 1(3-4),
     18703 (1990)
[16] Otto, K.N., Antonsson, E.K.: Design parameter selection in the presence of noise. Re-
     search in Engineering Design 6(4), 234–246 (1994)
[17] Antonsson, E.K., Otto, K.N.: Imprecision in engineering design. Journal of Mechanical
     Design 117(B), 25–32 (1995)
[18] de Neufville, R., Scholtes, S.: Flexibility in Engineering Design, 1st edn. The MIT
     Press, Cambridge (2011)
[19] Zadeh, L.A.: Fuzzy sets. Information Control 8, 338–353 (1965)
[20] Moore, R.E.: Interval analysis, 1st edn. Prentice Hall, New Jersey (1966)
[21] Cao, L., Rao, S.S.: Optimum design of mechanical systems involving interval parame-
     ters. Journal of Mechanical Design 124(3), 465–472 (2002)
[22] Nikolaidis, E., Chen, S., Cudney, H., Haftka, R.T., Rosca, R.: Comparison of probabil-
     ity and possibility for design against catastrophic failure under uncertainty. Journal of
     Mechanical Design 126(3), 386–394 (2004)
[23] Mourelatos, Z.P., Zhou, J.: Reliability estimation and design with insufficient data based
     on possibility theory. AIAA Journal 43(8), 1696–1705 (2005)
[24] Dubois, D., Prade, H.: Possibility theory, 1st edn. Plenum Press, New York (1988)
[25] Aughenbaugh, J.M., Paredis, C.J.: The value of using imprecise probabilities in engi-
     neering design. Journl of Mechanical Design 128(4), 969–979 (2006)
[26] Walley, P.: Statistical reasoning with imprecise probabilities, 1st edn. Chapmal Hall,
     London (1991)

Capturing Variability in Model Based Systems Engineering

Cosmin Dumitrescu, Patrick Tessier, Camille Salinesi,
Sebastien Gérard, Alain Dauron, and Raul Mazo

Abstract. Automotive model-based systems engineering needs to be adapted to the industry specific needs, in particular by implementing appropriate means of representing and operating with variability. We rely on existing modeling techniques as an opportunity to provide a description of variability adapted to a systems engineering model. However, we also need to take into account requirements related to backwards compatibility with current practices, given the industry experience in mass customization. We propose to adopt the product line paradigm in model-based systems engineering by extending the orthogonal variability model, and adapting it to our specific needs. This brings us to an expression closer to a description of constraints, related to both orthogonal variability, and to SysML system models. We introduce our approach through a discussion on the different aspects that need to be covered for expressing variability in systems engineering. We explore these aspects by observing an automotive case study, and relate them to a list of contextual requirements for variability management.

Cosmin Dumitrescu · Alain Dauron
RENAULT
1 Av. du Golf
78288 Guyancourt, France
e-mail: {cosmin.dumitrescu,alain.dauron}@renault.com

Patrick Tessier · Sebastien Gérard
CEA, LIST
Laboratory of Model Driven Engineering for Embedded Systems
Point courrier 174 F91191 Gif Sur Yvette, France
e-mail: {patrick.tessier,sebastien.gerard}@cea.fr

Camille Salinesi · Cosmin Dumitrescu · Raul Mazo
Université Paris I Panthéon-Sorbonne
90 rue de Tolbiac, 75013 Paris, France
e-mail: {camille.salinesi,raul.mazo}@univ-paris1.fr

M. Aiguier et al. (eds.), *Complex Systems Design & Management 2013*,
DOI: 10.1007/978-3-319-02812-5_10, © Springer International Publishing Switzerland 2014

1   Introduction

Product diversity has an impact on all organization levels. However, one of the areas that could benefit and improve in respect to the management of variability in automotive, is model based systems engineering (MBSE). Our purpose is to introduce a variability modeling technique in MBSE, that would bridge the gap between vehicle features and component specifications, by introducing variability management on an intermediate level. This would enable engineers to design systems for reuse. For instance, reuse modules, components, and also specifications, requirements, or documented system architectures. We aim to draw on the valuable experience provided by software product line engineering and either adopt or adapt modeling techniques for model based systems engineering context and existing product variety practices at Renault.

1.1   Motivating Examples

The problem we are facing can be analyzed from multiple perspectives: from an information systems point of view (structure, storage, accessibility of data, compatibility with legacy systems), modeling techniques and extension of systems engineering meta-models, processes and specific practices. We need to adopt a research protocol which enables us to gradually refine our research problem in respect to each of these aspects, as explained in Section 4.

However, in order to provide some insight on the context and modeling issues, we present some simple variability modeling examples.

Issue 1: Need for flexibility and detail. The Constraint Satisfaction Problem (CSP) formalisms are suited for a great variety of problems. For instance, they can be used for product configurations, and the representation of the catalog or range of products, at Renault [2]. This representation provides an efficient mean of interacting with customers through online configurators in order to create custom vehicle configurations.

For example, let's consider the following boolean variables:

1. *Door opening system*: SOPA, SOPB, SOPC (three types of door opening systems);
2. *Window lifts*: FMWL (front manual window lifts), EFWL (electric front window lifts), FAPWL (front anti-pinch window lifts), EIWL (electric impulse window lifts), FIDEP (front electric impulse for driver side and electric for passenger side);

and the constraint:

$$SOPC \Rightarrow (EFWL \vee EIWL \vee FIDEP) \tag{1}$$

Constraint 1 imposes that variant *SOPC* of the vehicle *door opening system* requires the use of only one of the three variants of window lifts: "electric front window lifts", "electric impulse front window lifts", or "front electric impulse for driver side and electric for passenger side window lifts".

In the case of Renault, the CSP formalism, implemented through the organization "documentary language", captures characteristics of already developed products. This is why it is only possible to take into consideration some of these constraints only once the product is fully developed and the appropriate variables and constraints were added to the lexicon (variables in the "documentary language" expressing vehicle characteristics). However, some of these constraints, as in the case of constraint 1, are technical by nature and are discovered only during system design. Furthermore, detailed requirements and design variability is not captured through the "documentary language", which would enable system engineers to apply configuration techniques for the reuse of requirements and specifications on any level of detail.

Renault's product range is very broad, leading to 10^{21} possible configurations for the Traffic van product family [2]. On the one hand, further increasing the number of variables, to capture more detailed variability for each vehicle system, would only create even more complex problems for online configurators. On the other hand, variability expression in MBSE needs to be compatible with legacy systems, although variability models could be local for each family of systems (vehicle subsystems).

Issue 2: Representation of "complex" constraints. One of the modeling techniques for variability, the orthogonal variability model (OVM), was introduced by Pohl et al. [16] to manage variability in the context of software product lines. The formalism, presented in Figure 1, enables the representation of constraints such as "requires" and "excludes" between variants and variation points. To express constraints 1 using OVM, we need to introduce an auxiliary variation point - *AuxiliaryVP* - as shown in the diagram[1] from Figure 1.

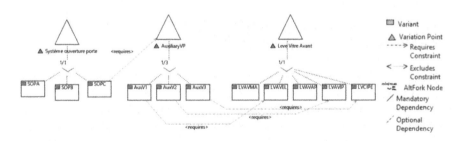

Fig. 1 Using an auxiliary variation point to express constraints using OVM notation

---

[1] Diagram realized using the VARMOD Prime Tool
(http://www.sse.uni-due.de/en/projects/varmod-prime)

Issue 3: Variability in system architectures. Another formalism for variability modeling is the Feature Model (FM) [13]. FMs have become the most widely accepted modeling technique for variability management in the field of software product lines . FMs have the advantage of being easy to understand and independent of specific domains. Figure 2 presents two possible ways for modeling the functional decomposition for the electric parking brake system (EPB): (a) using the FM hierarchy to represent common and optional functions[2] and (b) applying the stereotype << *variableElement* >> [26] to distinguish common and optional functions in SysML(UML).

In the case of system architectures, hierarchies are already introduced through decomposition, such as the functional or physical structures, which means we can imagine two scenarios to make use of FMs. In the first case, a partial feature structure may be deduced from variability in the system architecture along with the constraints imposed by the semantics of the SysML system model. However, characteristics defined in the "documentary language" are not structured in the same hierarchical way. In a second case, the system engineer would create hierarchies for both variability and architectures. However, this would result in redundant information and increase in the model complexity.

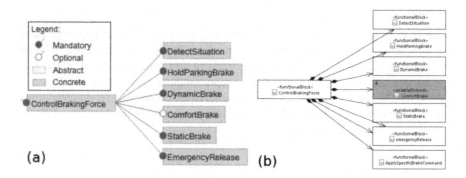

Fig. 2 Electric Parking Brake system (partial) functional decomposition (a) using the FM notation; (b) using stereotyped "optional" SysML block elements

We propose in this article to adopt an extended orthogonal variability model Co-OVM (Constraint Oriented Orthogonal Variability Model) to cover specific needs for systems engineering at Renault. Co-OVM is implemented as a UML profile in the SysML modeler Papyrus as an extension to Tessier's Sequoia plug-in [26].

The article is structured in 7 parts. Following the introduction, we present in Section 2 an overview of existing variability modeling techniques. Section 3 develops the subject around different aspects that should be taken into account to introduce variability management in our MBSE framework. Section 4 introduces the

---

[2] Diagram realized using FeatureIDE

http://wwwiti.cs.uni-magdeburg.de/iti_db/research/
featureide/

requirements for a variability management approach in the context of systems engineering at Renault and the meta-model based on the identified concepts. Section 6 explains the implementation and results regarding MBSE variability management tools, based on the modeler Papyrus. Finally, Section 7 concludes the paper by some perspectives to our work.

2 Related Work

Many variability modeling techniques exist [13][16][26][22]; however, the simple implementation of such an approach in a large organization would have some shortcomings and challenges, as pointed out by Filho et al. [10]. They would have to be adapted to the specific context of the organization.

Variability models diverged into numerous dialects through extension, adaptation or through application in general purpose modeling languages.

By drawing on comparisons and classifications [7][6][1][23] for an overview of variability modeling techniques, we propose the following synthesis: (a) distinct conceptual models for variability; (b) embedding variability in existing modeling languages or artifacts; and (c) realization techniques [23], enabling technologies (e.g., off-the-shelf solvers for constraints). A short summary presenting the classification of these techniques is presented in table 1, which do not take into account aspect oriented modeling for variability, as we considered this to be a software specific technique.

Our aim is to embed variability concepts in SysML models, while providing compatibility to the company "documentary language" and support for our MBSE framework [5].

Table 1 Classification of variability modelling techniques

VM technique class	Standalone variability models	Applied/ embedded variability	Enabling techniques
Feature Model based	FODA[13] , FeatuRSEB, FORE, FOPLE, GP, CBFM, Forfamel	UML Integration of features [17]; Requi-Line (requirement engineering) [14]	
Variation Point based	Orthogonal Variability Model (OVM)[16], Gomma and Weber, Quality Aware OVM [19], COVAMOF[23], VSL[3]	UML integration of OVM: Halmans and Pohl [12]; von der Maßen and Lichter [14]	
Constraint Based		Sequoia (Tessier et al.)[26], Feature Modelling Constraint Language (FMCL) [11]	Salinesi et al.[22], Streitferdt et al. [24]

3 Identification of Concepts for the Representation of Variability in Systems Engineering

System models alone are not able to completely ensure traceability from requirements to the implementation when variability is present. For example, variability traceability needs to be ensured in relation to the project or business context and system assets, answering to the questions: who specifies variability?, what is the rationale for increasing variety?, to which context does a particular expression of variability apply? Based on our industrial experience, we can claim that variability is present not only on the client offer level, but also on the solution level through design alternatives and replaceable components.

The system model needs to be complemented with relevant information for capturing variability. Several aspects are essential for the representation of variability in model based systems engineering, which shall be discussed in this section.

Types of Variability. The central concept, implicit in all variability languages is *the choice* [15]. The act of performing *choices* can be regarded either as creating a configuration (e.g., online configurators) or as decision making, involving specific engineering domain information, with impact on the final derived system model.

Indeed, approaches such as DOPLER [18] refer to choices as decisions, and the "decision oriented" model represents a link between the domain of assets and the derived system. "Features" capture high level variability in requirements, exposed to the customer (or stakeholders) and enable system configuration from a higher level of abstraction, without focus on decisional impact.

In respect to the design space, a decision oriented derivation would be directed from the problem to the solution space, whereas a non-decision oriented configurations would not take into account any particular workflow. To our opinion all these approaches are useful at certain phases in the lifecycle of the system. We can thus identify: *design* decisions with impact on the system model, characteristics that contribute to products *diversity* and replaceable system parts (e.g., *components* from different suppliers, and COTS).

Sources of Variability. Because existing vehicle level variants further impact the development each subsystem, it is important to trace their origin, where they first occur. For example, variations for the gear box (automatic or manual), indirectly require the development of alternative behaviors for the electric parking brake (EPB) system. Since not all vehicle models propose both automatic and manual gearboxes, it is important to trace the origin of variability and the targeted configurations.

The source may be one of the system levels of decomposition, abstraction level (operational scope, behavior, physical), technical context (enabling systems), environment (climate, country external conditions etc.), stakeholder viewpoints, or the "documentary language". When the system model does not provide sufficiently fine grained information for traceability, it is possible to add more detailed information in respect to the variation source, to point to specific model elements. Once introduced, the variation may further impact lower levels of abstraction. (e.g., functional variability may induce the same variations of a physical level). One interesting aspect of automotive systems is that there is much more physical variability in systems - related to alternative or shifting technologies, and component suppliers - rather than functional variability.

Legacy Variability Specification. Product variety is not a new concern in the automotive industry. new. As systems engineering takes into account variability, both the tools and existing knowledge for mass customization need to remain compatible and provide a base for new MBSE tools.

The "documentary language" enables the description of variability on the vehicle level. Astesana et al.[2] describe the way variety (or "diversity") is expressed and exploited at Renault: as constraint satisfaction problem variables.

From a representation of data point of view, variables are grouped by the corresponding variation subject (e.g., type of fuel : diesel, gasoline etc.), which brings us closer to the modeling techniques known in product line engineering. In the case of the documentary language, there is no distinction between "mandatory" or "optional" variables and all defined variables need to be assigned a value [2]. Other semantic properties (like commercial packs of options, default options, preferences etc.) are specific to marketing and commercial activities. They are neither specific to systems engineering, nor crucial for compatibility.

Constraints, The Core of Product Line Engineering. Existing vehicle features (at Renault) are represented as Boolean expressions. At the core of product line engineering practices is constraint programming [21][15][4][2], which enables both the representation and the analysis of product line models. This is indeed the aspect addressed by all variability models, which is essential for the derivation of a single system model. At Renault, constraints are introduced through the commercial offer and stakeholder requirements, and also due to technical, architectural dependencies or supplier constraints. Technical constraints are the result of dependencies to variable resources from within the system or from external enabling systems. Sometimes technical constraints need to be rendered visible for the commercial offer (e.g., the technical constraint *a navigation system requires the presence of a CD player* needs to be visible for the commercial offer). Finally, we need to be able to express constraints both in variability models and in system architecture, to asses impact of variant selection.

Variability in System Architecture Models. Tseng & Jiao [27] define *variety generation* as a "way in which the distinctiveness of product features can be created". They also define the basic methods in which variety can be generated: attaching (*presence*), swapping (*replacement*), and scaling (*parametrization*). More complex *variety generation* techniques can be created based on the repeated use of these basic methods. The purpose of these techniques is to introduce *variety* through components, by proposing different configuration items which satisfy system requirements, for the conception of the system. These concepts are also present in the Common Variability Language (CVL) [25], described with slightly different vocabulary: existence, value assignment, substitution.

The variability model is linked to the architecture model through mapping relationships. We considered an *"n to n"* relation for the mapping of variant choices on system model elements (binding). In this case, the semantics of the mapping relationships should be specified by the user as follows: (i) *at least one selection* of the linked variants validates the presence of the system element, or (ii) *all selections* of the linked variants validate the presence of the system element. This type of relationships can be represented as constraints, which are specific to the architecture of the family of systems.

Multiple Viewpoints. It is useful to take into account existing system viewpoints to navigate through the available variability, and relate variability choices to the system analysis being performed. The definition of additional viewpoints on variability can broaden the scope of the system analysis (e.g., taking into account supported variability for suppliers may influence decisions for component selection). Furthermore, in our case, we also rely on the definition of viewpoints to partially define the flow of choices for product line derivation [9]. Successive partial derivations enable the engineer to gradually reduce the solution space and generate specific deliverables for each project milestone.

Customer Oriented Variability. The customer is represented within the organization by the "product department", which manages, structures and proposes the commercial offer, by defining the vehicle *options*, *packs of options* and *levels of equipment*. From a systems engineering point of view we consider the commercial offer as a part of the stakeholder requirements and we choose not to have any additional representations for this type of variability.

4 Research Approach and Requirements for the Description of Variability in Renault MBSE

The research protocol we used to define the model is based on the development of engineering scenarios for families of systems[8] as well as an ontology for variability. The ontology served as a mean of aggregation of concepts relative to engineering practices from within the organization, and from relevant literature on product lines. This approach is represented in Figure 3. At the same time the variability management ontology provides the specifications for developing tool support for MBSE, based on Papyrus SysML modeler.

The use of ontologies for variability has several benefits in our context. Ruiz & Hilera [20] define the objectives for using ontologies in software design and development. In particular they mention the following applications: specification to specify the requirements and components definitions or the system functionality; knowledge acquisition to rely on ontologies as a guide for the knowledge acquisition process.

Experimenting with the example of the Electric Parking Brake System (EPB) allowed us to better understand and confirm the issues that we identified through interviews and exploration of the company practices.

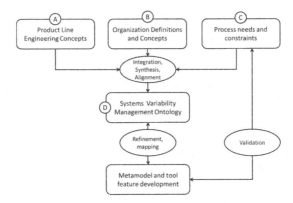

Fig. 3 Approach for the discovery of variability concepts

While we have identified requirements in a more exhaustive manner, we present briefly the core requirements that the method for variability management should satisfy:

R#01. The approach shall provide a representation for reuse, in extension to the system modeling paradigm (used at Renault).
R#02. The approach shall be compatible/integrated to existing legacy tools and processes in respect to variability.
R#03. The approach shall support current reuse strategies.
R#04. The approach shall be scalable, applicable to multiple levels of system decomposition.

In Table 2 we relate these requirements to the aspects regarding the representation of variability for systems from the Section 3, without regarding the requirement refinement details.

5 An Orthogonal Variability Meta-model for the Context of Renault Systems Engineering

Choosing a variation point based model for the description of variability, means that we can use a similar description for the different types of variability (design decisions, system characteristics etc.) and at the same time we can rely on the existing "documentary language" model which is in use in legacy systems. It is also a convenient representation when taking into account viewpoints, since the simple variation point based structure does not pose consistency problems.

The implementation is based on the the modeler Papyrus[3] and it extends the plug-in for product lines, Sequoia [26]. Sequoia already supports expression of variability

---

[3] http://www.eclipse.org/papyrus

Table 2 Overview of the covered aspects of variability

Aspect	Specific Concepts	Requirements
Types of variability	Diversity, DesignDecision, ComponentChoice	R#01
Sources of variability	FunctionalScope, PhysicalScope, TechnicalContext, OperationalScope, SystemEnvironement, Stakeholders, DocumentaryLanguage	R#01
Legacy variability specification	Object, Criteria, MTC	R#02
Constraints	Dependencies, VariantState, Seqouia VariationGroup[26]	R#01
Impact of choices on system elements	Binding, VariationForm	R#01
System model variability visibility	VariabilityImpact	R#01
Methodology support	DiversityContext, DiversityUseCase, Target (Vehicle) Project	R#01, R#01, R#04
Customer oriented variability	Vehicle Options, Packs of Options, Levels of Equipment etc.	not included

by using constraints between optional SysML/UML model items. However, we need to adapt it's functionalities in respect to the specific MBSE context at Renault.

Variability concepts were defined as a UML profile, that extends current notions present in SysML and in the SysML-R profiles [5]. It is worth pointing out, that variants are implemented as stereotypes on use-cases, which was also proposed by Halmans et al. [12], and by von der Maßen and Lichter [14]. This, along with the existing Sequoia profiles for modeling constraints, allowed us to adapt the use-case diagrams for modeling constraints between variants. To the model to be orthogonal, variability is stored separately from the system model and may be reused across different system models. We also take advantage of the existing Sequoia representation of constraints in UML to capture constraints issued by the definition of the product line through the representation of variability dependencies in the variability model, but also constraints issued from the propagation of variability on the system's architecture. The concepts for representing variability are presented in Figure 5.

6   Modeling Tool Example

We have used two simple systems to experiment with our models : the electric parking brake (EPB) and the automatic lighting system. They were exploratory case studies simple enough to provide us with valuable information through examples and observation.

The EPB model we considered, contains variations on all abstraction levels, and allowed us to exercise several scenarios [8] in respect to the sequence of choices that lead to different configurations :

(a) reuse of two alternative designs (based on a single DC motor and a puller cable, or caliper mounted electrical actuators);

(b) reuse of problem definition and introducing new solution alternatives (based on reduced power design, and relying on complementing braking force from the hydraulic brakes);

(c) separating or combining optional system functions thorough allocation alternatives (i.e., providing of the "hill start assistant" function through the hydraulic or electric parking brakes)

In Figure 4 we represent the context diversity for the EPB system: variants that contribute to defining the scope of the system, relative to stakeholder requirements and the system environment (e.g., the system is design only for Temperate Climate, EU countries). Usually engineers design a system for particular ranges of vehicles, meaning that they target from the start of the project certain configurations. Variants and constraints for a particular vehicle project my also be represented as individual diagrams. The design should be realized with the largest possible scope of re-usability, within budget, performance and time constraints, but these initial configurations have to be satisfied in the first place and represent the focus of the system design.

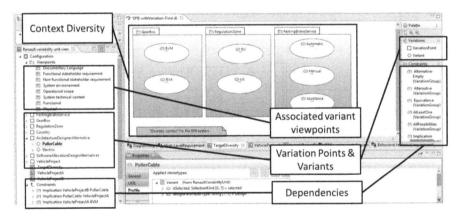

Fig. 4 Papyrus modeler including extensions for variability modeling Co-OVM

The EPB system model contains 21 variation points with 46 related variants representing system requirements (characteristics), design decisions and choices related to components. The model also contains 36 constraints directly related to variants, and 44 constraints issued from binding to the system model and from propagation of optional elements. It is often the case that in practice, both system and constraint models are far more complex. While in the EPB case the Co-OVM model satisfied our requirements, we still have to experiment with larger system models.

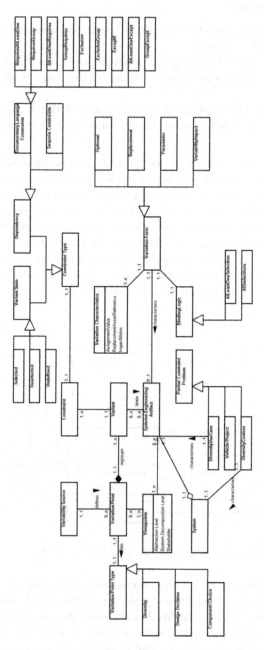

Fig. 5 Variability concepts in our MBSE framework

7 Conclusion and Perspectives

We proposed the adoption of a model for system variability in model based system engineering, based on concepts coming from the software product line literature and existing organization practices.

Our purpose was to:

- Provide explicit representation of variability during system conception for reuse of requirements and model elements across families of systems.
- Early documentation of configuration information for vehicle components in respect to the "documentary language".
- Facilitate the discovery and modeling of technical constraints for introducing them in the "documentary language".

We intend to further exploit the representation of variability to allow the engineer to obtain valuable information about the family of systems : calculate the coverage of the requirements of the family of systems for partial product configurations, calculate the coverage of the vehicle range for a family of systems, calculate the coverage of all possible system configurations containing a given component, and system variability or commonality index.

References

[1] Ahn, H., Kang, S.: Analysis of software product line architecture representation mechanisms. IEEE, 219–226 (2011), doi:10.1109/SERA.2011.22

[2] Astesana, J., Cosserat, L., Fargier, H.: Constraint-based modeling and exploitation of a vehicle range at renaults: Requirement analysis and complexity study. In: Workshop on Configuration, p. 33 (2010)

[3] Becker, M.: Towards a general model of variability in product families. In: van Gurp, J., Bosch, J. (eds.) Workshop on Software Variability Management, Groningen, The Netherlands, pp. 19–27 (2003),
http://www.cs.rug.nl/Research/SE/svm/
proceedingsSVM2003Groningen.pdf

[4] Benavides, D., Segura, S., Ruiz-Corts, A.: Automated analysis of feature models 20 years later: A literature review. Information Systems 35(6), 615–636 (2010)

[5] Chalé Góngora, H.G., Dauron, A., Gaudré, T.: A commonsense-driven architecture framework. part 1: A car manufacturers (nave) take on mbse. In: INCOSE 2012 (2012)

[6] Czarnecki, K., Grünbacher, P., Rabiser, R., Schmid, K., Wsowski, A.: Cool features and tough decisions: a comparison of variability modeling approaches. In: Proceedings of the Sixth International Workshop on Variability Modeling of Software-Intensive Systems, pp. 173–182 (2012)

[7] Djebbi, O., Salinesi, C.: Criteria for comparing requirements variability modeling notations for product lines. In: Fourth International Workshop on Comparative Evaluation in Requirements Engineering, CERE 2006, pp. 20–35 (2006)

[8] Dumitrescu, C., Salinesi, C., Dauron, A.: Towards a framework for variability management and integration in Systems Engineering. In: 22nd Annual INCOSE International Symposium, Rome, Italy, pp. 12–13 (2012a)

[9] Dumitrescu, C., Tessier, P., Salinesi, C., Grard, S., Dauron, A.: Flexible product line derivation applied to a model based systems engineering process, Paris (2012)

[10] Filho, J., Barais, O., Baudry, B., Le Noir, J.: Leveraging variability modeling for multi-dimensional model-driven software product lines. In: 2012 3rd International Workshop on Product Line Approaches in Software Engineering (PLEASE), pp. 5–8 (2012), doi:10.1109/PLEASE.2012.6229774

[11] Gmez, A., Ramos, I.: Automatic tool support for cardinality-based feature modeling with model constraints for information systems development. In: Pokorny, J., Repa, V., Richta, K., Wojtkowski, W., Linger, H., Barry, C., Lang, M. (eds.) Information Systems Development, pp. 271–284. Springer, New York (2011), http://dx.doi.org/10.1007/978-1-4419-9790-6_22, doi:10.1007/978-1-4419-9790-6_22

[12] Halmans, G., Pohl, K.: Communicating the variability of a software-product family to customers. Software and Systems Modeling 2(1), 15–36 (2003), doi:10.1007/s10270-003-0019-9

[13] Kang, K.C., Cohen, S.G., Hess, J.A., Novak, W.E., Peterson, A.S.: Feature-oriented domain analysis (FODA) feasibility study. Tech. rep., DTIC Document (1990)

[14] von der Maßen, T., Lichter, H.: RequiLine: a requirements engineering tool for software product lines. In: van der Linden, F.J. (ed.) PFE 2003. LNCS, vol. 3014, pp. 168–180. Springer, Heidelberg (2004)

[15] Mazo, R., Salinesi, C., Diaz, D., Djebbi, O., Lora-Michiels, A.: Constraints: The heart of domain and application engineering in the product lines engineering strategy. International Journal of Information System Modeling and Design (IJISMD) 3(2), 33–68 (2012)

[16] Pohl, K., Böckle, G., van der Linden, F.J.: Software Product Line Engineering: Foundations, Principles and Techniques. Springer-Verlag New York, Inc., Secaucus (2005)

[17] Possomps, T., Dony, C., Huchard, M., Rey, H., Tibermacine, C., Vasques, X.: A UML profile for feature diagrams: Initiating a model driven engineering approach for software product lines. In: Journe Lignes de Produits, pp. 59–70 (2010)

[18] Rabiser, R., Grunbacher, P., Dhungana, D.: Supporting product derivation by adapting and augmenting variability models. In: 11th International Software Product Line Conference, SPLC 2007, pp. 141–150 (2007), doi:10.1109/SPLINE.2007.22

[19] Roos-Frantz, F., Benavides, D., Ruiz-Corts, A., Heuer, A., Lauenroth, K.: Quality-aware analysis in product line engineering with the orthogonal variability model. Software Quality Journal (2011), doi:10.1007/s11219-011-9156-5

[20] Ruiz, F., Hilera, J.R.: Using ontologies in software engineering and technology. In: Calero, C., Ruiz, F., Piattini, M. (eds.) Ontologies for Software Engineering and Software Technology, pp. 49–102. Springer, Heidelberg (2006), http://dx.doi.org/10.1007/3-540-34518-3_2, doi:10.1007/3-540-34518-3_2

[21] Salinesi, C., Mazo, R., Diaz, D., Djebbi, O.: Using integer constraint solving in reuse based requirements engineering. In: 2010 18th IEEE International Requirements Engineering Conference (RE), pp. 243–251 (2010), doi:10.1109/RE.2010.36

[22] Salinesi, C., Mazo, R., Djebbi, O., Diaz, D., Lora-Michiels, A.: Constraints: The core of product line engineering. In: 2011 Fifth International Conference on Research Challenges in Information Science (RCIS), pp. 1–10 (2011), doi:10.1109/RCIS.2011.6006825

[23] Sinnema, M., Deelstra, S.: Classifying variability modeling techniques. Information and Software Technology 49(7), 717–739 (2007), doi:10.1016/j.infsof.2006.08.001

[24] Streitferdt, D., Riebisch, M., Philippow, K.: Details of formalized relations in feature models using OCL. In: Proceedings of the 10th IEEE International Conference and Workshop on the Engineering of Computer-Based Systems, pp. 297–304 (2003)

[25] Svendsen, A., Haugen, O., Moller-Pedersen, B.: Using variability models to reduce verification effort of train station models. In: 18th Asia Pacific Software Engineering Conference, APSEC 2011, pp. 348–356 (2011), doi:10.1109/APSEC.2011.21

[26] Tessier, P., Servat, D., Gerard, S.: Variability management on behavioral models. In: VaMoS Workshop, pp. 121–130 (2008)

[27] Tseng, M.M., Jiao, J.: Mass customization. In: Salvendy, G. (ed.) Handbook of Industrial Engineering, 3rd edn., pp. 684–710. Wiley Interscience in cooperation with Institute of Industrial Engineering (2001)

Simulation Methods in the Healthcare Systems

Andrey Khudyakov, Camille Jean, Marija Jankovic, Julie Stal-Le Cardinal, and Jean-Claude Bocquet

Abstract. Healthcare systems can be considered as large-scale complex systems. They need to be well managed in order to create the desired values for its stake-holders as the patients, the medical staff and the industrials working for health-care. Many simulation methods coming from other sectors have already proved their added value for healthcare. However, based on our experience in the French heath sector (Jean et al. 2012), we found these methods are not widely used in comparison with other areas as manufacturing and logistics. This paper presents a literature review of the healthcare issue and major simulations methods used to address them. This work is design to suggest how more systematic creation of so-lutions may be performed using complementary methods to resolve a common is-sue. We believe that this first work can help to better understand the simulation approaches used for health workers, deciders or researchers of any responsibility level.

Keywords: System modeling, Systemic tools, Healthcare sytem.

1   Introduction

For governments around the world, the main priority for the success and prosperi-ty of society became support enhancements of the existing healthcare systems (HCS) (WHO, 2000). Statistics suggest that in most countries, a significant frac-tion of public money is allocated to the healthcare sectors (OECD 2011). Despite these huge investments, HCS have not yet delivered all the expected improve-ments. While healthcare systems are complex and include many interconnected elements, the rules and heuristics generating managerial decisions are too simplis-tic to cope with the complexity involved in such systems (Lebcir 2006). The use

Andrey Khudyakov · Camille Jean · Marija Jankovic · Julie Stal-Le Cardinal ·
Jean-Claude Bocquet
Ecole Centrale Paris, Paris, France
e-mail: andrey.khudyakov@student.ecp.fr,
     {camille.jean,marija.jankovic,
     julie.le-cardinal,jean-claude.bocquet}@ecp.fr

M. Aiguier et al. (eds.), *Complex Systems Design & Management 2013*,   141
DOI: 10.1007/978-3-319-02812-5_11, © Springer International Publishing Switzerland 2014

of model simulation getting wider and wider distributed in the areas of health, but nevertheless not widely used in comparison with other areas, such as manufacturing, logistics and military applications (Sanchez et al. 2000). This paper aims to help health workers, deciders or researchers to understand the relation between the existing modeling approaches and to suggest how more systematic creation of solutions may be performed using complementary methods for the same issue.

Section 2 talks about the general definitions of HCS and briefly describes the nature and goals of HCS. Section 3 contains a review of literature of simulation methods used in HCS and answers the question why modeling approaches is important in this sector. Section 4 provides analyses and results and attempts to set a context for the need of generic framework for HCS which provides guidance for researcher, practitioners, and decision makers interested in utilizing different approaches for healthcare issues.

2   Context

2.1   The Healthcare Systems (HCS)

In this part of the paper, first we describe HCS. Secondly, we define their goals and structure. Thirdly, we briefly establish the existing problems in HCS.

The World Health Organization (WHO) has defined health systems as "the sum of the people, institutions and resources arranged together (in accordance with relevant policies) to maintain and improve the health of the population they serve" (WHO, 2000). A health system consists of all organizations, people and actions whose primary intent is to promote, restore or maintain health (WHO, 2000). This includes efforts to influence determinants of health as well as more direct health-improving activities. A health system is therefore more than the pyramid of publicly owned facilities that deliver personal health services. It includes, for example, a mother caring for a sick child at home; private providers; behavior change programs; vector-control campaigns; health insurance organizations; occupational health and safety legislation. (WHO, 2007).

The goals of HCS are a good health, responsiveness to the expectations of the population and fair financial contribution (WHO, 2007). But it is also good quality, efficiency, acceptability and equity (Brody 2007).

2.2   The Healthcare Environment

To understand the use of simulation techniques in HCS we have to understand what the healthcare environment represents and classify the types of existing problems. The key purpose of the framework is the promotion of common understanding of what a health system is and what constitutes health systems strengthening (OECD 2011). These are based on the functions defined in (World health report, 2000). The building blocks are: service delivery; health workforce; information; medical products, vaccines and technologies; financing; and leadership and governance (stewardship) (OECD 2011).

The WHO report summarizes the features of health systems oriented to population health and health equity as follows it has defined a single framework of HCS with six building blocks (WHO s. d.):

- Service delivery: It means good health services which deliver effective, safe, quality personal and non-personal health interventions to those that need them, when and where needed, with minimum waste of resources;
- Health workforce : it is a well-performing health labor team that works in ways that are responsive, fair and efficient to achieve the best health outcomes possible, given available resources and circumstances (i.e. there are sufficient staff, fairly distributed; they are competent, responsive and productive);
- HCS Information Systems : it is a well-functioning health technology system that ensures the production, analysis, dissemination and use of reliable and timely information on health determinants, health system performance and health status;
- Medical products, vaccines and technologies is a well-functioning health system that ensures equitable access to essential medical products, vaccines and technologies of assured quality, safety, efficacy and cost-effectiveness, and their scientifically sound and cost-effective use;
- Financing : it means a good health financing system that raises adequate funds for healthcare, in ways that ensure people can use needed services, and are protected from financial catastrophe or impoverishment associated with having to pay for them. It provides incentives for providers and users to be efficient;
- Leadership and governance (stewardship) involves ensuring strategic policy frameworks exist and are combined with effective oversight, coalition building, regulation, attention to system-design and accountability.

2.3 The Definition of Existing Issues in HCS

In Healthcare system there are both types of problems: complicated problems and complex problems. The complicated problems can be solved in a linear fashion using straightforward, reductionist, repeatable, sequential techniques. They are amenable to traditional project management approaches and they introduce limited/known/manageable consequences and no unintended consequences. A complicated problem is well defined; its solution is clear and can be given to a designer to create detailed specifications and project manager to implement (Robertson-Dunn 2012). Complex problems are problems, whose size, dependence on context, variety of elements and interdependence, make them unable to be fully predictable and therefore controllable. They tend to be non-linear, difficult to understand and their solutions can lead to other problems and unintended consequences. All problems involving new technology, new development environments or new applications should be considered to be complex. This kind of problem cannot be solved by reductionist or sequential approaches but more by using the systemic approach (Le Moigne 1985).

In the previous sections of this chapter the relevance and importance for understanding HCS was established. A deeper discussion of information related to approaches for solving the HCS problems can be found in the literature review in Chapter 3.

3   The Simulation Approaches for HCS

3.1   Why Simulation Is Important

The simulation techniques are now rapidly increasing in HCS modeling (Jun et al. 2009). The simulation can be defined as an imitation of a system. For examples, computer aided design (CAD) systems provide imitations of production facility designs and business process. There is a difference between the concepts of a static simulation, which imitates a system at a point in time, and a dynamic simulation, which imitates a system as it progresses through time (Law 2006). Simulation is the process of designing a model of a real system and conducting experiments with this model for the purpose either of understanding the behavior of the system or of evaluating various strategies (within the limits imposed by a criterion or set of criteria) for the operation of the system (Shannon 1975).

In the following segments we propose the literature review of five existing simulation approaches which have been developed in HCS.

3.2   The Literature Review, Reliability of Existing Approaches and Those Examples Studies

The objectives of this research are to gain insight into approaches and solved issues in HCS, and to suggest how more systematic creation of solutions may be performed in HCS. In the following segments we describe what we found to be five major approaches. We propose the literature review of approaches with their motivations of usage and case applications.

3.2.1   Markov Models

Markov models are stochastic models (Hidden Markov model; Markov chain; Markov chain Monte Carlo; Markov decision process; partially observable Markov decision process) describing a sequence of possible events in which the probability of each event depends only on the state attained in the previous event. Hospital managers need to make defensible resource allocation decisions driven by hospital inpatient inventory predictions (Broyles et al. 2010). Accurate predictions of hospital inpatient inventory can help evaluate strategic decisions such as policy changes (Marshall et al. 2005). Harper (2002) argues that most existing models of the complex hospital inpatient inventory system are application specific and cannot be generalized to all hospital inpatient inventory systems. Hospital planning and management must consider hospital patient flow for decisions on hospital inpatient staffing and capacity. Hospital inpatient inventory is a "primary driver of resource use in hospitals" and affects demand of many ancillary service including laboratory, pharmacy, physical therapy, radiology, housekeeping, and surgical services (Littig & Isken 2007). Accurate hourly and daily hospital inpatient inventory prediction is critical to short term hospital operations improvement (Littig & Isken 2007), staffing costs (Ledersnaider & Channon 1998), ambulance diversion (Schull et al. 2003), hospital bed management (Mackay, 2001), in the areas of inpatient placement (Clerkin et al.1995). It have been used by (Broyles et al. 2010) in order to decrease the delay on the part of customers waiting for service, for delivery, operating cost reduction, lead time reduction, faster plant changes, capital cost reduction and improved customer service and for not arriving as promised ineffective management of chronic illness (Broyles et al. 2010).

3.2.2   Discrete-Event Simulation (DES)

Discrete Event Simulation (DES) is modeling the operation of a system as a discrete sequence of events over time. Each event occurs at a particular instant in time and marks a change of state in the system (Zeigler et al. 2000). It have been used to spread and containment of hospital acquired infections (Hagtvedt et al. 2009); planning for disease outbreaks (Aaby et al. 2006); determining bed requirements (Griffiths et al. 2010); investigating emergency departments (Paul et al. 2010); and determining appropriate ordering policies in the blood supply chain (Katsaliaki & S. C. Brailsford 2007). It is use for planning, scheduling, reorganization and management of healthcare and hospital services, communicable diseases, bio-terrorism, screening, costs of illness, economic evaluation (comparing alternative healthcare interventions), policy and strategy evaluation (Fone et al. 2003). And also for risk reduction, operating cost reduction, lead time reduction, faster plant changes, capital cost reduction and improved customer service (Robinson et al. 2012).

3.2.3   System Dynamic Modeling (SD)

System dynamic modeling (SD) is an approach to understanding the behavior of complex systems over time. It deals with internal feedback loops and time delays that affect the behavior of the entire system. It is described as an analytical modeling approach whose roots could be said to lie in the general systems theory approach (Bertalanffy 1993). System dynamics is a computer-based simulation modeling methodology developed at the Massachusetts Institute of Technology (MIT) in the 1950s by Jay Forrester as a tool for managers to analyze complex problems (Forrester 1999). The word "dynamic" implies continuous change and that is what dynamic systems do - they continuously change over time. Their position, or state, is not the same today as it was yesterday and tomorrow it would have changed yet again. Using system dynamics simulations allows us to see not just events, but also patterns of behavior over time. Sometimes the simulation looks backward, to historical results. At other times it looks forward into the future, to predict possible future results. Understanding patterns of behavior, instead of focusing on day-to-day events, can offer a radical change in perspective. It shows how a system's own structure is the cause of its successes and failures. This structure is represented by a series of causally linked relationships. The implication is that decisions made within an organization have consequences, some of which are intentional and some are not. Some of these consequences will be seen immediately while others might not be seen for several years. System dynamics simulations are good at communicating not just what might happen, but also why. This is because system dynamics simulations are designed to correspond to what is, or might be happening, in the real world. As example, we can cited the work about long waiting lists (Siciliani 2006); ineffective management of chronic illness epidemics of obesity; heart disease (Jones & Homer 2006) and decision-making in resources allocation for telehealth (Jean et al. 2012).

3.2.4   SD+

SD+, Brailsford describe it as a form of mixing methodologies of SD and DES and due to fundamental differences, mixing methodologies is quite challenging (S. Brailsford 2008). There are problems which exhibit elements which require both SD and DES, and there are interactions between them. In those scenarios accurate analysis demands to capture those interactions. It has been argued in literature SD-, where SD and DES are integrated symbiotically, will provide more insight and accurate analysis of such problems with fewer assumptions (Chahal et al. 2013). Also for them this it means that the same thing has been represented in both models but this representation does not have same face value. Mostly SD represents the aggregated version of the variable which is disaggregated in DES and can be represented in DES either by single or group of variables which holds value equivalent to SD variable. It has the potential to provide a more complete way of dealing with the complexity of the real world (Chahal et al. 2013)..

3.2.5   Agent-Based Approach and Multi-agent Systems

Agent-based approach & Multi-agent systems is presented, aiming to minimize the waiting time of patient as well as the cost of care within emergency department (Daknou et al. 2008). Artificial Intelligence and knowledge based systems are assuming an increasingly important role in medicine for assisting clinical staff in making decisions under uncertainty (e.g. diagnosis decisions, therapy and test selection, and drug prescribing (Huang et al. 1995). In France (Daknou et al. 2008) were successfully adopted the software agents to provide the means to accomplish such real-time application for taking care of all arrival patients at the emergency department and to improve the quality of care.

4   Analysis Approaches in Relation to the Problematic Element

In this part of the analysis we propose to look into the relation between the different approaches and the problem element(s). The development activities and perspective of simulation techniques is now rapidly increasing in healthcare systems modeling ((Barnes et al. 1997; Jun et al. 2009). We understood that usually for the more realistic picture in complex issues, as in HCS issues, it cannot be used just one of the approach, very often it have to be mixed. For example, for right medical decision making and making decisions under uncertainty we can use agent-base approach or multi-based approach. To simulate the existing system, it is appropriate to use the Markov chain, for simulation the system over time, the interaction between the elements of system and understanding of avocation in relation to the element of problem complexity it is better to use SD, DES or (SD+).

Table 1 Decision Level

Decision Level	Markov Model	System dynamic modelling	Discrete event simulation (DES)	SD+	Agent-based approach
Short term decision	1				1
Mid term decision	1		1		1
Long term decision		1	1	1	

Also, it is necessary to look into the effectiveness of methods in making the best solutions (decisions) in time. The table 1 describes and classifies these decisions according to the classical operations of decision framework, which would consist of three levels: short-term decisions; mid-term decisions; long-term decisions. Short-term decisions: would consist of one or several determination of process, rules, short planning staff, use of equipment, human resources. Mid-term decisions: would consist to reorganization or optimization of allocation of resources, human resources or staff in fix (known) location. Long term decisions: would consist of one or several objectives of modeling, human resources, medical equipment, as well as the identification of new location (allocation of resources), bases regions, capacity of system.

5   Conclusion

In this paper we have attempted to set a context for the need of generic framework for HCS which provides guidance to the prospective researcher, practitioners, decision makers interested in utilizing different approaches for healthcare system issues. Research project focused on approaches to solve one of problems developed in the context of HCS. HCS problems are understood as any problems that has mainly based in HCS. Based on the literature it was recognized that generic framework should be capable of providing guidance to prospective users with regards to mapping between different approaches.

It has been argued in literature that the investment and effort involved in development of integrated "hybrid modeling" (decision making method which includes both qualitative and quantitative techniques; using two or several approaches for the same problem) is wasted if the problem does not require "hybrid modeling".

It is expected that this work will encourage those engaged in simulation and modeling (e.g., researchers, practitioners, decision makers) to realize the potential of cross-fertilization of the several modeling approaches.

References

Aaby, K., et al.: Montgomery County's Public Health Service Uses Operations Research to Plan Emergency Mass Dispensing and Vaccination Clinics. Interfaces 36(6), 569–579 (2006)

Barnes, C.D., et al.: Success Stories in Simulation In Health Care. In: Proceedings of the 1997 Winter Simulation Conference, pp. 1280–1285 (1997)

von Bertalanffy, L.: Théorie générale des systèmes, Dunod (1993)

Brailsford, S.: System dynamics: What's in it for healthcare simulation modelers. In: Proceedings of the 2008 Winter Simulation Conference, pp. 1478–1483 (2008)

Brody, W.: What's Promised Waht's Possible? (2007)

Broyles, J.R., Cochran, J.K., Montgomery, D.C.: A statistical Markov chain approximation of transient hospital inpatient inventory. European Journal of Operational Research 207(3), 1645–1657 (2010)

Chahal, K., Eldabi, T., Young, T.: A conceptual framework for hybrid system dynamics and discrete event simulation for healthcare. Journal of Enterprise Information Management 26(1/2), 50–74 (2013)

Daknou, A., et al.: Agent based optimization and management of healthcare processes at the emergency department. International Journal of Mathematics and Computers Simulation 2(3), 285–294 (2008)

Fone, D., et al.: Systematic review of the use and value of computer simulation modelling in population health and health care delivery. Journal of Public Health 25(4), 325–335 (2003)

Forrester, J.W.: Industrial dynamics. Pegasus Communications (1999)

Griffiths, J.D., et al.: A simulation model of bed-occupancy in a critical care unit. Journal of Simulation 4(1), 52–59 (2010)

Hagtvedt, R., et al.: A Simulation Model to Compare Strategies for the Reduction of Health-Care–Associated Infections. Interfaces 39(3), 256–270 (2009)

Huang, J., Jennings, N.R., Fox, J.: Agent-Based Approach to Health Care Management. Applied Artificial Intelligence 9(4), 401–420 (1995)

Jean, C., et al.: Telehealth: towards a global industrial engineering framework based on value creation for healthcare systems design. In: Proceedings of the 12th International Design Conference DESIGN 2012, pp. 949–958 (2012), http://hal.archives-ouvertes.fr/hal-00714273 (consulté le juillet 3, 2012)

Jones, Homer: Understanding Diabetes Population Dynamics Through Simulation Modeling and Experimentation. American Journal of Public Health (2006)

Jun, G.T., et al.: Health care process modelling: which method when? International Journal for Quality in Health Care 21(3), 214–224 (2009)

Katsaliaki, K., Brailsford, S.C.: Using Simulation to Improve the Blood Supply Chain. Journal of the Operational Research Society 58(2), 219–227 (2007)

Law, A.: Simulation Modeling and Analysis with Expertfit Software, 4th edn. McGraw-Hill Science/Engineering/Math. (2006)

Lebcir, R.: Health Care Management: The Contribution of Systems Thinking. University of Hertfordshire (2006)

Ledersnaider, D.L., Channon, B.S.: SDM95–reducing aggregate care team costs through optimal patient placement. The Journal of Nursing Administration 28(10), 48–54 (1998)

Littig, S.J., Isken, M.W.: Short term hospital occupancy prediction. Health Care Management Science 10(1), 47–66 (2007)

Marshall, A., Vasilakis, C., El-Darzi, E.: Length of Stay-Based Patient Flow Models: Recent Developments and Future Directions. Health Care Management Science 8(3), 213–220 (2005)

Le Moigne, J.-L.L.: La théorie du système général. Presses Universitaires de France, PUF (1985)

OECD, Health at a Glance, OECD Indicators (2011)

OMS (World Health Organisation), Rapport sur la santé dans le monde, 2000 – Pour un système de santé plus performant (2000),
http://www.who.int/whr/2000/en/whr00_fr.pdf

Paul, S.A., Reddy, M.C., Deflitch, C.J.: A Systematic Review of Simulation Studies Investigating Emergency Department Overcrowding. Simulation 86(8-9), 559–571 (2010)

Robertson-Dunn, B.: Beyond the Zachman Framework: Problem-oriented System Architecture (2012),
http://www.academia.edu/1863616/
Beyond_the_Zachman_Framework_Problem-
oriented_System_Architecture (Consulté le avril 20, 2013)

Robinson, S., et al.: SimLean: Utilising simulation in the implementation of lean in healthcare. European Journal of Operational Research 219(1), 188–197 (2012)

Sanchez, S.M., et al.: Emerging issues in healthcare simulation. In: Proceedings of the 2000 Winter Simulation Conference, pp. 1999–2003 (2000)

Schull, M.J., et al.: Emergency department overcrowding and ambulance transport delays for patients with chest pain. Canadian Medical Association Journal 168(3), 277–283 (2003)

Shannon, R.E.: Systems simulation: the art and science. Prentice-Hall (1975)

Siciliani, L.: A dynamic model of supply of elective surgery in the presence of waiting times and waiting lists. Journal of Health Economics 25(5), 891–907 (2006)

WHO. Everybody business: strengthening health systems to improve health outcomes (2007)

Zeigler, B.P., Kim, T.G., Praehofer, H.: Theory of Modeling and Simulation: Integrating Discrete Event and Continuous Complex Dynamic Systems, 2nd edn. Academic Press Inc. (2000)

Unifying Human Centered Design and Systems Engineering for Human Systems Integration

Guy A. Boy and Jennifer McGovern Narkevicius

Abstract. Despite the holistic approach of systems engineering (SE), systems still fail, and sometimes spectacularly. Requirements, solutions and the world constantly evolve and are very difficult to keep current. SE requires more flexibility and new approaches to SE have to be developed to include creativity as an integral part and where the functions of people and technology are appropriately allocated within our highly interconnected complex organizations. Instead of disregarding complexity because it is too difficult to handle, we should take advantage of it, discovering behavioral attractors and the emerging properties that it generates. Human-centered design (HCD) provides the creativity factor that SE lacks. It promotes modeling and simulation from the early stages of design and throughout the life cycle of a product. Unifying HCD and SE will shape appropriate human-systems integration (HSI) and produce successful systems.

1 Introduction

The International Council on Systems Engineering (INCOSE) defines systems engineering (SE) as "an interdisciplinary approach and means to enable the realization of successful systems." The SE approach includes early definition of customer needs, documentation of the desired functionality and the technical requirements before design activities are begun. SE spans all stages of the life cycle

Guy A. Boy
University Professor,
Director of the Human-Centered Design Institute,
Florida Institute of Technology,
and IPA Chief Scientist for Human-Centered Design at
NASA Kennedy Space Center

Jennifer McGovern Narkevicius
Jenius LLC, and Chair of the Human-Systems
Integration Working Group of INCOSE

M. Aiguier et al. (eds.), *Complex Systems Design & Management 2013*,   151
DOI: 10.1007/978-3-319-02812-5_12, © Springer International Publishing Switzerland 2014

of a product including design, manufacturing, operations, training and support, disposal and costs (Haskins, 2011). Even with this broad approach to system development, there remains much to improve to take into account and define people's needs wholly. Achieving human-centered SE requires a mindset that differs from current technology-centered and finance-driven practices. We observe consistent program and project failures despite adherence to the very holistic SE approach. These failures range from small and accommodatable (e.g., annoying "quirks" in handheld devices) to the truly spectacular (the failure of a power grid). The source of these failures can commonly be found in inadequate requirements that are technology focused, poorly specified, unmeasurable, and unachievable. The failure to incorporate a human-centered focus also results in the lack of these requirements driven by organizations and people, which are the most flexible but unrepresented interfaces in the system. SE methods and tools impose rigid processes that end in failure because our technology-based world evolves very quickly making flexibility in solutions essential.

Human-centered design (HCD) has developed across diverse technical disciplines drawing on human-computer interaction, artificial intelligence, human-machine systems, social and cognitive sciences. It is based on the same objective as SE, that is developing a holistic approach to support fuller integration of Technology, Organizations and People but HCD brings different concepts and terminology (cf. the TOP model: Boy, 2013). This broader, trans-disciplinary language supports a new, more comprehensive approach to systems design. In HCD, "systems" are commonly denoted as "agents" which can be people or software-based artifacts. Agents are functionally defined as displaying cognitive functions, defined by their roles, the contexts of validity and supporting resources (Boy, 1998). The roles can be expressed in terms of objectives, goals and/or purposes. Minsky (1985) defined an agent as a "society of agents" and by analogy, a cognitive function is a society of cognitive functions, which can eventually be distributed among different agents.

It is interesting to compare this HCD definition of an agent with the SE definition of a system. INCOSE Systems Engineering Handbook defines a system as an "integrated set of elements that accomplish a defined objective" (Haskins, 2011). Note the correspondence of Minsky's definition of an agent to the current definition of SE community's System of Systems (SoS) (Clark, 2009). The association of the concepts of systems and agents highlights the parallel paths of HCD and SE. SE processes include technology and people. This approach takes the functionality of people into account as a central part of systems. HCD includes agents (both the cognitive functions of people and technology). It is time to merge these two approaches to achieve a better definition of Human-System Integration (HSI).

HSI complexity can be analyzed by explaining the emergence of (cognitive) functions that arise from agent activity. This expansion of agents results in a need for function allocation approaches that supersede the classical MABA-MABA[1] model (Fitts, 1951), where functions are not only allocated deliberately, but as

---

[1] Men Are Better At - Machine Are Better At.

they emerge from agents' activities. Our socio-technical world is dynamic and the concept of static tasks should also be superseded by the concept of dynamic activities. A task is what we prescribe; an activity is what we effectively do. Activities are characterized by their variability and should be modeled by a dynamic non-linear model, not as signal and noise. This is where the difficulty lies - systems are no longer only mechanically complicated, they are highly interconnected and, therefore context-dependent and complex. People generally handle variability well, engineered systems cannot. Engineered systems are programmed (handling procedures only); people are flexible and creative. Engineered systems are excellent at deduction; people are unique at handling induction (Harris et al, 1993) and, more importantly, abduction (Peirce, 1958).

Conventional linear methods and techniques only work in the short term and are not useful for handling the complexity of HSI. It is imperative to develop and use methods and techniques built on complexity science that study non-linear dynamical systems, attractors, bifurcations, fractals, catastrophes, and more. Instead of avoiding complexity and working in the artificial world of the Gaussian "bell" curve, where we remove the parts that we do not understand or do not fit our hypotheses or a priori conclusion, we need to model the overall complexity of HSI and benefit from it. Design can only succeed by restoring the necessary flexibility of control, management and accountability of systems being designed and developed today. It is essential to understand why SE fails, why context matters, why HSI requires complexity analysis, and why HCD generates a path directly toward HSI.

2   Why Does Systems Engineering Fail?

Software Is Embedded in Almost all Systems. Until recently, the concept of machine was mainly associated with hardware and mechanical entities. Engineering cars was the kingdom of mechanical engineers. Electrical and computer engineering have progressively penetrated this industrial sector and current cars are full of wires and electronic circuits. It is no longer possible to repair a car without complex diagnostic systems. The next step, represented by the Google Car, is drastic. It promotes software engineering, human-computer interaction and artificial intelligence to lead the design of the entire car. This approach is backward - instead of putting electronics into a thermal engine and a car body, wheels are stuck onto a computer! We are designing tangible objects in which software will make itself as invisible as possible, working undetectably supporting the various tasks people have to perform.

Large industrial programs require organization. This significantly contributed to the development of SE. However, industrial organizations are still designed with hardware in mind. Software is still often considered as an add-on. Hardware engineering and software engineering should be integrated in a human-centered way (i.e., from purpose to means and not the other way around). This organization is essential to assure that the engineering work is supported and consistent, allowing good work by both hardware and software engineering. It is important to make a distinction between design and manufacturing. When design is well done, manufacturing can be easily accomplished using current SE approaches. However, when design is rushed or pressured by outside influences (including local politics or program management goals) the SE processes are aborted or abandoned resulting in "patches" and constant re-engineering of products. We then need to better use creative, flexible and effective software engineers cooperating with human-centered designers to do conceptual models and adapted human-in-the-loop simulations that lead to effective requirements for manufacturing.

Interconnectivity Grows Exponentially. Software is no longer only embedded locally; it enables interconnections among a variety of systems considered as agents. Internet is a great example of such interconnectivity. People are connected to other people and systems, anywhere and anytime, synchronously and asynchronously. Such interconnectivity creates communication with no delays; but new kinds of behaviors and properties emerge which cause new kinds of problems. This is why organizational models are so important. These models can be descriptive, prescriptive or predictive, and should be developed in that order. As a first step, they need to describe the various outcomes of SE use in industry and governmental agencies. Descriptive models require the development of appropriate syntax (e.g., nodes and links) and semantics (e.g., contents and meaning of these nodes and links). The military organizational model or architectural framework has been in use for a long time. It is based on pyramidal hierarchies with primarily vertical information flow. This model is still in use in most large organizations. However, in these organization information runs transversally (via mobile phone, email, chat, and other typically internet based communication). Several new types of communities have emerged from the use of this growing interconnectivity including communities of practice organized through social media. These changes in interconnectivity increases the organizational dynamics and may cause the emergence of events that would not have been possible before.

These emerging organizational models in a highly interconnected world must be analyzed. The Orchestra model (Boy, 2009 and 2013) was proposed to describe workers' evolution from old-time soldiers (i.e., executants) to a new model, musicians (creative, flexible experts). The main difference between executants and experts is their level of autonomy. The more agents become autonomous, the more coordination rules they will need to interact safely, efficiently and comfortably in society. The composers of the Orchestra model are needed to create, coordinate and fine-tune the musicians' scores. This is the first level of coordination, prescriptive coordination, the task level. During a performance, the conductor is required to coordinate musicians in real-time. This second level, effective coordination, is the activity level.

Rapid and Constant Socio-technical Evolution. Our socio-technical world is changing fast and drastically and interconnectivity is the major contributor. Flexibility is required to enable the necessary rapid adaptation. Three main issues continue to be relevant: complacency, situation awareness and decision-making. Automation is a contributing cause of complacency. Many systems are highly automated and human operators do not have much to do in nominal situations. Consequently, they do nothing or they do something else. In either case, they are barely in the control loop of the process they are supposed to monitor and manage. Issues arise when unanticipated events occur, when human operators have little time to jump back into the control loop, often leading to major issues. Recent commercial aircraft accidents have been attributed to this problem. To avoid this in systems design we need to look at the bigger picture of Technology, Organizations and People (the TOP model) to answer this question (Boy, 2012). The answer is not necessarily technological; it can be organizational, training, change of practice, or some combination of these.

This may be the first time in the history of humanity that remembering things is not the best way to keep them handy almost anytime, anywhere. The Internet gets the information that we want in a few clicks. However, having information handy does not guarantee that we understand it - the meaning is crucial (Boy, 2013). Pilots may face all kinds of relevant and timely information, but if they do not understand it in context, it is neither useful nor usable. This is why situation awareness progressively emerged as a fundamental concept. The TOP model points to appropriate solutions to this problem.

Procedures and rules impose rigidity, thwarting the flexibility required to solve emergent problems. The unexpected is crucial today because of our highly procedural world (Boy, 2013). Procedures can be imposed organizationally or embedded in automated systems (machines) but when something unexpected occurs, solutions are not handy to make appropriate decisions, and socio-cognitive adaptation becomes an issue. In such situations, decision-making is handicapped by lack of situation awareness, educated and embodied practice of problem solving and risk taking (Boy & Brachet, 2010).

3   Context Matters

People Issues. SE provides a framework that is often context-free. However, every project or program involves many different contexts that often lead to ad-hoc planning and operations, and adherence to SE procedures may establish roadblocks. Even if SE processes work well, people must articulate them. These people need to understand SE, organizations, contexts, human resource management and the space between these where major social issues emerge when SE processes are very complex. One of the main flaws in SE is to think that people are systems in the same way machines are. People are far more complex than engineered systems (i.e., artifacts). In addition, when people interact with simple artifacts, complexity emerges from the various possible usages of artifacts and other interactions, which are difficult to anticipate exhaustively.

People have creativity, complexity, flexibility, inductive and abductive cognition. Modern artifacts equipped with sophisticated software can also display complex cognition. People easily forget what they have to do; well-programmed software does not. People are unique inventors and creative creatures; machines are not. The set of capabilities and limitations that people and machines bring the system solution are broad and complementary – if they are considered early in design. Context variations are a natural variability that needs to be modeled and understood to generate appropriate decisions and actions to develop and design successful systems.

People have their own motivations and systems engineers are no different. Work is always better done when workers are more autonomous and trusted. In a new world where the Orchestra model can be a reference, workers have to be able and agree to play the same symphony by articulating their own personal production. They alternatively need to be leaders and followers. For example, astronauts learn both leadership and *followship*. Depending on context, they may change their roles. This imposes excellent skills and deeper knowledge of the domain and organization where work must be accomplished. Motivation is a matter of respect and engagement. When SE processes are too rigid and constraining without apparent reasons, workers get less motivated because they feel that they are not respected. They may think that they are treated as chattel and robots (e.g., as a spreadsheet box with a dollar number in it). Engagement is also a crucial factor that embraces dedication, completion of work toward goal satisfaction, and pleasure to perform. It is important to include *design thinking* into engineering and management practices (Boy, 2013).

Organizational Issues. Organizations are set up with rules and procedures. The global market economy and interconnectivity afford large organizations the opportunity to involve contractors distributed across many countries. This distribution of work requires modification in work practices including the need to divide the work into parts that can be produced independently, and later be integrated at the top level. This dichotomization involves standardization that cannot be static because of rapid and constant social and technical evolution. Top-level integration involves technical, social and organizational competence that may not be entirely present today in organizations that are almost exclusively led by finance managers and stakeholders. This absence of socio-technical leadership influences many factors including personnel motivation, delays and customers satisfaction.

Culture is another emerging property of socio-technical systems. Organizations develop their own culture, which depends on a tradeoff between order and freedom. Organizational culture is autopoietic, that is, the culture is incorporated into systems being developed and in their usages (Maturana & Varela, 1980). An autopoietic system or organization is defined as a network of processes that can regenerate, transform or destroy components that interact among each other. We can see that for a given culture the creation and refinement of products will be inevitably based on this culture (e.g., if the culture is technology-centered, the product will also be technology-centered). A human-centered culture in systems design and engineering will result in a human-centered organization and human centered products.

Knowledge Management. SE has become a way of thinking and a standardized international engineering practice. Process quality is improved continuously. Protocols are now systematic and there are prescriptive strategies for many situations. Incremental experience feedback feeds this process of improvement and this quality approach contributes to reduced variability. However, this is only true for well-specified problems and known situations. Unknown unknowns are still problematic.

Tools have been developed to support SE practice however this results in abundant generation of documents. Model-based SE was a response to this document-centered approach (Brown, 2011). For example, modeling tools are now used to capture and identify requirements, system element structure, traceability, verification, and configuration management. However, people are not taken into account as they should be.

Knowledge management is difficult because (discrete) documents cannot replace (continuous) human expertise. Storing information does not guarantee that people reading, hearing or interacting with it will understand the content, what the document makers wanted to report, or the knowledge that was meant to be captured. Documentation is contextual but context is difficult to capture. It requires document makers to have an external view of the (current) context and imagine the (future) context in which document readers will be.

4   Human-Systems Integration Requires Complexity Analysis

Moving from Cartesian Philosophy to Complexity Science. The Capacity Maturity Model (CMM) proposes a sequential set of processes (Paulk et al., 1995). Organizations become certified for the maturity of their processes but CMM does not guarantee product maturity nor does it promise that a product will be sustainable; maturity of practice is also necessary (Boy, 2005 and 2013).

Quality is a matter of tests of a technology within its environmental, organizational and individual use domain (the TOP model again). Tests are predicated on a Cartesian approach of successive parsing, moving away from the complexity of the whole to the simplicity of the atomic. Interconnectivity and rapid, constant socio-technical evolution, makes these tests more difficult. Linear local tests are not sufficient and sometimes not relevant. We need to investigate new methods for non-linear holistic tests. We need to look for emerging behavioral patterns, attractors and bifurcations, in the complexity science sense (Mitchell, 2008). We must move from Cartesian positivist methods, which are constraining and limiting toward phenomenological approaches.

Applying the Orchestra model to what we can observe in our industrial companies and governmental agencies, it is obvious that we have not moved from the military model to an emerging orchestra-like model. This older military model is formal and highly structured while the new emerging model is informal, more flexible, and leverages individual creativity as a strength that contributes to the performance of the whole. There is a disjunction between the way organizations are still structured and the way people behave in them, which causes consistency and synchronization problems. Practices have already evolved within the Orchestra model, and they are imposed in the old military model. The SE V model is very linear and deliberately imposed as a rigid prescriptive model. Unfortunately, not everything can be predicted and we do not have predictive models that can be accurate and robust enough to anticipate variations. The role of people is to handle these variations. Since such variations may happen anytime anywhere in the organization, the old model is very limited. Thinking of the organization in a positivist sense (i.e., the whole is equal to the sum of the parts) does not work anymore; complexity is too high and emergent properties require a gestalt. SE often fails because of the lack of a holistic model.

Discovering Emerging Phenomena. Understanding complex systems behavior starts by looking for emerging phenomena. For example, the use of new technology may typically generate surprises and face human operators with unexpected events. Instead of considering these events as noise, it is important to model the overall human-system activity by considering the parameters involved in observed emerging phenomena. Consequently, the concept of emergence is crucial in SE, especially when addressing SoS.

Stacey defines emergence as "… the production of global patterns of behavior by agents in a complex system interacting according to their own local rules of behavior, without intending the global patterns of behavior that come about. In emergence, global patterns cannot be predicted from the local rules of behavior that produce them. To put it another way, global patterns cannot be reduced to individual behavior." (1996).

It is interesting to watch the emerging behavior of termites. Each of these creatures looks very simple, but when they are working together they can produce amazing things including thermo-regulated galleries and habitats. We need to think in terms of phenomena, and we need to study them. Biologists and physicists have already developed models to study convection and diffusion phenomena. They use integrated partial differential equations to produce these emergent phenomena. Analog phenomena could be modeled for engineering processes.

More specifically, going from the old military model to an orchestra model will require the definition of new types of management. Composers, conductors and musicians in general know about emerging musical patterns. They can adjust their own contribution to participate in the making of these musical patterns. In highly interconnected, rapidly changing and diversity-rich complex systems, we necessarily need to study management issues. Performing SE locally in a homogeneous group of people is drastically different from doing it in a multi-cultural distributed environment.

Designing for Flexibility. Choosing complexity science as a support for managing SE imposes a functional approach and a structural approach to system design. The main problem is that classical "a priori" function allocation does not work since some functions, perhaps the most interesting ones, will emerge from operations. The utilization of knowledge, gleaned from experience to develop virtual prototypes to be iterated is essential. These prototypes can be easily designed, developed and tested with people in the loop, which may result in discovering emergent functions. This is the only way to test human-system activity before anything has been manufactured.

Agent-based architectures should then be developed beyond the prototypes while incorporating the emerging features. These architectures are flexible functional networks that can be easily modified. Beyond design purposes, this kind of flexibility can be used at operations time to improve HSI in a continuous manner.

5   Human-Centered Design Is a Clear Way to Achieve HSI

Distinguishing Task and Activity: Cognitive Function Analysis (CFA). We have seen that SE is a matter of SoS and can be modeled using the agent representation, supported by cognitive function analysis (Boy, 1998). A cognitive function is defined as an entity transforming a task into an activity (i.e., from something prescribed into something effective) and can be applied to people or software artifacts. Functions can be deliberately defined, but they are really understood at operations time. When defined a priori, they are directly associated with tasks. Cognitive functions can only be defined when an agent produces the related activity. This is why task analysis is important as a first resource to the definition of functions, but it should be complemented by an associated activity analysis. This is not possible without simulation-based or real-world operations.

Comparison of the task and activity reveals the variability. Variability should be analyzed using the TOP model to determine if variability is due to technology, organization or people. Tasks are typically described using hierarchical decompositions. Activity is more difficult to describe because we usually capture events and actions traces that need to be interpreted and categorized into cognitive functions.

CFA consists in incremental construction and refinement of roles, contexts of validity and resources. Resources can be physical or cognitive functions. CFA is based on scenarios (stories) that represent the "real world" in which the tasks are to be performed as well as simulated or real-world traces of activities of the various agents involved. CFA is typically carried out on possible scenarios (what ifs) that make sense to be tested. It departs from classical causal analyses that are purely based on the continuity of the past. Prediction is typically short term because it is a derivative extrapolation (i.e., we cannot predict too far in time). It is event-driven. CFA is fundamentally goal-driven and creative. CFA contributes to create, design and refine cognitive functions based on experience and exploration of possible futures. Consequently, CFA can support HCD.

Departing from Short-Term Prediction to Testing Possible Futures. Risk taking is an issue in industry and government projects. We seek to decrease risks or at least understand them as much as possible. Predictive models have been proposed and used, but the predictions of today's technologies came from the creativity of science fiction writers (*cf.* How William Shatner Changed the World, 2005). Jules Verne (1873) wrote "Around the World in Eighty Days" more than a century before we were capable of going around the world in 90 minutes in the International Space Station. A focus on creativity and methods that enable the test of possible futures is necessary and preferable to trying to predict in the short term.

However, creativity alone will not achieve our engineering aims. Rather creativity must be melded with engineering practice to succeed. HCD is based on cognitive engineering, complexity analysis, organization design and management, human-computer interaction although we have focused on modeling and simulation (M&S) in this paper. HCD provides a methodology with depth and breadth to support development of design throughout the SE process.

In other words, conceptual thinking must be based on virtual prototypes, usability and usefulness tests on advanced simulators, to define appropriate requirements. This step is mandatory if we want to understand the complexity of HSI. It may be considered as more costly than what is done today in requirements engineering, but it will certainly be much less costly for the life cycle of a product. The current short-term, linear view of SE should be transformed into a longer-term nonlinear view incorporating HSI, with testing done on both virtual prototypes and real mockups, considering issues, capabilities and constraints of technology, organizations and people simultaneously.

Promoting Modeling and Simulation. Traditionally, engineering focused on designing and developing machines that were further tested by human factors and ergonomics specialists. Incorporating HSI into SE resulted in identifying and incorporating human capabilities and limitation in system development. HCD brings a methodology and philosophy to leading the design process prior to engineering and feeding forward solutions that can be successfully introduced into the society and organizations that will use the system. Requirements should be human-centered, but the main asset of modern technology is to provide M&S from the very beginning of a design project. We now have the tools to do so.

The software evolution that was initially seen as a difficult problem is now a resource for HCD. Human-centered designers can create, shape, refine and test new possible systems in a virtual world, and deduce appropriate requirements for the later manufacturing stages of the life cycle of a product. Dassault's Falcon 7X was entirely modeled and simulated using software for example. It can also be used during the whole life cycle to support risky decisions and operations while reducing risk. In the Rail domain, the Cab Technology Integration Lab (CTIL) has developed to test not only new integrations of controls, displays, and decision aids but also to explore new automation technologies and their HSI impacts to SE purpose (Jones et al., 2010). Simulation provides a venue to explore outcomes in a low risk cost effective methodology. In addition, simulations can be orchestrated, linking models together across domains to examine previously unknown design spaces. It is through the linking of these various simulations that more is learned about the dynamical properties of the whole.

6 Conclusion

Software is great at solving problems that are well stated, but cannot state (or solve) the messy problems people can. Therefore, the real issue of stating problems and imperatively taking abduction into account requires the art and technique of forecasting, projecting possible futures and evaluating their validity. This goal-driven approach is a fundamental alternative to the current event-driven short-term approach. Solutions are necessarily context-dependent and should emerge from a concurrent consideration of the TOP model.

Function allocation, using the strengths of each component is key and more productive than designing a system, which relies on the human flexibility to overcome flaws in architecture or design. We need to better understand human and systems roles, contexts of validity of these roles, and appropriate resources they can use. Interconnectivity involves numbers of nodes and links between these nodes. Complexity science tells us that behavioral attractors emerge from interactivity between these nodes therefore we need to discover these attractors and their emerging properties. Using HCD, risks can be mitigated by using M&S from the very beginning of design projects, and continuously during the life cycle of a technology. HCD introduces creativity and design thinking that SE urgently needs.

This paper promotes unification of HCD and SE towards better HSI. We need to acknowledge that engineering requires creativity as well as solid technical knowledge and skills. If we want to create a livable future, HCD will need to come first, anticipating engineering practice. Even engineering as it is thought today will evolve. Having good M&S, people will be able to develop systems because they will have support for both their functional and structural parts.

Money spent wisely during the early stages of a product minimizes technical, cost, and schedule risks, minimizing continuous costly repairs caused by design flaws once it is delivered. There will always be a tradeoff between procedural linear SE leading to rigid practices, and creative non-linear HCD providing flexibility and motivation. We propose an integration of HCD and SE where socio-technical leadership should return at the top of organizations, irrigating motivation and collective pride.

References

Boy, G.A.: Cognitive Function Analysis. Greenwood/Ablex, CT, USA (1998) ISBN 9781567503777

Boy, G.A.: Knowledge management for product maturity. In: Proceedings of the International Conference on Knowledge Capture (K-Cap 2005). Banff, Canada. ACM Digital Library (October 2005)

Boy, G.A.: The Orchestra: A Conceptual Model for Function Allocation and Scenario-based Engineering in Multi-Agent Safety-Critical Systems. In: Proceedings of the European Conference on Cognitive Ergonomics, Otaniemi, Helsinki area, Finland (2009)

Boy, G.A., Brachet, G.: Risk Taking. Dossier of the Air and Space Academy, Toulouse, France (2010) ISBN 2-913331-47-5

Boy, G.A.: Orchestrating Human-Centered Design. Springer, U.K. (2013) ISBN 978-1-4471-4338-3

Brown, B.: Model-based systems engineering: Revolution or evolution?Thought Leadership White Paper. IBM Rational (December 2011)

Clark, J.O.: System of Systems Engineering and Family of Systems Engineering from a Standards Perspective. In: 3rd Annual IEEE International Conference on Systems Engineering, Vancouver, BC, Canada (2009) ISBN 978-1-4244-3462-6

Fitts, P.M.: Human engineering for an effective air navigation and traffic control system. Ohio State University Foundation Report, Columbus (1951)

Harris, S.D., Ballard, L., Girard, R., Gluckman, J.: Sensor fusion and situation assessment: Future F/A-18. In: Levis, A.H., Levis, I.S. (eds.) Science of Command and Control, Part III Coping with Change. AFCEA International Press, Fairfax (1993)

Haskins, C.: International Council on Systems Engineering (INCOSE) Systems Engineering Handbook v. 3.2.2.A Guide for System Life Cycle Processes and Activities. INCOSE-TP-2003-002-03.2.2 (2011)

Jones, M., Jones, M., Olthoff, T., Harris, S.: The cab technology integration lab: A locomotive simulator for human factors research. Proceedings of the Human Factors and Ergonomics Society Annual Meeting 54(24), 2110–2114 (2010)

Maturana, H., Varela, F.: Autopoeisis and cognition: the realization of living. In: Cohen, R.S., Wartofsky, M.W. (eds.) Boston Studies in the Philosophy of Science, vol. 42, D. Reidel Publishing Co., Dordecht (1980) ISBN 90-277-1016-3

Minsky, M.: Society of Mind. Simon and Schuster, New York (1985)

Mitchell, M.: Complexity: A Guided Tour. Oxford University Press, New York (2008)

Paulk, M.C., Weber, C.V., Curtis, B., Chrissis, M.B.: The Capability Maturity Model: Guidelines for Improving the Software Process. SEI series in software engineering. Addison-Wesley, Reading (1995) ISBN 0-201-54664-7

Peirce, C.S.: Science and philosophy: collected papers of Charles S. Peirce, vol. 7. Harvard University Press, Cambridge (1958)

Shatner, W.: How William Shatner Changed the World. Directed by Julian Jones. Handel Productions, Montreal (2005)

Stacey, R.: Complexity and creativity in organization. Berrett-Koehler, San Francisco (1996)

Verne, J.: Around the World in Eighty Days. Série Les Voyages extraordinaires. J. Hetzel, Paris (1873)

A Product Development Architecture with an Engineering Execution Representation of the Development Process

Gregory L. Neugebauer

Abstract. Successful development projects share common characteristics and attributes documented in project management curriculums and professional societies. This report gives tribute to those common elements as well as advancements in Systems Engineering to provide guidelines for system life cycle processes and activities.

Closely guarded by many corporations is the actual product development "process" which includes the steps, critical decisions, and roadmaps that enable a development team to efficiently evolve products from conceptual design to delivery and support. This paper addresses the question "can a single tailorable product development architecture be formulated and selectively optimized to address specific needs of a commercial or military application?" It will also guide the reader through a case study of creating a product development architecture and the benefits derived from the journey.

This approach will take a different perspective than the management, planning, and control viewpoints of product development and engage the topic with an engineering execution representation.

1 Introduction

The waste in traditional Product Development programs, results from a number of causes: craft mentality of engineers, poor planning, ad hoc execution, and poor coordination and communication cultures. (B. Oppenheim, 2008)

Gregory L. Neugebauer
Sandia National Laboratories,
Albuquerque, New Mexico, United States
e-mail: glneuge@sandia.gov

M. Aiguier et al. (eds.), *Complex Systems Design & Management 2013*,
DOI: 10.1007/978-3-319-02812-5_13, © Springer International Publishing Switzerland 2014

Figure 1 below provides evidence from a study of Product Development (PD) efforts which establishes a challenge and opportunity for improved methods and tools.

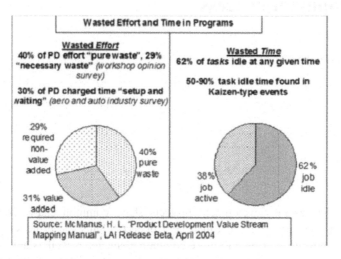

Fig. 1 A Product Development Opportunity

Acknowledgment must be given to the term *product development*. While it may be convienient to refer to these initiatives as products, the final deliverable is often a *system*, typically distinguished by the level of complexity.

While attempting to present a comprehensive view of a product development architecture, there are constituents that are not addressed in this research which must be inscribed within the product development architecture or as a component of portfolio management which include marketing analysis, product pricing, and end user testing.

Models and representations of the product development process may use the terms "architecture" and "framework" inter-changeably, although the use of "framework" is frequently associated with portrayals such as DoDAF, MoDAF, and Zachman. The term "product development architecture" used throughout this paper will refer to a substructure within the enterprise architecture.

2   Motivating Factors

The fact that most projects still fail in some manner, suggests that conventional project management does not meet current business needs. Although the conventional project management body of knowledge forms a good foundation for basic training and initial learning, it may not suffice for addressing development of complex systems. (Shenhar & Dvir, 2007)

Furthermore, to acknowledge the demand for a product development architecture, we must recognize the vast difference between the stable world of repetitive manufacturing and the high-variability world of product development. A product development process must thrive in the presence of variability. (Reinertsen, 2009)

From an engineering execution perspective, this will include:

- Learning from waste
- An antithesis of the earned value approach
- Critical early stage enablers, engagement of stakeholders, and building team confidence
- Application of Systems Engineering principles
- Recognition of an inefficient development process
- Maximization of agile concepts

In product development, a cross-functional team is brought together because its members have collective knowledge that cannot be held efficiently by any individual. This collective knowledge is not present by definition when the team is merely assembled, it is only potentially present. (Madhaven & Grover, October 1998) This condition begs for a product development architecture that catalyzes team members into a common goal and purpose.

3   Timeline

As a formal discipline, project management as we know it was born in the middle of the twentieth century. The Manhattan Project, which built the first atomic bomb during World War II, exhibited the principles of organization, planning, and direction that influenced the development of standard practices for managing projects. During the Cold War, large and complex projects demanded new approaches. In programs such as the U.S. Air Force intercontinental ballistic missile (ICBM) and the Navy's Polaris missiles, managers developed a new control procedure called program evaluation and review technique (PERT). This approach evolved simultaneously with the critical path method (CPM) which was invented by DuPont for construction projects. (Shenhar & Dvir, 2007)

The premier project management organization, the Project Management Institute (PMI) was founded in 1960 and has since done a remarkable job in building the guide to the Project Management Body of Knowledge (PMBoK), which has become the de facto standard of the discipline. (Shenhar & Dvir, 2007)

The modern discipline of Systems Engineering (SE) was developed in the ballistic missile program by Si Ramo and Dean Woldridge in 1954, with the first formal contract to perform "Systems Engineering and Technical Assistance

(SETA)." Under this contract, Ramo and Wooldridge developed some of the first principles for SE and applied them to the ballistic missile program—considered the most successful major technology development effort ever undertaken by the U.S. Government. (Jacobson)

"Total Quality Management" (TQM) which swept the industry by storm in the early 1980s was led by Deming [1982]. It was an attempt to adopt the successful Japanese industrial management methods to the U.S. industry. A strong message of TQM was that pursuit of higher quality is compatible with lower costs. (B. Oppenheim, 2008)

An important component of Concurrent Engineering (CE) was multifunctional design teams, sometimes called Integrated Product Teams or IPTs, which included representatives from the subsequent phases in the upfront engineering design. CE when effectively implemented with electronic design tools and workforce training led to dramatic reduction in design rework and, consequently, cost and schedule. (Hernandez, 1995)

Originating at Motorola several years later and relying on rigorous measurement and control, Six Sigma focused on systematic reduction of process variability from all sources of variation: machines, methods, materials, measurements, people, and the environment. (Murman, et al., 2002)

The term Lean as an industrial paradigm was introduced in the United States in the bestselling book, *The Machine That Changed the World, The Story of Lean Production* published by the MIT International Motor Vehicle Program [Womack, Jones, and Roos, 1990], and elegantly popularized in their second bestseller *Lean Thinking*, [Womack and Jones, 1996]. The authors identified a fundamentally new industrial paradigm based on the Toyota Production System. The paradigm is based on relentless elimination of waste from all enterprise operations, involving the continuous improvement cycle that turns all front-line workers into problem solvers to eliminate waste. (B. Oppenheim, 2008)

4   Transformation of Product Development Architectures

Research into the formulation of product development processes results in a deductive conclusion that many architectures or structures are the result of "accidental architectures" which evolved from sensible initiatives (TQM, Concurrent Engineering, Six Sigma...) or synthesized business practices. It is not uncommon for companies to treat internal processes as intellectual property, and when examined closely require non-disclosure agreements for release.

The construction of an architecture offers a unique opportunity for many companies. As will be explored later in this paper, product development can be considered a form of knowledge management.

Within any organization there exists a reluctance to change. A carefully constructed product development architecture which takes on an engineering approach with value based principles will find an eager audience of supporters. The method and approach used to develop the architecture will likely uncover strengths, weaknesses, opportunities and threats (SWOT) that otherwise would remain unacknowledged. Leveraging these conditions will provide the organization a pathway to improved marketability and profitability.

Benefits realized by the shear process of creating a product development architecture may be enough to initiate and justify this endeavor.

A key outcome of the architecture may be the mental transformation from viewing the deliverable as a present day product to a future state system. The rewards of this transformation will be realized many times over as subsequent systems are delivered.

The presence of a product development architecture presents a compelling argument for improved and consistent supply chain management. All suppliers will understand the techniques, tools, and processes used by the integrating entity. Outsourcing portions of the product development will burden less risk as all parties are acutely aware of their responsibilities. Research and technology entities both within and outside the organization will have direct line-of-sight to the key goal of delivering a product to the customer.

Another benefit of adopting a product development architecture is the agility to move technical personnel between projects and organizations. It is difficult to overestimate the importance of having an agile workforce that can move between assignments without the encumbrance of a new product development style, and it will drive ideal behaviors. Using a common methodology, multiple interpretations of business practices are significantly reduced. A common product development architecture will also enable projects to ease staffing transitions as transfers, promotions and retirements take place.

Productivity will be enhanced by accelerating each project teams' transition phase by moving through "forming, norming, and storming", allowing the team to reach the "performing" stage (Psychologist Bruce Tuckman's phrase) in a shorter time. Team members will understand their roles and feel secure about their presence on the project.

Monitoring of projects will improve and manifest decisions to increase funding of promising work, allowing this process to be data driven. A sense of urgency, early in the project will also be established, avoiding the late stage heroics often associated with product delivery.

The Diamond Approach to Project Management

To measure if a tailorable product development architecture can be developed, Shenhar and Dvir provide a useful examination of the many types of product development campaigns that may exist [*Reinventing Project Managment, The Diamond Approach to Successful Growth and Innovation*]. By categorizing the project, a unique footprint can be visualized. See Figure 2.

The diamond approach looks at the four dimensions of novelty, technology, complexity, and pace of a project. Novelty, as used by the authors, involves the uncertainty of the goal that is to be achieved and is an indicator of how clear or unclear the project goal and requirements are. Is the project about developing a new derivative, building upon an existing

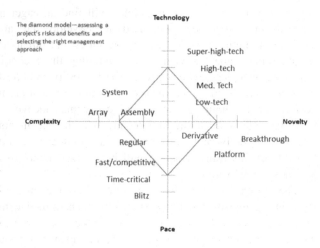

Fig. 2 NCTP Model (Shenhar & Dvir, 2007)

platform, or creating a breakthrough product or service? Technology describes the level of technological uncertainty involved and ranges from low to super high-technology. The complexity of a project, however, involves both the complexity of the product and of the organization involved. The measures of complexity range from an assembly of components to a complex system of systems. Finally, pace is the urgency required for the project, or its time frame for completion. (Mulenburg, 2007)

From Embedded to Embodied Knowledge

Using the notions of tacit and distributed cognition as a basis, Madhaven and Grover have proposed a model that links team members' and leaders' cognitive attributes and the team's process attributes to the efficiency and effectiveness with which the potential knowledge, resident within the team, is realized as a new product. (Madhaven & Grover, 1998). Their belief is that approaching product development as knowledge management will enhance the understanding of this critical process.

Participants in the Systems Architecture Forum identified Reference Architecture as a knowledge repository which facilitates knowledge transfer and communication. A Reference Architecture aids the understanding of the basic architectural and design principles. A Reference Architecture can also serve as a lexicon of terms and naming conventions as well as structural relationships within a company, industry or a domain. (Cloutier, Muller, Verma, Nilchiani, Hole, & Bone, 2010)

While the success of a system or product is ultimately determined by conformance to requirements and end use performance, failures may occur due to dysfunctional project teams. Thus, addressing the needs of project team members can be a critical success factor for a project.

Existence of a formalized product development architecture serves the development teams by building trust between team members. As described by Madhaven and Grover, the aggregate level of trust in team orientation will be related positively to the efficiency and effectiveness that embedded knowledge is converted to embodied knowledge. This principle applies equally to the team members' trust of one another.

The presence of an architecture also invites front loading of the project. This strategy has been found to be a key indicator of project success and allows program managers the visibility to determine if development funding should be continued or curtailed. The earned value approach in contrast, is better suited for serial type initiatives that have been previously executed. As will be discussed in the next section, incremental innovation is a possible candidate for applying earned value constraints to a project.   Innovative or significant changes to the interfaces or use-application of a product must be managed differently than incremental improvements to a product with the similar interfaces.

Architectural Innovation

Henderson and Clark assert that innovation can be categorized as incremental or radical and that each has competitive consequences requiring different organizational capabilities.

Architectural innovation is characterized as a change in the way constituent components are linked together in the system. Radical innovation is identified when a new dominant design emerges for the organization and creates significant challenges since it extends the usefulness of existing capabilities. It needs new modes of learning and a unique skill set of developers. See Figure 3.

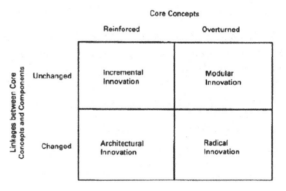

Fig. 3 A Framework for Defining Innovation (Henderson & Clark, March 1990)

Under this construct, tailorable product development architectures are needed to address the complexities and demands of various levels of innovation or radical challenges.

A product development architecture with focus on an engineering representation could prove to be an extremely useful tool to bridge the boundaries of innovation.

If the architecture were tailorable, the potential time-savings in understanding the variable focus areas that must be addressed is considerable. Incremental innovation would be streamlined and rely on the development efforts of previous systems where possible. Radical innovation would necessarily assume more risk, so additional validation and verification measures may be necessary.

Architectural innovation would benefit from a product development architecture by thorough evaluation of interfaces and modifications necessitated by the emergent new environments that the product or system would experience.

Mikkola provides an image to aid architects to evaluate the modularity of products that will be driven by the architecture. See Figure 4. The work outlines testable propositions

	Customizable	Non-Customizable
Standard	n_{STD-C} • off-the-shelf components • detail-controlled components	n_{STD-NC} • carry-over components • supplier-proprietary components
New-to-the-Firm	n_{NTF-C} • new materials • new versions of upgradable components • modular innovations	n_{NTF-NC} • unique components • product-specific components

Fig. 4 Classification of Components

and discusses the managerial and theoretical implications for the modularization function. (Mikkola, 2006)

Meyer and Dalal observed that high levels of reuse generally indicate that a product family was developed with a platform discipline. Upon analysis, one product family showed substantial platform discipline, emphasizing a common architecture and processes across specific products within the product line. The other product family was developed with significantly less sharing and reuse of architecture, components, and processes. The conclusion offered that the platform-centric product family outperformed the latter along a number of performance dimensions over the course of the decade under examination. (Meyer, 2002)

Shingo Model

The Shingo Prize is awarded to organizations that demonstrate a culture where principles of operational excellence are deeply embedded into the thinking and behavior of all leaders, managers, and associates. (Huntsman Business School; Utah State, 2012)

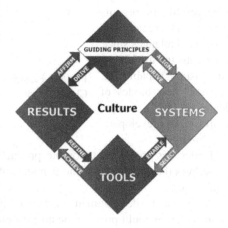

Fig. 5 The Shingo Transformational Process

The Shingo Prize for Operational Excellence is named after Japanese industrial engineer Shigeo Shingo. Dr. Shingo distinguished himself as one of the world's thought leaders in concepts, management systems and improvement techniques that have become known as the Toyota Business System. (Huntsman Business School; Utah State, 2012) A model for the transformational process is shown in Figure 5.

Principles associated with operational excellence can be summarized.

1. There is a clear and strong relationship between principles, systems, and tools.
2. Operational excellence requires focus on both behaviors and results.
3. Business and management systems drive behavior and must be aligned with correct principles

Although structured for operational excellence the Shingo model provides a strategic departure point for augmentation of new product development architectures.

5   Perspectives of Architecture

A comprehensive product development architecture initiative must recognize that an integrated effort with top-down and bottom-up construction, utility, and deployment will be needed.

In optimized cases the product development architecture will interface with the corporate reference architecture as shown in Figure 6. (Muller & Hole, 2007)

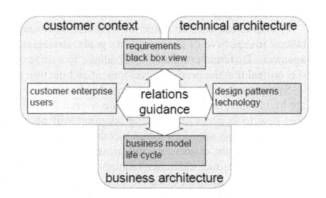

Fig. 6 Domains of the Reference Architecture

In practice, business architectures and customer context are often missing from the fully described reference architecture. (Rosen, 2002) Techniques using a SysML taxonomy or the Boardman Conceptagon potentially offer a systematic and critical thinking approach to address the complexities of integrating a technical architecture into a larger reference architecture. (Boardman & Sauser, 2008)

Conceptually, the technical architecture will scope the boundary, interior and exterior conditions. This paper will address these constituents, using a derivative of the Shingo model to address:

- Culture, Principles, Subsystems, Results (interior)
- Critical Tasks (interior)
- Ideal Behaviors (interior)
- Tools (interior)
- Links to critical resources and compliance with corporate business practices. (exterior)
- Communication Strategy (boundary)

Culture, Principles, Subsystems, Results

Culture is what we do when no one is looking and is behavioral based. Behavior is something that we can see and observe. Rich personal interaction, consisting of direct, frequent, and essential information exchange among team members, will influence the trust in the teaming commitment of other members positively. (Madhaven & Grover, October 1998)

NASA describes the Systems Engineering culture as a pervasive mental state and bias applied to problem solving across the development lifecycle and at all levels of enterprise processes. (NASA, 2007)

Principles are used to guide an organization throughout its life in all circumstances, irrespective of changes in its goals, strategies, type of work, or top management. Different companies will adhere to a different set of core principles and it is critical that the product development architecture reflect these principles.

Systems can be considered an organized collection of parts (or subsystems) that are highly integrated to accomplish an overall goal or defined objective. Companies new to systems engineering adoption will necessarily need to emphasize a systems thinking approach to product development.

Results are the observable and measurable outcomes of a system and evidence that the system met customer requirements. Best practices identified by NASA include visible metrics, effective measures and visible supporting data for better decisions at each organizational level. (NASA, 2007)

Critical Tasks

Critical tasks further refine the ideal behaviors into industry standards or corporate determined best practices. The tasks must support the ideal behaviors and delineate the actionable elements into manageable portions.

Ideal Behaviors

With behaviors now characterized as observable activities, a product development architecture will detail activities to a level that assures customers, stakeholders and participants that critical tasks are conducted with appropriate rigor, responsibility is properly delegated, and that the scope or breadth of work expected is appropriately communicated.

Tools

Tools are devices or process that aid in accomplishing a task. Tools can be a point solution, or a generic set of applications that can be applied to solving a problem or developing a solution.

Product Development tools, may be thought of as being formalized means for aiding product developers to carry out their jobs, and often summarized by terms such as Methods, Approaches, Diagrams, Guidelines, Models, Working Principles, Procedures, Representations, Standards, Steps, Techniques, and Methodologies. (Araujo & Duffy, 1996)

An important complexity in product development architectures is examination of the question "do tools drive architecture or does architecture drive tools?" At one abstraction, the product development architecture can be viewed as a tool itself. At another, the architecture may be represented by a comprehensive and integrated tool suite.

Links

Many organizations find themselves replete with business practices, quality assurance guides, and corporate policies and governance. Many of these regulations should flow-down into the product development architecture to insure that product launch is not encumbered by technical, supervisory or administrative delays.

Architects may consider these business procedures as exterior elements of the product development architecture with internal coupling.

Communication Strategy

Inherent in the implementation of any architecture, a training and communication strategy must be developed and supported by both key stakeholders and staff or user community. Pilot projects may be used for testing the architecture on a limited scale, but eventually the architecture will be broadcast to a larger user base.

The extent of the shared models will be related to the efficiency with which embedded knowledge is converted to embodied knowledge. Shared models will also have a relationship with the effectiveness with which the embodied knowledge is converted. (Madhaven & Grover, October 1998)

6   An Approach

An examination of an initiative at Sandia will demonstrate the utility of an engineering perspective. Figure 7 depicts the Principles and Ideal behaviors as foundational elements, while the Phases distinguish the maturity of the product as it traverses the development cycle. Tasks, Behaviors, Tools and Links provide a second tier of detail discussed in later sections.

Creating the Architecture

The process of creating the case study architecture was launched with formal sponsorship from executive management. As with most projects, a vision evolved from a decomposition of the "as-is" state into the conception of the future state. This vision matured into a mission statement with objectives and strategies. The constraints and assumptions were discussed and acknowledged. Fundamentally, desire to change along with foundational knowledge and a team of motivated professionals were the primary ingredients.

The purpose of many product development activities is to produce information that increases certainty about the ability of the design to meet requirements— i.e., these activities decrease performance uncertainty and assure a manufacturable product. Since risk is the product of uncertainty and consequences, reducing uncertainty equates to reducing risk in many cases. The objective of the product development activities is to drive this index to an acceptable level. Product development activities add value when they contribute towards this objective. (Browning, 2003)

Eventually the primary consideration of guiding principles and constructs of the architecture evolved. The recent enterprise adoption of integrated phase gates for all projects was apparent. In keeping with systems engineering principles outlined by INCOSE and corporate governance, these phase gates became the backbone of the architecture.

Further decomposition was approached using parallel information gathering and critical thinking approaches, in both focused forums and working meetings. Phases were broken down into tasks which were further resolved into behaviors. All through this process, management remained involved as active listeners or barrier-busters to furnish resources and offer perspective.

Checklists proved to be an important credential for the architecture, found in corporate quality, business, and technical procedures. Input and output criteria for each phase gate were examined and the critical activities to support these requirements identified and turned into actionable statements.

Only after the behaviors were narrowed did the architects begin to evaluate tools and links.

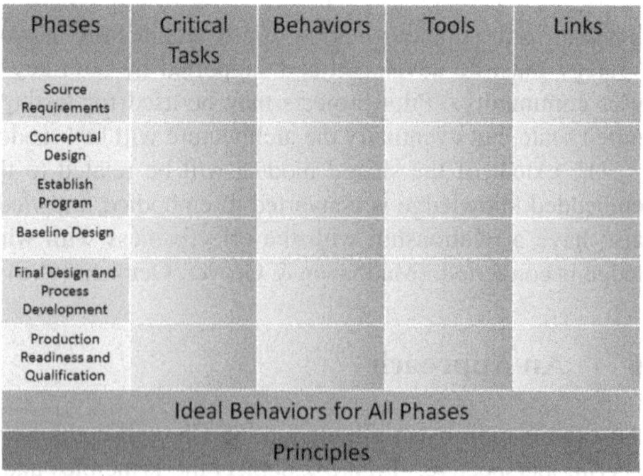

Phases	Critical Tasks	Behaviors	Tools	Links
Source Requirements				
Conceptual Design				
Establish Program				
Baseline Design				
Final Design and Process Development				
Production Readiness and Qualification				
Ideal Behaviors for All Phases				
Principles				

Fig. 7 An Enginnering Execution Perspective

Ultimately the architecture was "rolled-out." Several training sessions were offered to introduce the architecture. Deployment was supported by the presence of an internal web-site that provided a hierarchical representation of the architecture and active links to tools and infrastructure systems supporting the initiative.

Presently, the web-site provides a vital communication utility. It offers the potential to consolidate project data, while integrating various intra and inter organizational functions and processes. Integration challenges of web-based product development tools described by Sethi et al. and decisions on scaling the information technology investment for appropriate sophistication were examined. (Sethi, Pant, & Sethi, 2003)

Principles

Perfection
We build on our learning.

We consider designing products as a family (rather than single products) to obtain multi-product synergy.

Pull
We consider with our customer a broad set of options and trade spaces.

We seamlessly transition to production manufacturable designs that meet customer requirements (performance, schedule), are cost effective, and are sustainment-friendly.

Flow
We establish a project baseline early in order to understand the impact of changes and communicate those to the customer.

We understand, manage, react, and communicate risk effectively.

We recognize and act on critical decisions based on data.

Value Stream
We have a process that translates customer requirements to a manufacturable design based on project management and system engineering principles.

Value from Customer Perspective
We have active, continuous, and ongoing engagement with the customer.

We put ourselves in our customers' shoes to understand their environments and how they measure our success.

Ideal Behaviors for All Phases

A) Clearly define and negotiate roles and responsibilities throughout the project with high emphasis on interfaces with the customer and production.

B) Utilize a decision making process (i.e. who makes decision, identify points in time where decisions should be made, how decision is made, constraints, etc.) that will minimize/avoid rework and stick to it!

C) Identify a core set of peer reviewers (external to the Integrated Product Team) who will support the team throughout the process.

D) Integrate peer reviewing as early as possible to the extent that value is added and document results.

E) Consider risk impact in all planning and decisions.

F) Complete the check/act cycle continuously in order to build on learning and measure the project against customer expectations and requirements.

G) Collaborate with the appropriate people early in each phase to minimize rework from the reviews by addressing matters early in the process rather than later.

H) Go to where the work is done (internal, external, and customers) to see the environments first-hand.

I) Apply good project management skills by balancing and communicating the three tradeoffs of cost, schedule, and performance.

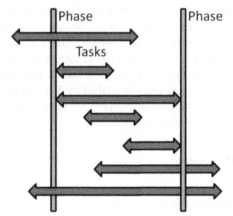

Fig. 8 Critical Tasks Muste Extend Beyond Phases

J) Consider next assemblies and subassemblies in the execution of this architecture.

Phases, Critical Tasks, and Behaviors

Prior to further examination of the case study a review of task structure is paramount. As cautioned by Reinertsen, tasks must be allowed to start when appropriate. Some may be constrained to a particular phase, or require passage of a milestone to begin. It is vital that the majority of these tasks remain flexible to start and complete as needed. The following sections describe these tasks; however they are presented as suggested starting points. See Figure 8.

Throughout the following section, Phases are shown in bold, Tasks are numbered and Behaviors designated by letter.

Assess Source Requirements Phase

1. Identify mission and scope of work (charter)
A) Question what is in scope and what is out of scope
B) Request and/or receive authorization for work
C) Identify stakeholders and customers
D) Clearly define measurable stakeholder/customer expectations/requirements for budget, schedule, deliverables, and constraints

2. Identify cross-functional team
A) Identify resources for roles (e.g. Financial, Quality, Design, Production, Customer, Management, Modeling, Supply Chain, Tooling, S&T, ES&H, Statistics, or any role that may be used in the product lifecycle).

B) Define responsibilities for roles defined in A

C) Consider all phases of the project when identifying roles

D) Negotiate roles& responsibilities with each team member and his/her manager (to meet both project needs and competency growth)

E) ID mentoring opportunities and resource/skills gaps (backups so no skill only one-deep), feeding those back to management for resolution

3. Draft project plan (resource plan, risk, budget, and schedule

A) Build a project plan collaboratively with the appropriate team members with a customer perspective

B) Determine project schedule strategy using a graded and phased approach (Don't over plan early!)

C) Plan for minimal fractionation of resources

D) Plan for team member co-location

E) Plan value-added tasks and decisions

F) Determine critical decisions that would preclude moving forward without substantial rework downstream

G) Integrate team building philosophy into project schedule (e.g. milestone celebrations)

H) Know the critical path

I) Request budget profile to start high and taper off to support planning activities before changes are expensive and time consuming

J) Strive to receive budget as planned/requested over the course of the program to minimize churn and rework

K) Consider Design, Qualification, Manufacturing, Sustainment, and Retirement aspects of lifecycle

L) Estimate budget for resources (people, facilities, equipment) for lifecycle

M) Understand the funding sources to avoid gaps in resource plan (i.e. who pays for what)

O) Work with management on make-buy decisions

Note: resource plan = budget (source and allocation), people, facilities, equipment, and procurement

4. Create risk management plan

A) Identify risks with corresponding plans to either Exploit, Mitigate, Avoid, Transfer, or Accept (Execute to EMATA)

B) Consider consequence, likelihood, and detectability (ability to detect risk occurrence) aspects in scoring each risk

5. Obtain & analyze source requirements

A) Distinguish requirements from background information/rationale/applicability

B) Question the requirement to understand the rationale and margin

C) Ensure the requirements are written using the SMART (specific, measurable, attainable, realistic, timely) guidelines

D) Consider all applications for the product as well as product capability and functioning with the next assembly

E) Obtain not only fit, form, and function requirements, but also consider cost and schedule

F) Consider the full life cycle of the product when developing requirements

G) Collaborate with stakeholders/customers to establish baseline and detailed requirements

H) Educate the customer about the options based on design, manufacturability, cost (e.g. costs of negotiable requirement verification) , risk, and schedule

I) Know what can and cannot be negotiated (musts vs nice-to-haves, know your tradeoffs)

J) Be open to any and all design options and associated risks

K) Consider how each requirement will be verified to enhance understanding of requirements

L) Understand and challenge assumptions that drive requirements

6. Negotiate requirements

A) Distinguish requirements from background information/rationale/applicability

B) Question the requirement to understand the rationale and margin

C) Ensure the requirements are written using the SMART (specific, measurable, attainable, realistic, timely) guidelines

D) Consider all applications for the product as well as product capability and functioning with the next assembly

E) Obtain not only fit, form, and function requirements, but also consider cost and schedule

F) Consider the full life cycle of the product when developing requirements

G) Collaborate with stakeholders/customers to establish baseline and detailed requirements

H) Educate the customer about the options based on design, manufacturability, cost (e.g. costs of negotiable requirement verification) , risk, and schedule

I) Know what can and cannot be negotiated (musts vs nice-to-haves, know your tradeoffs)

J) Be open to any and all design options and associated risks

K) Consider how each requirement will be verified to enhance understanding of requirements

L) Understand and challenge assumptions that drive requirements

7. Identify requirement option trades

A) Be open to any and all design options and associated risks

B) Challenge the team to keep thinking of potential requirement trades

C) Establish acceptable margin

D) Identify critical performance and quality parameters and their weight/priority per customer perspective

Complete Conceptual Design Phase

8. Create design concepts
A) Weigh design options against requirements (musts and nice-to-haves)
B) Diversify and challenge the team to keep thinking of options until multiple options exist
C) Draft physical and functional architecture considering interfaces (see system engineering terminology for further definition)
D) Review previous and existing designs and identify commonality and improvement opportunities
E) Create scalable designs based on requirement space (i.e. that vary across systems or vary overtime)
F) Balance variations in design attributes to meet your requirements
G) Identify options that can be modeled/simulated
H) Engage R&D subject matter experts to understand leading edge technology options

9. Analyze and begin down select design options and document
A) Understand design relationship with manufacturing options
B) Use modeling to analyze design options
C) Consider margin and maturity of technology when down selecting designs
D) Quantify design options against requirements to prioritize options
E) Establish and weigh down select criteria
F) Consider existing manufacturing and product capabilities
G) Analyze risk for lifecycle cost, schedule, and performance
H) Consider full lifecycle costs when down selecting
I) Carry forward as many options that passed down select criteria as feasible

10. Identify manufacturing options and capabilities
A) Collaborate with Design and Manufacturing personnel on manufacturing options corresponding to multiple design options
B) Compare existing manufacturing capabilities to manufacturing options to identify maturity gaps with a goal of commonality
C) Consider any and all manufacturing options and associated risks
D) Be open to design options that currently may lack manufacturing capability
E) Obtain or predict cost/benefit, maturity, process capability, facilities, and capacity for proposed manufacturing processes
F) Begin to identify critical manufacturing parameters
G) Consider the number of process steps, piece parts, and time
H) Consider make/buy and supplier trades and options
I) Consider lifecycle impacts (flow and future use) within manufacturer and supplier
J) Identify manufacturing constraints (e.g. maturity, floor space, etc.)

11. Draft system qualification plan
A) Define constraints and strategy for the qualification activities
B) Set a strategy such that rework is minimized to accomplish qualification (e.g. plan, requirement mapping, demonstration of each requirement, data collection system, and final report)
C) Determine how to demonstrate each requirement
D) Consider and plan all resource needs to execute qualification plan (e.g. assets, facilities, people, etc.)
E) Consider multi-system use in qualification planning
F) Consider both product and process in the Qualification Plan
G) Consider next assembly and subassemblies in your strategy
H) Glean lessons learned by reviewing legacy qualification documentation

12. Begin down selecting manufacturing options and document
A) Use modeling to analyze options
B) Consider margin and maturity of technology when down selecting
C) Quantify manufacturing options against design options to prioritize options
D) Establish and weigh down select criteria (e.g. manufacturability, Manufacturing Readiness Levels)
E) Carry forward as many options that passed down select criteria as feasible
F) Use data collected in previous step (i.e. analyze and begin down select design options and document) to down select manufacturing options

13. Conduct Conceptual Design Review
A) Capitalize on previous collaboration with stakeholders to conduct the Conceptual Design Review to minimize rework

Establish Product and Program Design Phase

14. Document conceptual manufacturing process flows
A) Decompose physical architecture into major process steps
B) Perform gap analysis to identify new process needs with goal of commonality across products
C) Develop the strategy and supporting schedule to address manufacturing needs concurrent with the design
D) Simulate flow and capacity which would include takt times, level loading, work cell configuration, process capability (goal of Cpk of 1.3), minimizing costs& risks, integration with other products, etc.
E) Incorporate quality into processes vs. inspection and testing at end of processes
F) Mistake proof processes
G) Model and minimize environmental impact and reduce waste streams, benefiting health and safety

15. Baseline project plan

A) Detail the schedule collaboratively with the appropriate team members with a customer perspective

B) Plan for minimal fractionation of resources

C) Plan value-added tasks and decisions

D) Determine critical decisions that would preclude moving forward without substantial rework downstream

E) Integrate team building philosophy into project schedule (e.g. milestone celebrations)

F) Know the critical path

G) Request budget profile to support lifecycle costs

H) Strive to receive budget as planned/requested over the course of the program to minimize churn and rework

I) Consider Design, Process capability, Qualification (process, product and supplier), Manufacturing readiness, On-going production, Sustainment, and Retirement aspects of lifecycle

J) Refine and submit budget for resources (people, facilities, suppliers, equipment, risk mitigation)

K) Understand the funding sources to avoid gaps in resource plan (i.e. who pays for which activity)

L) Place project under formal change control

Note: resource plan = budget (source and allocation), people, facilities, equipment, and procurement

Create Baseline Design Phase

16. Perform detailed design

A) Capture interfaces and functional requirements into product definition

B) Calculate margins to understand tradeoffs between designs

C) Approach design characterization in a cost effective manner (minimize redundancy, rework, and by utilizing modeling capabilities)

D) Optimize your design via trade-off between product stability, cost, and design margin

E) Design once and apply in many different uses

F) Design for: sustainment, testing, assembly, acceptance, manufacturability, state of health monitoring, mistake proofing, etc

17. Develop & execute test plan

A) Iteratively execute Plan, Do, Check, Act

B) Use Design of Experiments to develop test plans

C) Correlate accelerated aging with requirements

D) Understand your risks associated with accelerated aging, HALT and HASS philosophies, and margin determination

E) Plan for test facilities in advance to ensure meeting schedules

F) Balance fidelity of testing against individual requirements with efficiency of enveloping

G) Test plan consists of mapping of requirements, pedigree of parts, risks of using different pedigrees, conformance vs acceptance tests, statistically significant sample sizes, assumptions, hypothesis, etc.

H) Understand the measurement uncertainties

I) Incorporate computational simulation to a level consistent with the fidelity of the models

J) Use historical data to augment testing

K) Balance margin with cost when down selecting to baseline design

L) Improve models based on test results

M) Update product definition based on test evaluation/results

18. Update and release qualification plan

A) Update qualification plan based on test evaluation/results

B) Update the requirement map to reference requirement demonstration

C) Update resource needs to execute qualification plan (assets, facilities, people, etc.)

D) Build and test to emulate any variability expected in production

E) Ensure the Qualification Plan is for product and process

F) Peer review the Qualification Plan and update as necessary

19. Conduct Baseline Design Review

A) Capitalize on previous collaboration with stakeholders to conduct Baseline Design Review to minimize rework

Finalize Design & Process Development Phase

20. Formalize product definition support drawings

A) Define the what's not the how's

B) Conduct tolerance analysis to allocate tolerance

C) Map requirements to product definition/critical features

D) Ensure that the product built matches the product qualified

E) Create product definition that can be exported directly to models (manufacturing and simulation)

21. Finalize manufacturing process flows

A) Detail and validate the capacity analysis which would include takt times, level loading, work cell configuration, process capability (goal of Cpk = 1.3), minimizing costs & risks, integration with other products, etc.

B) Create a risk-based assurance plan noting which data collection points are required before final development and after qualification.

C) Map design requirements to manufacturing processes

D) Execute schedule and monitor and build improvement plans including mistake-proofing methods

E) Define critical manufacturing parameters based on criticality to product performance

F) Design processes that allow for quick change out between processes (flexible and agile manufacturing)

G) Design for co-processing for multiple products

H) Use COTS equipment and tooling

I) Document process flows and environmental safety and health waste streams

J) Collaborate with Production to verify final development build readiness

K) Develop and implement a transition plan from development to production (i.e. transfer of staff, training, etc.)

22. Conduct Final Design Review

A) Capitalize on previous collaboration with stakeholders to conduct Final Design Review to minimize rework

Complete Production Readiness & Qualification Phase

23. Support process qualification and first production build   activities

A) Partner with Production to prove in processes utilizing continuous communication

B) Conduct process walk through

C) Collect evidence for process qualification

D) Perform real-time monitoring of processes to ensure Cpk and yield goals are met

E) Complete process qualification reports based on final development lot data collected

F) Optimize processes as needed

G) Evaluate product qualification readiness (joint evaluation between Development and Production)

H) Complete the development to production transition plan

24. Verify requirements & issue complete & qualified engineering release

A) Verify and document that all requirements are met

B) Review the risk based assurance plan and modify data collection points as appropriate for full scale production

25. Support product submittal to customer

A) Continue to collaborate and support full scale production and field surveillance programs

Tools

Today's engineering departments typically standardize on particular modeling, analysis, configuration management and authorization tools. Additional tools for decision analysis, reliability estimations, requirements documentation and tracking can be ad-hoc or simply recommendations. The identification of these tools and the timeframe of when to use them can be a powerful addition to a product development architecture.

Standardization on tools is very important and will provide new team members with a list of approved applications, reference databases and general guidelines.

Tools may also include suggested or required training courses. Inclusion of this information will

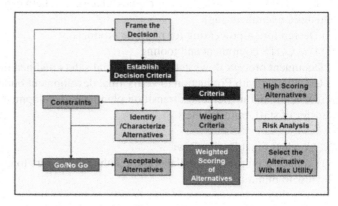

Fig. 9 Kepner Tregoe Decision Process

provide the product development architecture architects an opportunity to work with internal and external training providers to create a technical skills development program for engineering product teams

A time honored methodology for choosing tools can be found in the Kepner-Tregoe Decision Making process. The Kepner-Tregoe process provides a suitable method for dealing with complex, qualitative, somewhat subjective information and converting this into semi quantitative information more useful for decision making. See Figure 9.

The process provides a reasonable basis for keeping track of many factors simultaneously as must be done in comparing several complex alternatives. This approach is reasonably transparent in that it provides high visibility to the key elements leading to the final results.

Links

The utility of comprehensive and tailorable product development architecture will become evident when linking behaviors and tasks to business practices and programmatic requirements.

In particular, links to required checklists for mandatory reviews (conceptual, prototype, baseline, and final) along with the required or suggested content, list of reviewers, peers, stakeholders, et cetera, and the actionable tasks will be a powerful addition to the architecture.

A secondary and perhaps even greater value for the organization will be the ability to link these critical tasks and behaviors with work package agreements or a work breakdown structure. Handoffs and interfaces with internal and external entities will be apparent as the roles and responsibilities are associated with the behaviors.

Communication Strategy

As presented by Madhaven and Grover, the path to embodied knowledge requires a significant effort to improve communications. The acceptance of a product development architecture by the organization or distinct product development teams will be significantly influenced by the degree to which this architecture is absorbed and retained by the participants.

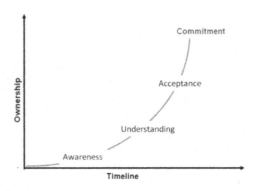

Fig. 10 Awareness to Commitment Curve

Project teams can be considered a special class of stakeholders for emerging product development architectures. There exists a life cycle to transitions: people involved in the transition must move through the Awareness-Understanding-Acceptance-Commitment curve shown in Figure 10. (Hahn, 2013)

A singular model that quickly communicates the fundamental elements of the product development architecture is vital. Visibility of this model in conference rooms and passage ways will provide further reinforcement of the strategic product development architecture.

Traditional top level architectures are useful for managers and program leaders but strand key staff on an engineering project thirsty for greater detail and guidance. This gap is further compounded when new team leaders are assigned who have not benefitted from the mentoring of an experienced leader. Effective product development architecture must address this gap if it is to be deployed and utilized.

Details can be absent; however the image of the architecture and the specifics of where to drill down for further information is one of the distinguishing characteristics of an engineering representation of the product development architecture.

Along with principles, and the earlier illustration of communication strategy, Figure 11 and the accompanying notes provide a useful example of how a strategy can be developed and communicated.

The backbone of the Product Development System is the Integrated Phase Gates (A thru F) which are based on system engineering principles.

The Product Realization Team (PRT) is a cross-functional team which enables seamless transition to manufacturable & sustainment-friendly designs that meet customer requirements.

Having customers at true north of the objectives will encourage active, continuous, and ongoing engagement. This enables putting the development team in the customers' shoes to understand their environments and how they measure our success.

The model promotes the team to recognize and act on critical decisions based on data through iterative Plan, Do, Check, Act cycles as prototypes and final product as *systems* advance through stages of technical and manufacturing readiness.

The funnel shape in the system image illustrates how we carry many options as long as possible to avoid discounting options too early. Our designs are

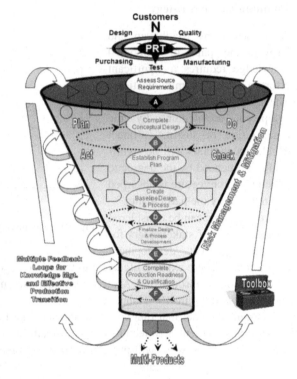

Fig. 11 Communication Strategy

set-based to allow us to extract an optimum solution from multiple options.

Feedback loops are not only employed to build on our learning, they are emphasized to illustrate the importance of an agile system. Sequential or "waterfall" styles of product development architectures may miss this agile opportunity at all stages of the program.

Naturally, the questions of architecture "enforcement" will arise. While there are several options ranging from management authority to voluntary adoption, persuasion is a model where the architect builds relationships and a position of authority. Persuasion requires clear communication. However, even when the architecture is clearly captured in diagrams and the architects have fully applied themselves; the outcome may not be as intended. In other words, initiating architecture is insufficient to enforce it. Monitoring and adaptation of architecture is also necessary.

The construction of a product development architecture will reap many benefits. The development of teams and their dynamics is vital to project success. This is enabled by a consistent reference to the adopted product development architecture.

Figure 12 illustrates the hierarchy and integration of the product development architecture relative to the enterprise model and distinguishes the parts and relationships to the whole.

This approach will lead to rich personal interaction, consisting of direct, frequent and informal interaction among team members, and will influence the team commitment positively. In today's competitive environment, every project is schedule-driven to an extreme, while attempting to address all aspects of product development with precious resources.

Expertise is converted appreciably to embodied knowledge. (Madhaven & Grover, October 1998)

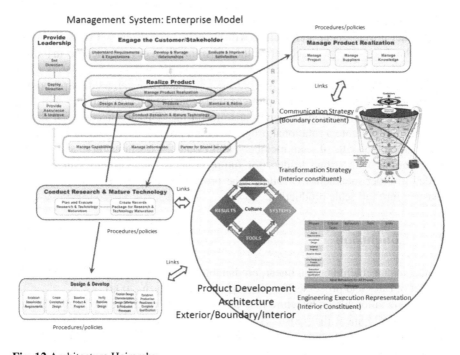

Fig. 12 Architecture Heirarchy

Future Work

Product development architectures encourage upfront communication, planning while setting expectations in accordance with budget, availability of resources, and the importance of maintaining open rapport. Architecture disciplines us to identify options early to prevent acute problem-solving dilemmas typical of schedule-driven projects.

This leads to a culture that is intended to reflect the activities we should be doing AND builds on each other's learning. Being grounded in an architecture gives the confidence to say "no" when "no" needs to be said...and that "no" is NOT a reflection of inadequacy.

7 Summary

Constructing a product development architecture will enable an organization to excel in the practice of engineering.

Architectures will strengthen management and engineering execution of programs by adopting or developing techniques and practices that enable a disciplined approach for the engineering of products and systems with assured quality by utilization of common product development practices.

In meeting this objective, organizations will provide customers and themselves with increased confidence that product or system development will reliably and consistently deliver on commitments, and will advance the state-of-the-art in the practice of engineering.

Product development architecture will provide a consistent set of tools, terms and technologies that will accelerate the transition from development to manufacturing. Manufacturing personnel will understand critical design parameters while design personnel will appreciate and consider alternative manufacturing proposals.

An architecture that requires participation from production partners will ease the difficult transition from design to manufacturing. This architecture can also be leveraged to promote a rotational program within product development to broaden and sharpen the skills of team members, abbreviate the learning curve and provide organizations with strong leaders for future projects. Opportunities may be afforded to other members to "follow" the product as it transitions from phase to phase or into full scale production.

8 Conclusion

Project management is not a linear, predictable process or a universal activity with one set of rules and processes for all projects. (Reinertsen, 2009)

A single tailorable architecture may not be possible to bridge the many types of product development that exist. A review of published architectures and execution of a sequence of steps – described in this paper, will lead the architects to a strategic product development architecture tailored and optimized for a general class of product or system development initiatives.

Product development architectures must not be viewed as static or perfect models. An effective architecture that can be tailored to the needs of the organization or the specific product under development will provide systematic agile execution of the product development cycle. An organization may choose to have a "reference" architecture, which is customized as needed to provide a point of departure for a product development activity. Additions or omissions should be defendable and provide feedback to the responsible architects improving the system.

As shown earlier, surveys indicate that 42% of product development effort is categorized as waste and 62% of tasks are idle at any given time. Given this possibility, architects should address the questions:

What types of "waste" are present in the current product development process?

What obstacles can be eliminated while developing new products that must be eliminated?

What embedded or embodied knowledge could reduce waste via the product development architecture?

Assessment should be an integral element of evaluating the product development architecture. Architecting is primarily a forward oriented activity, where by pro-active analysis and synthesis, a solution is shaped. Assessments are reactive, architecture assessments tend to detect issues that have not been covered sufficiently by the architecting effort. All processes, tools, and checklists that are proposed for architecture assessments are presumably also beneficial for the forward architecting activity. (Muller & Hole, 2009)

Finally, stimulating the conversation on new product development architectures and identification of the fundamental characteristics of both unsuccessful and proven implementations will undoubtedly contribute to richer and more useful models. The scalable nature of this work is allied to the dilating scope of architectures, and best practices of systems engineering.

Architecting is recognized as a discipline that connects customer needs and constraints with feasible technology-based solutions. The exchange of experience between practitioners from different domains may provide an overview of the status of systems architecting. (Muller & Hole, 2005) The strength of a product development architecture is its ability to be adaptable and provide an engineering execution model of the process!

Acknowledgements. The author would like to thank Eirik Hole, faculty advisor Stevens Institute of Technology for topical guidance, challenge and defense; Carla Busick, Dee Dee Griffin, Tina Hernandez, and Gary Pressly for their collective contribution to the Product Development Initiative; Cliff Renschler and Neil Lapetina for management support, review, and encouragement.

Sandia National Laboratories is a multi-program laboratory managed and operated by Sandia Corporation, a wholly owned subsidiary of Lockheed Martin Corporation, for the U.S. Department of Energy's National Nuclear Security Administration under contract DE-AC04-94AL85000.

References

Araujo, C.S., Duffy, A.: Product Development Tools. In: III International Congress of Project Engineering, Barcelona (1996)

Oppenheim, B.W., Murman, E.M.: Lean Enablers for Systems Engineering. Wiley InterScience (2008)

Boardman, J., Sauser, B.: Systems Thinking, Coping with 21st Century Problems. CRC Press, Boca Raton (2008)

Browning, T.: The Journal of Systems Engineering 12(1), 69–90 (2009)

Huntsman School of Business, Utah State University: The Shingo Model. The Shingo Prize: http://www.shingoprize.org/model-guidelines.html (retrieved February 8, 2013)

Cloutier, R., Muller, G., Verma, D., Nilchiani, R., Hole, E., Bone, M.: The Concept of Refernce Architectures. Systems Engineering 13(1) (2010)

Hahn, H.A.: Managing the Project Team as a Special Class of Stakeholder for Enterprise Transformation Projects. Presented at INCOSE Enchantment Chapter Meeting, Los Alamos, New Mexico, USA (March 2013)

Henderson, R.M., Clark, K.B.: Architectural Innovation: The Reconfiguration of Existing Product Technologies and the Failure of Established Firms. Administrative Science Quarterly 9 (March 1990)

Hernandez, C.: Challenges and Benefits to the implementation of IPTs on large military procurements. SM Thesis. MIT Sloan School, Cambridge (June 1995)

Jacobson, C.: JDAM. TRW, Cleveland

Madhaven, R., Grover, R.: From Embedded Knowledge to Embodied Knowledge: New Product Development as Knowledge Management. Journal of Marketing 62, 1–12 (1998)

Meyer, M.H.: Managing platform architectures and manufacturing processes for nonassembled products. Journal of Product Innovation Management 19(4), 277–293 (2002)

Mikkola, J.H.: Capturing the Degree of Modularity Embedded in Product Architectures. Journal of Product Innovation Management 23(2), 132 (2006)

Mulenburg, G.: Standish Group International, Inc. (2007)

Muller, G., Hole, E.: The State-of-Practice of Systems Architecting: Where are we heading? Architecting Forum, Helsinki, Finland, October 4-5 (2005)

Muller, G., Hole, E.: Reference Architecture; Why, What and How. Architecting Forum, Hoboken, NJ, USA, March 12-13 (2007)

Muller, G., Hole, E.: Architecture Assessments; Needs and Experiences. Architecting Forum, Washington, DC, USA, April 14-15 (2009)

Murman, E.M., Allen, T., Bozdogen, K., Cutcher-Gershenfeld, J., McManus, H., Nightingale, D., et al.: Lean enterprise value: Insights from MIT's Lean Aerospace Initiative. JDAM, Paggrave (2002)

NASA, Key Enablers for Successful Programs in Aerospace. NASA Pilot Benchmarking Initiative, Washington, DC (2007)

Reinertsen, D.G.: The Principles of Product Development FLOW. Celeritas Publishing, Redondo Beach (2009)

Rosen, M.: Enterprise Architecture Trends 2007: The year ahead. Cutter Executive Report (September 2002)

Sethi, R., Pant, S., Sethi, A.: Web-Based Product Development Systems Integration and New Product Outcomes. Journal of Product Innovation Management 20, 37–56 (2003)

Shenhar, A.J., Dvir, D.: Reinventing Project Management. Harvard Business School Press, Boston (2007)

Passenger's Transport between Platform and Train within the Metro in Paris

Jérôme Amory

Abstract. Transport systems are complex and in constant evolution. The number of passengers carried every year requires that the operator of this system analyses in terms of dangers and risks, the various stages of a passenger trip within the transport system. The transfer of passengers "platform – train" and "train – platform", is a crucial sequence since it is the interface between two journeys, passenger's trip and train's trip. The transfer of passengers generates - among others - risks of passengers falls on the platform, in the train and even on the track.

Keywords: systems' complexity, risks control, functional analysis, defense in depth, methods, systemic.

1   Introduction

The passengers transport system managed by the RATP, Autonomous Operator of Parisian Transports, the transport is a complex one. Reasons for this complexity are numerous and from different natures. They include technical and regular aspects, but also a constant evolution in order to adapt to other transport modes and its environment. In order to manage risks generated by the system, the RATP has developed a global and systemic approach called "safety system". This approach relies on rigorous and structured methods and integrates the concept of defense in depth. A reliable risks control does not exist without a complete understanding of the system to manage, the RATP completes its approach by a deliberate enhancement of the system knowledge and its environment using a group of graphical representations and referentials which contributes to control

Jérôme Amory
Railway Safety, Executive Management,
RATP (Régie Autonome des Transports Parisiens),
Paris, France
e-mail: jerome.amory@ratp.fr

M. Aiguier et al. (eds.), *Complex Systems Design & Management 2013*,
DOI: 10.1007/978-3-319-02812-5_14, © Springer International Publishing Switzerland 2014

risks, to re-establish a transversal coherence, to create synergies between actors of security and safety but also with decisions makers, optimizing a level of security for each step of transport on a daily basis.

2   Relevance and Approach of the Defence In Depth (DID)

RATP aim is to maintain and perpetuate a high level of performances on overall criterias which characterize transport. The safety of passengers and its staff is obviously the priority criteria. The risks control generated by the system and its environment translates in different approaches and complementary. The methodological monitoring has led the RATP to have an interest on the concept of defense in depth and to capture this concept on the field of transport.

2.1   Basic Notions

This old concept, erected in dogma on the military field by roman armies has gradually been adopted in other industrial fields such as the nuclear, chemistry, oil and information systems. The appropriations are different depending on the field of interest but a number of common notions are emerging, such as:

- lines of successive defences,
- each line of defence is autonomous but participates to the global defence,
- a unitary processing of each threat, taking into account their diversity and their difficulty in forecasting.

The defence in depth is focused on final effects. The question asked in terms of risks management is not "what do we undesired event?" but "what do we want to protect, from whom, of what, and at what level?" The level of risk acceptability defines defence requirements.

2.2   Relevances of the Defence in Depth Approach

The relevance of the appropriation of this concept of defence in depth and its modeling which follows, lies in several aspects:

- a global and systemic approach: The concept of defence in depth approach relies on the systemic approach adopted by the RATP. This global approach implements system engineering's principles take into account the whole transport system.
- a perennial and structured approach: In the classical management of risks, the various experts often debate on how to define with accuracy the undesired context, the studied undesired event and initiating events for a given situation. The DID approach ensures the singleness of these definitions.

- a complementary concept to existing approaches: This approach completes and feeds classical approaches of analyzing risks, of safety, of mechanisms and experiences feedbacks. These three levels of defence (and the notions of undesired contexts / events) have an impact on the build up of Preliminary Analysis of Risks (PAR). The PAR takes, generally, into account few actions post undesired events or undesired contexts. The section "measures of reduction" should be structured according to three levels of defence: prevention, protection and safeguard.
- a powerful modeling: The logic of the proposed modeling allows to identify or to conceive elements of various natures which contribute to finalities of defence and to understand their interactions, in order to control the relevance and efficiency. It gives to the experts opportunity to share a common vision of the system.
- a contribution of the notion of in depth: The distinction of lines of prevention, protection and safeguard roots the RATP commitment, with respect to the safety of transport systems, in order to avoid an unacceptable final effect.

The structured analysis of undesired contexts and of defence elements supplies a stringent basis for the determination of indicators of efficiency of the defence system, and the identification of precursors.

The concept of defence in depth is often associated to the term « barrier » which, if it illustrates well this notion of separation between the aggressor and the sensitive element to an aggression, is nevertheless constraining and often considered as equipment. The concept of defence in depth will allow to create a real typology of this defence element.

2.3   System of Defence in Depth

In associating engineering system with defence in depth, it is possible to identify a system of defence, the overall organized resources aimed at maintaining the level of security required for the transport system, which fulfills functions of defence to which are tied in security demands.

In terms of structure of the defence in depth system, engineering brings its stringency of representation. The functional analysis is a resolutely systemic approach and its relevance is founded for all models on perennial references with aims and system functions.

The functions of defence and its requirements are directly linked to dangers emanating from the transport system environment, of choices of principles, from components (technical and human) and from failures or mistakes.

A « system of defence in depth » (or SDD) is a coherent group of arrangements and of organized resources, and structures on several lines of defence, contributing to the control of the final effects susceptible to be created by all forms of aggressions on sensitive elements (man, system, company and/or environment)

The functions of defence express in view of the entities involved (figure 1), namely:

- of the three present entities: potential aggressors, aggressive flows produced by these aggressors and sensitive elements which may be damaged by these aggressions,
- direct final effects or combined (which we want to control) according to the types of sensitive elements and this field of sensitivity,
- final effects embody all the domains: services expected from the transport system, security of people, protection of the environment, company financial durability, company image,
- the notion of acceptability integrates a multi-criteria reasoning, inherent to all decision processes. These levels of acceptability are independent of undesired situations and events which may lead to the final considered effect. They strictly rely upon the upheld value systems and clearly depend on the choice of level "company policy".

The defence "target" chosen at the policy level offers the framework of a referential of requirements to satisfy for all system conception, managing or improvement.

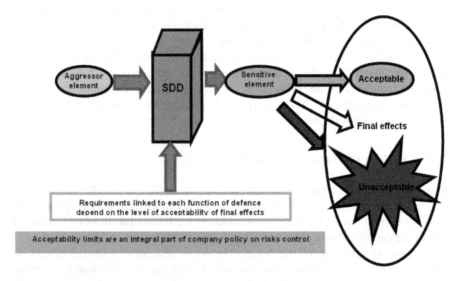

Fig. 1 Acceptability of final effects

2.3.1 A Structuring by Lines of Defence, in Response to the Finalities of Defence

The proposed structuring states each finality of defence while specifying under three complementary principles, which correspond to strategic action choices aimed at satisfying requirements characterized previously:

- a principle of prevention, which consists of acting on the probability of an undesired event taking place,

- a principle of protection, which aims at preserving final effects within the acceptable limits defined into policy of defence when an undesired event could not be avoided,
- a principle of safeguard, aimed at limiting the magnitude of consequences in cases where the accident can not be avoided, and acts simultaneously on the seriousness of the final effect and the uncombination with other final effects.

These three principles infer a system architecture set in three sub-systems or "defence lines", with a fourth sub-system of leading the group (which can be partially integrated to each line).

2.3.2   An Identification of Elements of Defence Making up Each Line of Defence (Figure 2)

Elements constituting each line of defence (whether they exist or projected) are defined through the determination of choices of principles of action in the face of aggression, and of interactions among elements of defence, of a same line or between lines.

The level of analysis, deemed equivalent, in system engineering, to the definition of a logical architecture of the defence system, requires that one justifies each element with respect to the relevance of the upheld action in order to satisfy the function of prevention, of protection or of safeguard:

- action on the aggressor,
- action on the sensitive element,
- or action on the aggressive flow.

The determination of this mode of action allows to define (or to validate) the performances expected from the element, within the context of its contribution to the line of defence under consideration.

2.3.3   A Definition of Each Element of Defence

Each element of defence can therefore be defined concretely through:

- means of actions deployed:
  - resources internal or external to the system, even mixed,
  - equipment or automatisms, or man, or a combination of these resources.
- modes of activation and of control of these resources,
- functions expected of the element of defence, and the associated requirements.

These definitions allow to model the technical architecture of the defence system.

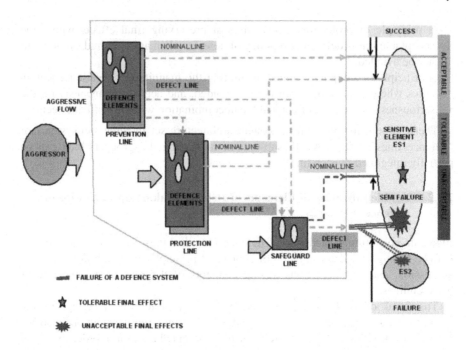

Fig. 2 Model of architecture of a defence system

3   Structured Methodology of the Diagnostic of System

The methodological plan articulates itself on the progress illustrated on figure 3.

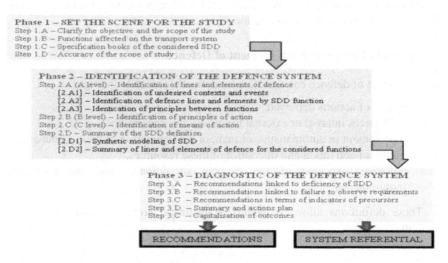

Fig. 3 Methodological plan of the diagnostic of a system

Note: Depending on the cases under consideration, from an incident or from the system function, the methodology steps will be totally or partially described.

4   Summary of the Application of the Methodology Following a Metro Passenger Fall during Transfer

4.1   Phase 1: Set the Scene for the Study

The study over the fall of passengers during transfer platform / train is linked to a function of the transport system. This study contribute to have available referential DID for the function "transfer platform / train" and to be able to diagnose the effectiveness of the existing SDD for this function. The input data comes from the observation of the existing system for the subway in nominal and degraded modes.

The study is centered on the transfer function for the functional analysis side and on the fall of passengers on the track side for the analysis of risks part. This study only makes sense within the context "operation with passengers" in nominal and degraded modes.

In referring to the transport system functional description according to the method APTE™ (figure 4) within the context of operation with passengers, we can identify the system function(s) directly linked to this study as well as choices of principle selected (figure 5).

These functions are set out as followed:

- SF1.A3.C1 – Insure to the groups of passengers an access to the train from the platform (transfers platform/train) for each point served by the considered line,

- SF1.A3.I1 – Insure to the groups of passengers an access to the platform from the train (transfers train/platform) for each point served (destination point) on the considered line.

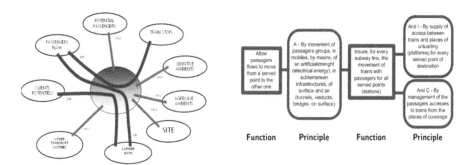

Fig. 4 Transport system during operation with passengers (octopus)

Fig. 5 Simplified functional breakdown of the system function to the transfer function

The drafting of the defence system transfer function specifications happens through the identification of entities: aggressors, aggressive flow and sensitive elements.

The potentially aggressive external environments of the defence system are the platform, the track and the train cars. The "platform passengers" and "train passengers" are the sensitive elements. The aggressive flow considered here is a mechanical flow mainly of reaction to a fall. The final unacceptable effect considered is the assault to the physical integrity of passengers. Figure 6 summarises these results.

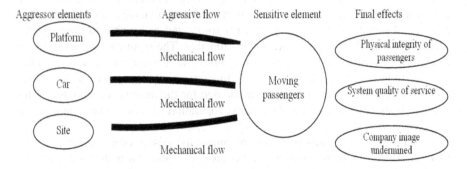

Fig. 6 Representation of the mechanical flow with aggressors and sensitive elements

The defence system must answer safety requirements of the transport system and it therefore uses components of the transport system and external components of the transport system (figure 7). In this case study, the defence system provides the functions which finalities are to insure the physical integrity of passengers in relation to aggressors' elements during the transfer platform / train.

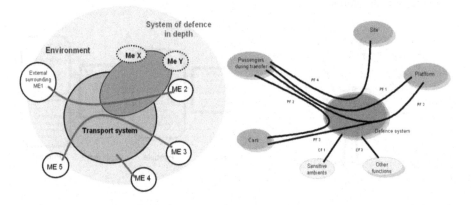

Fig. 7 General model of a defence system   **Fig. 8** Expression of main functions and constrained functions of a defence system

The main functions (figure 8) of the defence system in the case of passengers falls are of insure the physical integrity:

- of passengers during transfer in relation to the platform,
- of passengers during transfer in relation to the car,
- of "platform passengers" in relation to the interface platform-car,
- of passengers during transfer in relation to the track.

To these principle functions one has to add constraining functions with respect to surroundings and other functions of the transport system.

On the specifications book is drafted the level of acceptability of the final effect. As an indicator, we can specify:

- for the acceptable level: no injuries requiring care,
- for the tolerable level: light injured, or temporarily incapacitated,
- for the unacceptable level: loss of life or permanently incapacitated.

As stated previously, the functions of the defence system must also take into account the degraded mode of the system function. Several cases must then be analysed, as an example:

- the train doors do not open,
- the train is not positioned as per passengers exchange conditions,
- the platform is partly taken by building work,
- the ambiance conditions (lighting) are not satisfactory.

4.2   Phase 2: Identify the System of Defence

From the functional description of the transport system, it is also possible to identify danger scenarios associated to the function "transfer".

4.2.1   Identify Lines of Defence

The identification of the undesired context and undesired events (danger scenario) is used as the reference in ascertaining the position of lines of defence (prevention, protection and safeguard). For reasons of sizing, this communication is only limited to the undesired context "fall of passengers on the track between two cars during transfer". The identification of undesired events happens in conjunction with the functional referential of the transport system such as on figure 9.

A danger scenario can be considered as the « negative » of an arm of the functional description. In a system engineering analysis approach, the functional analysis is the link between the need expressed and the technical solution meant to fulfill the functions expressed. All failure of the technical system has an impact in the realization of one or several functions. These failures are initiatory events of a chain of incident.

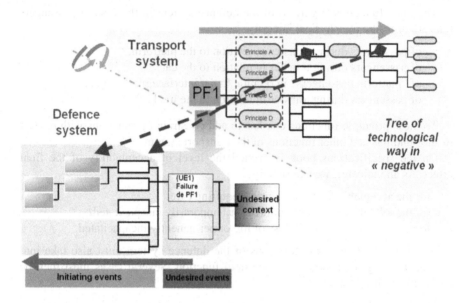

Fig. 9 Representation of the transport system functional failures making up the undesired events and initiating events of the danger scenario

The fall of « platform passengers » on the track between two cars is one of the dangers that the defence system, if well sized, must avoid. This undesired context results partly from choices of selected principles during the conception of the transport system namely:

- a raised platform surface relative to the track,
- trains made up of several coupled cars (mechanically and electrically) and leaving a space in between the cars.

This undesired context may become reality from undesired events which illustrate failures of the system functions. Failures of functions can be caused by defaults of conception, of realization, of use or of maintenance. Nonconformity to the requirements creates a failure of the concerned function. On this diagram "butterfly tie knot" of undesired events can also be created by initiating events of deliberate origins. In applying this approach, the undesired events and initiating events are identified automatically and found results are illustrated on figure 10.

On this type of diagram "butterfly tie knot", it could be interesting to position defence lines using finalities defined previously. Figure 11 illustrates these defence lines.

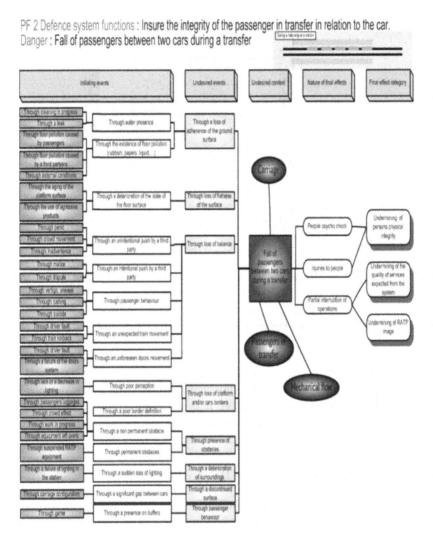

Fig. 10 Diagram « butterfly » of functional defects or diagram causes effects in relation to passengers fall between two cars during transfer

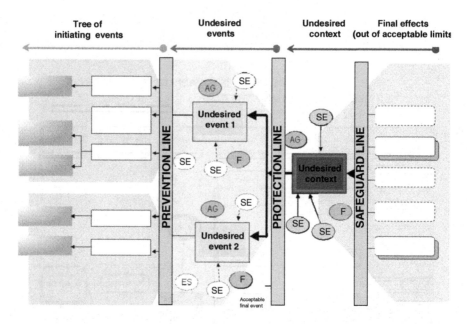

Fig. 11 Positioning of defence lines in relation to initiating events, undesired events and the undesired context (SE: Sensitive element, AG: Aggressor, F: Aggressive Flow)

4.2.2   Identify Defence Elements Thanks to Action Principles

To each undesired event is associated a sub function of defence where action principles are stated according to the aggressor, the aggressive flow and the sensitive element in order to get different defence elements. The defence sub function PF2.2.1 "Insure the passenger integrity during transfer in relation to passenger fall on the track between two cars following a loss of adherence on the platform surface" is validated by experts and illustrated on table 1.

4.2.3   Identify Means of Actions of Defence Elements

For each element of defence, means of internal or external action of human type, equipment, automatism or mixed are associated logic of execution. An extract of the identification of means of actions for the sub function PF 2.2.1: "Insure the integrity of passenger during transfer in relation to a passenger fall on the track between two cars by loss of adherence of the platform surface" is illustrated on table 2.

Table 1 Extract of the identification of defence elements of the sub function defence PF 2.2.1

Defence sub function		PF 2.2.1 : insure the passenger integrity during transfer with respect to the passenger fall on the track between two cars by loss of adherence on the ground surface		
Undesired context	Choice of actions principle	PREVENTION *Avoid (restrict the probability of) loss of passengers balance linked to the adherence of the platform surface*	PROTECTION *In case of passengers loss of balance, preserve final effects within acceptable limits*	SAFEGUARD *Control the consequences of passengers fall on the track between two cars and avoid other final effects*
Passengers fall on the track between two cars during transfer	Aggressor : Track + Interface Platform - Train	- Conception : Defines the ground characteristics (adherence,...) - Realisation : Ground meets defined characteristics - Conception : Allows passengers to preserve the cleanness and clarity of the ground - Conception : Design a draining system. - Operation : Control platform adherence... - Maintenance : Clean the ground outside passenger service operations...	Conception : trains with a small space between cars	
	Sensitive element : passenger on a platform	- Conception : Design handrails on the platform - Conception : ..... - Operations : Passengers awareness	- Conception : Put in place recovery means on the platform (handrail,...) - Conception : Put in place recovery means on the train (rails, nets,..)	- Conception : Organise tools of communication with internal and external rescue services... - Operation : ensure a staff presence (driver or in station). - Maintenance : Preserve the tools of communication with the internal and external rescue services
	Aggressive flow : track reaction		- Conception : Put in place a floor which acts as a partial or total shock absorber to a fall	
				- Automatic shutdown of traction current by presence detection on the track

Table 2 Extract of the identification of actions means of the defence function PF 2.2.1

DEFENCE ELEMENTS LINE OF PREVENTION	ACTION PRINCIPLE		ACTIONS MEANS	EXECUTION LOGIC
Conception : Define floor (platform) characteristics	Aggressor: Track + Interface Platform-Train	Avoid making the platform slippery during passenger service	H.2 : INTERNAL GROUP OF MEN : agents of conception	The platform adherence is defined according to the station ambient (humidity, temperature, traffic...)
Conception : Realise a floor (platform) conform to defined characteristics	Aggressor: Track + Interface Platform-Train	Avoid making the platform slippery during passenger service	He.2 : EXTERNAL GROUP OF MEN : external companies	The companies carry out a floor covering of platforms made up mainly in asphalt with a studied power of adherence
Conception : Allow passengers to preserve cleaness and clarity of floor (platform)	Aggressor: Track + Interface Platform-Train	Avoid making the platform slippery during passenger service	Ei.2 : EQUIPMENT : equipment specific to storing rubbish	The platforms are fitted with specific equipments to gather rubbish (bins, gutters ...)
Conception : Devise a drainage system	Aggressor: Track + Interface Platform-Train	Avoid making the platform slippery during passenger service	Ei.2 : EQUIPMENT : platform slightly biased and water drainage system	The stations are designed with platforms slightly bent to facilitate water drainage via the gutters

4.3 Phase 3: Diagnose and Recommended Defence System

Diagnose the System Defence: Over the illustration of the passenger fall on the track between 2 cars during transfer by loss of adherence on the surface of the platform, devices aimed at reducing the risk, organized by finality and analyzed through the global effectiveness of defence are estimated as "robust". In fact, the defence elements of the prevention and safeguard type are numerous and efficient in terms of conception, operations and maintenance. However the protection line is deemed as a slightly weak. Finally, an analysis of accidents since 2003 depicts that the adherence loss on the floor surface is not a significant cause of passengers fall over the track between 2 cars during transfer.

Recommended: Over this illustration of passenger fall on the track between 2 cars during transfer by adherence loss on the floor surface, the quantitative analysis and the qualitative analysis of the effectiveness of defence elements allow for the consideration of the risk (couple occurrence / gravity) as acceptable, that is why no complementary action is recommended.

5 Conclusion and Perspectives

The concept of defence in depth is rich and powerful in dealing with complex systems such as a transport system.

The will to structure the defence system in defence line, in defence elements, in principle of actions and in means of actions increase the knowledge of defence devises by their contribution to a finality of defence. The characterization of defence elements, true typology, paves the way to improve the qualitative judgment of the effectiveness of the defence system for a given function.

The appropriation such as established by the RATP also favour dialogue among the various actors of the conception, of operations, of maintenance and of risks control, by using models and representations. It also facilitates decision taking in cases of improvements of the defence system.

Furthermore, the use of a methodology based on functional analysis makes it possible to build the stable referentials on the transport system and its defence system. These referentials also allow for a thorough inventory of generic dangers linked to a transport system. In this, the approach of defence in depth contributes to a capitalization of knowledge.

References

[1] INERIS, Analyse des risques et prévention des accidents majeurs (DRA-07) (Juin 2001)
[2] Planchette, G., Valancogne, J., Nicolet, J.L.: Et si les risques m'étaient comptés Editions (Octarès 2002)
[3] INERIS, Eléments importants pour la sécurité (DRA 35) (Mai 2003)

[4] SGDN/DCSSI, La défense en profondeur appliquée aux systèmes d'informations (Juillet 2004)

[5] Cointet, A.: Méthodologie d'identification des éléments de défense des systèmes de transport. Communication 14 (Octobre 2004)

[6] Foinant, G., Cointet, A.: Défense en Profondeur et Précurseurs de dangers. Communication ATEC (Février 2006)

[7] Laval, C., Cointet, A.: Défense en Profondeur et Référentiel Outillé. Communication NTIC (Mai 2006)

[8] Laval, C., Cointet, A.: Défense en Profondeur et ingénierie système : une synergie au profit d'une approche globale des risques et des responsabilités. Communication 15 (Octobre 2006)

[9] Laval, C., Cointet, A.: Défense en Profondeur et ingénierie système : un outil d'aide à la décision face à la complexité des choix en maîtrise des risques d'entreprise. Communication 16 (Octobre 2008)

[10] Cointet, A., Amory, J.: , Maîtrise des risques d'un système complexe. Communication 18 (Octobre 2012)

Specifying Some Key SE Training Artifacts

David Gouyon, Fabien Bouffaron, and Gérard Morel

Abstract. Systems Engineering (SE) training is increasing in academic curriculum to satisfy the growing need for engineers aware of systems perspective. This training of a systems perspective is achieved by the formalization of selected best practices into a mature corpus with the expected objective to make SE a recognized engineering discipline for any specific curriculum. However, one major training difficulty is to infuse multi-disciplinary views of a system as a whole, beyond teaching the related standardized engineering processes. Our training practice leads us to formalize the specification process as one basic driver to rationally guide both teaching and learning SE basics.

1 Introduction

There is a growing interest to make Systems Engineering (SE) [Haskins et al. 2011], the discipline of engineering a system as a whole [Von Bertalanffy, 1968]. The objective of SE is to address a project's increasing complexity caused technically by multiple system component architectures as well as behaviors caused by multiple interacting agents. Classical engineering disciplines have to broaden their respective theoretical and technical domains to meet this crosscutting challenge in order to define, develop, deploy, and sustain a system satisfying stakeholder requirements.

For SE to be recognized as a standardized engineering discipline by the industrial and academic worlds, the BKCASE[1] international project (Body of Knowledge and Curriculum to Advance Systems Engineering) promotes a worldwide knowledge repository composed of two major deliverables. The first deliverable is

David Gouyon · Fabien Bouffaron · Gérard Morel
Centre de Recherche en Automatique de Nancy
Université de Lorraine, CNRS
Boulevard des Aiguillettes, BP 70239, 54506 Vandoeuvre les Nancy, France
e-mail: {david.gouyon,fabien.bouffaron,
       gerard.morel}@univ-lorraine.fr

[1] http://www.bkcase.org

M. Aiguier et al. (eds.), *Complex Systems Design & Management 2013*,
DOI: 10.1007/978-3-319-02812-5_15, © Springer International Publishing Switzerland 2014

the SEBoK (Systems Engineering Body of Knowledge) [Pyster et al. 2012a], which purpose is to "provide a widely accepted, community based, and regularly updated baseline of SE knowledge". It is available as a wiki to describe the precepts of Systems Science, System Thinking, and largely about Systems Engineering. The second deliverable is the GRCSE (Graduate Reference Curriculum for Systems Engineering) [Pyster et al. 2012b], as a set of "recommendations for the development and the implementation of a systems-centric professional master's degree program in Systems Engineering", compliant with the SEBoK.

Having the same objective as the SEBoK, the French chapter[2] of INCOSE[3] proposes a SE reference book [Fiorèse & Meinadier 2012]. The chosen approach is to describe the basic SE concepts ("with what?", "why?", "what?"), before describing SE processes, methods, and tools ("how?"). Eight years of experience teaching a master's degree program in Complex Systems Engineering[4] has convinced us that yes indeed, it is important for students to first understand how SE basic precepts are structured prior to teaching SE processes.

In this paper, we prescribe that the specification rationale based on Jackson's works [Jackson, 1997] is a key artifact for teaching in order to understand the concurrent interactions between the SE pivotal domain and the system-of-interest domain as well as the concurrent specialist engineering domains (Fig. 1).

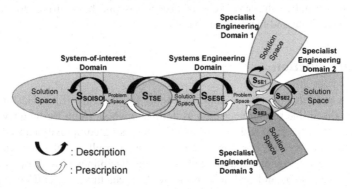

Fig. 1 Collaborative specification process applied to SE training [Bouffaron et al. 2012]. Teachers are in charge of the specification between System-of-interest and Systems Engineering domains. Students are in charge of the specification between Systems Engineering and specialist engineering domains.

Note that this specification rationale relies to system requirements as well as to functional and physical architecture requirements. Some specification facts are logically formalized using Object Role Modeling[5] in order to associate the

---

[2] http://www.afis.fr

[3] http://www.incose.org

[4] http://formations.univ-lorraine.fr

[5] http://www.ormfoundation.org/. Note that Object Role Modeling (ORM) enables to verbalize our specified training facts.

relevant standardized SE processes. These artifacts are then assessed in a teaching-learning case-study, supported by SysML [OMG 2012] as example of a Model-Based Systems Specification process.

2   SE Training Problem Statements

SE teaching is increasingly being implemented in some schools and universities, but practices as well as processes outcomes are not well unified. SE is mainly taught by experienced engineers, who are trained by years of practices on large projects, and still few professors rather autodidacts to be trained. These large engineering projects pushed practicing systems engineers to cope with different types of problems, to find solutions, and to formalize their "best practices" for implementation on future projects. Consequently, SE approaches mostly based on feedback require more scientific basis to be recognized by academia. Furthermore, in function of teachers' experience, some complementary points of views can be followed, with advantages and disadvantages, to teach SE: ensuring compliance with SE standards, following Model-Based Systems Engineering (MBSE) methods, and applying project management concepts.

Indeed, a major effort has been completed to standardize SE processes (ANSI/EIA 632 [ANSI/EIA 1999], IEEE 1220 [IEEE 2005], ISO/IEC 15288 [ISO/IEC 2008], ISO/IEC TR 24748-2 [ISO/IEC 2011], ISO/IEC 26702 [ISO/IEC 2007]). As a result, agreement, enterprise, project and technical processes within a system life cycle are highlighted and well defined. These processes enable the systems engineer, or student, to realize which processes are to be applied, and what outcomes are expected. In this way, students can verify if they have applied SE processes well in order to "solve the problem right", but the main drawback is that the compliance to the SE processes does not guarantee that they have met "the right problem".

This traditional document-centric approaches usability limit [Pyster et al. 2012 a] is being answered by MBSE [Estefan 2008] based on the de facto SysML standard. But in practice, if students are not well guided in the application of MBSE methodology, the result is a set of diagrams representing aspects of the system, which can present inconsistencies. In reality students can have only "drawn" rather than "modeled" the system. In both cases, students must be guided by sound rationales.

Even if the management of SE processes cannot be strictly disassociated from project management [PMI 2008], we consider in our teaching/learning situation that it is pragmatic for students to first practice engineering in a managed context (Fig. 2b), in order to avoid overlapping (Fig. 2c) or dissociation (Fig. 2a) between capacities to manage a project or to engineer a system.

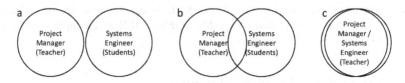

Fig. 2 Overlap of project roles, adapted from [Pyster et al. 2012]: teacher and students share project processes

This interoperation relationship between teacher and students (Fig. 3) must be also contextualized in a common domain-of-interest, from which the system-of-interest must be specified. This system-of-interest must have a scale factor and a complexity justifying that various disciplines (system, specialties...) are involved as well as that different roles are allocated to students.

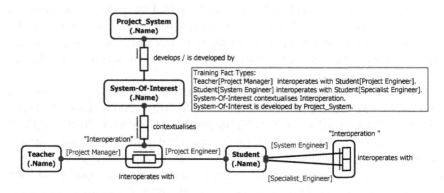

Fig. 3 Interoperations between teachers and students within SE training projects

As example of training context, the CISPI lab platform[6] reflects some requirements of the large-scale Connexion Project[7] aiming to innovate in control system architectures for nuclear power plants in France and abroad. CISPI system-of-interest reflects the main principles of a steam generator Auxiliary Feedwater System (AFS). The purpose of such an AFS is to cool the primary circuit, in case of emergency. Its mission is to feed steam generators with water in order to establish the necessary conditions for the use of a Shutdown Cooling Heat Exchanger (SCHE). The main objective of AFS is to maintain a sufficient water flow rate into the steam generators to ensure heat transfers in safe conditions. Student training projects impact modifications of the current CISPI platform related to an integrated Control, Maintenance and technical Management System [Morel et al. 2009] as well as fluid circuits.

---

[6] http://safetech.cran.univ-lorraine.fr/
[7] https://www.cluster-connexion.fr

3   Specification as a SE Training Driver

The problem statements we pointed out in the previous section bring to formalize the contractual specification relationship between teachers and students, as a pivotal driver to logically organize SE artifacts within training projects.

3.1   Problem-Solution Spaces Interoperation

In Fig. 3, Teacher[Project_Manager] and Student[Systems_Engineer] can be associated to *problem spaces*. Student[Project_Engineer] and Student[Specialist_Engineer] can in turn be associates to *solution spaces*. Indeed, problem space *describes* the *problem* and the solution space *prescribes* a *solution to the problem*. If the solution proposed solves the problem, this solution is seen as a *specification result*. In this sense, as underlined by [Hall et al., 2002] the interoperation between problem space and solution space is seen as a *specification process* producing this specified result. This is consistent with previous work [Bouffaron et al., 2012] wherein the specification process is seen as a descriptive-prescriptive interoperation relationship between problem space and solution space. This formalization (Fig. 4) allows clarifying the use of the "specification" word (which is used for referring to the *specification process* as well as the result of this process [Van Lamsweerde, 2000]) and the emergence of a key SE artifact which is the separation of problem and solution spaces.

Thus, **Student[Systems_Engineer] and Student[Specialist_Engineer] would not be the same person**. Therefore, to avoid role confusions, **students would be specialized in either systems engineering domain, or in specialist domains.**

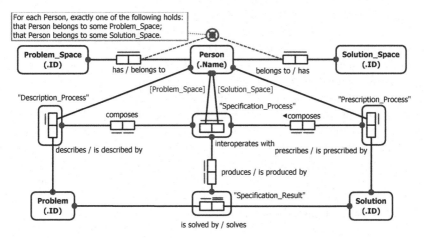

Fig. 4 Problem-solution spaces interoperation artifacts (note that constraints in ORM diagrams express some knowledge issued from SE best practices)

3.2 Source-Sink Objects Interoperation

In a more general way, the interaction between two spaces highlights that the specification process treats stakeholder requirements (Problem) as *source* objects, and to specify system requirements (Solution) as *sink* objects. We noticed that the specification process evolution is clocked by the different roles of the objects handled by the processes composing the specification process: an object produced by a process with a sink role will have a source role when it will be consumed by another process (Fig. 5). A main interest in the Source-Sink artifact is that it can be useful to perform traceability between source objects and sink objects during process execution. Thus, **teacher and student would trace all actions performed during process specification execution to ensure traceability between problem and solution.**

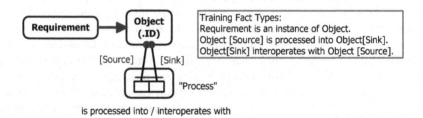

Fig. 5 Source-sink objects interoperation artifacts

3.3 Optative and Indicative Moods Interoperation

The process formalization (Fig. 6) can be put in relation with the works by Jackson in the software engineering domain, who formalized the concepts of optative and indicative moods [Jackson, 1997]. During the specification process, *optative objects* (representing requirements at different levels of abstraction) undergo several transformations performed by *processes* according to *indicative objects* belonging either to the problem space or to several specialist solution spaces. *Optative object* expresses conditions over the problem space that have to become true. *Indicative objects* represent the known properties (skills) of a domain which are validated by experts regardless of the behavior or given properties of the solution. Thus, domains can act as solution spaces if they have skills to solve a problem, or as problem space if they don't [Czarnecki, 1998]. This highlights that **some students would be involved at the system level, and some other students would be specialized in engineering domains.**

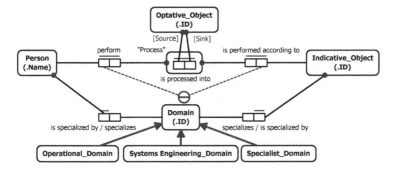

Fig. 6 Optative and indicative moods interoperation artifacts

3.4 Verification and Validation Processes Interoperation

The relationship between problem and solution spaces appears as a contractual process involving validation and verification [Pyster et al., 2012a] (Fig. 7). "Validation is used to ensure that one is working the right problem, whereas verification is used to ensure that one has solved the problem right" [Martin, 1997]. In this sense, we propose a formalization of the verification and validation processes by interpreting Jackson's works [Gunter et al., 2000] about optative and indicative moods, and considering the predicate: $W \wedge S \Rightarrow R$, where the specification S (Optative object Sink) must satisfy the requirement R (Optative object Source) considering the domain knowledge W (Indicative Object). Thus, the verification process performed by solution space consists in the satisfaction of the predicate:

$$W_{solution\ space} \wedge Optative\_object_{Sink} \Rightarrow Optative\_object_{Source}$$

In a similar way, the validation process performed by problem space has to satisfy the predicate:

$$W_{problem\ space} \wedge Optative\_object_{Sink} \Rightarrow Optative\_object_{Source}$$

This shows that **students would not valid their solution by themselves, their solution would be evaluated by the corresponding problem space**. The validation can be done either by the teacher for the system level, or by other students for specialists.

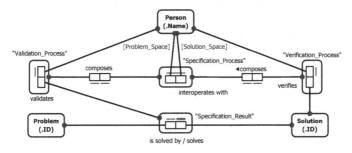

Fig. 7 Verification and Validation processes interoperation artifacts

4   SE Training Solution Assessments

To illustrate the potential benefits of the recommendations expressed in the previous section (in bold), we focus on the specification of the CISPI platform modifications. In order to specify them, a set of projects are proposed by relevant
teachers (Systems Engineering – problem space) to student teams (Specialist Engineering – solution spaces), as parts of their SE training curriculum.

The first set of observations we make is that, with such formalizations and explanations, students easily understand SE concepts and the relative positioning of
SE processes. They are aware of their respective roles within the problem and solution spaces, and establish contractual interoperations between themselves, and
the teachers. Students clearly make the difference between requirement definition
and analysis processes, and the verification and validation processes.

The second set of observations concerns the application of the recommendations on MBSE, more specifically with SysML [Holt & Perry 2008]. We proposed
to students to interpret SE recommendations into SysML modeling rules, in
order to facilitate diagram authoring, and to improve SysML semantics
[Ober et al. 2011]:

- The problem / solution spaces partitioning of SysML models can be done using
  packages: "problem space package" and "solution space package". This has a
  main interest for requirement definition and analysis, for example to clearly
  separate stakeholder requirements and system requirements;
- The source & sink concept is closed to UML customer & supplier roles used in
  dependencies to ensure traceability between objects. Given that SysML is a
  UML profile, we propose to use dependencies for the traceability during the requirement specification process, as presented in Table 1. Note that during the
  requirement analysis process, we have identified 4 types of transformations: refinement, decomposition, composition and induction [Bouffaron et al., 2012];

Table 1 SysML dependencies used to model source and sink requirement relations

Process	Source	Sink	SysML dependencies
Description	Stakeholder requirement – Problem space	Stakeholder requirement – Solution space	trace
Prescription	System requirement – Solution space	System requirement – Problem space	trace
Requirement Analysis	Requirement level n	Requirement level n+1	Refine, Derive, Requirement containment relationship ...
Validation	Stakeholder requirement	System requirement	satisfy

- As transformations rely on skills, it is very important to trace their use in models. We propose to include skills into models using SysML "Rationales". Such rationales can be linked to dependencies between requirements to justify and trace the transformation performed. Moreover, this enables to trace students reasoning, in a formative evaluation purpose.
- Although verification and validation processes can be executed according to skills, which can be traced using SysML rationales as presented before, they are usually performed according to SysML "Test Cases". Problem and solution spaces do not have necessarily similar test cases to verify and validate requirements. Thus we propose to link solution space test cases to requirements with a "verify" dependency as the indicative mood for the verification process, and to create the "validate" dependency for the validation process in order to clearly make the difference between verification and validation.

To illustrate this second set of observations, we focus on the following stakeholder requirement, extracted from a student project: "CISPI-AFS shall produce a sufficient water flow rate for the return of the primary circuit to SCHE conditions". Considering the work of students who are not guided by the recommendations made in section 3, the resulting requirement diagram is very poor, presenting only a system requirement traced to a stakeholder requirement (Fig. 8).

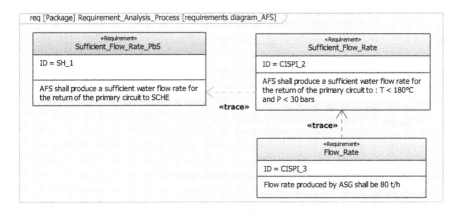

Fig. 8 Requirement diagram made by students without rules application

Considering now the work done by students guided by section 3 recommendations (Fig. 9), the approach is the following: once expressed in the *teacher problem space* (represented by a package), the stakeholder requirement is described to the *students' system level solution space* (also represented by a package). To transform (refine) this requirement into system requirements, as well as to verify the requirements they produce, students need skills that they can have or that can be required within the specialist engineering solution spaces. These skills are then traced in models with rationales. The validation of system requirements is done by the teacher in the problem space using a test case which is traced in models.

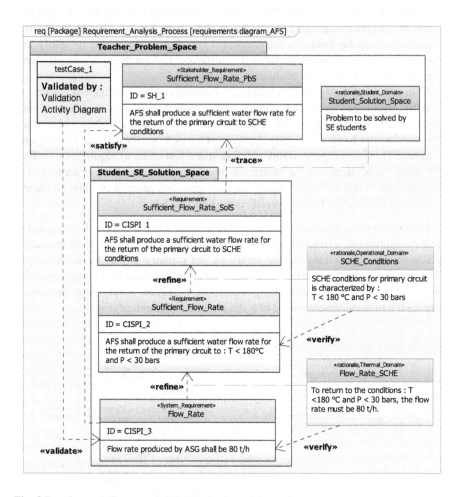

Fig. 9 Requirement diagram made by students applying rules

The comparison of the requirement diagrams of Fig. 8 and Fig. 9 shows that training recommendations improve the traceability because requirements are clearly classified into packages, and because skills used for requirement transformations and verification are traced into models. The verification of student work is also simpler, which is important in a learning context in which teachers can be the contracting authority and support student assessments at the same time. Detailed dependencies links (verify, validate…) helps the mutual understanding of models and accelerates the specification process by decreasing the number of iterations required before converging on validated system requirements.

Note that this training process is based on a reference specification process to rationally extend the project structure of SysML based tools as for example the one[8] used for the CISPI project, as well as their metamodel, for the relevant use of required SE artifacts.

5   Conclusions

In order to cope with the large amount of "best-practices" required to face any SE project with the limited amount of training resources (time, platforms, teachers' skill and knowledge, students' knowledge heterogeneity...), we share between teacher and students a common model of understanding the process of engineering a system as a whole. This as a prerequisite to logically train the technical processes and their related SE artifacts based on the specification process considered as a key SE driver.

ORM diagrams presented in this article are parts of a metamodel under development for training and for engineering purposes. This SE specification-based metamodel will be used as a pivotal reference to map systems modeling languages, and tool artifacts, with best SE artifacts.

References

ANSI/EIA, ANSI/EIA-632. Processes for engineering a system. Electronic Industries Alliance, Government Electronics and Information Technology Association Engineering Department, EIA Standard (1999)

Bouffaron, F., Gouyon, D., Dobre, D., Morel, G.: Revisiting the interoperation relationships between Systems Engineering collaborative processes. In: INCOM 2012, 14th IFAC Symposium on Information Control Problems in Manufacturing, Bucharest, Romania (2012)

Czarnecki, K.: Generative programming: Principles and techniques of software engineering based on automated configuration and fragment-based component models. PhD thesis. Technical University of Ilmenau, Germany (1998)

IEEE, 1220. IEEE standard for application and management of the systems engineering process. IEEE Computer Society (2005)

Estefan, J.A.: Survey of model-based systems engineering (MBSE) methodologies. Incose MBSE Focus Group 25 (2008)

Fiorèse, S., Meinadier, J.-P.: To discover and understand Systems Engineering. Cépaduès (2012) (in French) ISBN: 978.2.36493.005.6

Friedenthal, S., Moore, A., Steiner, R.: A practical guide to SysML: the systems modeling language. Morgan Kaufmann (2011) ISBN: 978-0123852069

Gunter, C.A., Gunter, E.L., Jackson, M., Zave, P.: A reference model for requirements and specifications. IEEE Software 17, 37–43 (2000)

---

[8] IBM® Rational® Rhapsody® supporting Model transformation Based Systems Engineering processes.

Hall, J.G., Jackson, M., Laney, R.C., Nuseibeh, B., Rapanotti, L.: Relating software requirements and architectures using problem frames. In: IEEE Joint International Conference on Requirements Engineering, Essen, Germany (2002)

Haskin, C., Forsberg, K., Krueger, M., Walden, D., Hamelin, R.D. (eds.): Systems Engineering Handbook, A guide for systems life cycle processes and activities, vol. 3.2.2. INCOSE (2011)

Holt, J., Perry, S.: SysML for Systems Engineering, vol. 7. Inst. of Engineering & Technology (2008)

ISO/IEC. ISO/IEC 26702. Systems engineering - Application and management of the systems engineering process (2007)

ISO/IEC. ISO/IEC 15288. Systems and software engineering - System life cycle processes (2008)

ISO/IEC. ISO/IEC TR 24748-2. Systems and software engineering - Life cycle management - Part 2: Guide to the application of ISO/IEC 15288 (System life cycle processes) (2011)

Jackson, M.: The meaning of requirements. Annals of Software Engineering 3, 5–21 (1997)

Martin, J.N.: Systems Engineering Guidebook: A process for developing systems and products. CRC Press (1997) ISBN: 978-0849378379

Morel, G., Pétin, J.-F., Jackson, T.L.: Reliability, Maintainability, Safety. In: Springer Handbook of Automation, pp. 735–747. Springer (2009) ISBN: 978-3540788300

Ober, I., Ober, I., Dragomir, I., Aboussodor, E.A.: UML/SysML semantic tunings. Innovations in Systems and Software Engineering 7(4), 257–264 (2011)

OMG, OMG Systems Modeling Language (OMG SysML), (v.1.3) (2012)

PMI. A guide to the Project Management Body of Knowledge (PMBoK guide), 4th edn. Project Management Institute (2008) ISBN: 978-1933890517

Pyster, A., Olwell, D., Hutchison, N., Enck, S., Anthony, D., Squires, A. (eds.): Guide to the Systems Engineering Body of Knowledge (SEBoK). Version 1.0.1. The Trustees of the Stevens Institute of Technology, Hoboken (2012)

Pyster, A., Olwell, D.H., Ferris, T.L.J., Hutchison, N., Enck, S., Anthony, J., Henry, D., Squires, A. (eds.): Graduate Reference Curriculum for Systems Engineering (GRCSE™). Trustees of the Stevens Institute of Technology, Hoboken (2012)

Van Lamsweerde, A.V.: Formal specification: a roadmap. In: Proceedings of the ACM Conference on the Future of Software Engineering, pp. 147–159 (2000)

Von Bertalanffy, L.: General System Theory: Foundations, development, applications, revised edn. George Braziller (1968)

An Engineering Systems Model for the Quantitative Analysis of the Energy-Water Nexus

William Naggaga Lubega and Amro M. Farid

Abstract. The *energy-water nexus* has been studied predominantly through discussions of policy options supported by data surveys and technology considerations. At a technological level, there have been attempts to optimize coupling points between the electricity and water systems to reduce the water-intensity of technologies in the former and the energy-intensity of technologies in the latter. To our knowledge, there has been little discussion of the energy-water nexus from an engineering systems perspective. A previous work presented a meta-architecture of the energy-water nexus in the electricity supply, engineered water supply and wastewater management systems developed using the Systems Modeling Language (SysML). In this work, models have been developed that characterize the various transmissions of matter and energy in and between the electricity and water systems.

1 Introduction

Water and electricity are inextricably linked, and as a consequence have to be addressed together[12]. Extraction, treatment and conveyance of municipal water and treatment of wastewater are dependent on significant amounts of electrical energy. Simultaneously, large volumes of water are withdrawn and consumed from water sources everyday for electricity generation processes. This *energy-water nexus*, which couples these critical systems upon which human civilization depends, has existed since the first implementations of the electricity, water and wastewater systems. The coupling, however, is becoming increasingly strained due to a number of global mega-trends[17]:(i) growth in total demand for both electricity and water

William Naggaga Lubega · Amro M. Farid
Masdar Institute of Science and Technology, Abu Dhabi
e-mail: {wlubega,afarid}@masdar.ac.ae

Amro M. Farid
MIT Technology Development Program
e-mail: amfarid@mit.edu

M. Aiguier et al. (eds.), *Complex Systems Design & Management 2013*,
DOI: 10.1007/978-3-319-02812-5_16, © Springer International Publishing Switzerland 2014

driven by population growth; (ii) growth in per capita demand for both electricity and water driven by economic growth; (iii) distortion of availability of fresh water due to climate change; and (iv) multiple drivers for more electricity-intensive water and more water-intensive electricity.

These trends raise concerns about the robustness of the electricity and water systems today and their sustainability over the coming decades. There is a risk that if the nexus is not optimally managed, scarcity in either water or energy will create aggravated shortages in both.

A number of discussions on the energy-water nexus have been published in recent years. The two approaches that have predominantly been taken in the literature to discuss this topic are: (i) discussions of various policy options [12, 13, 15, 16, 17, 18, 20]; and (ii) evaluation of the electricity-intensity of water technologies and the water-intensity of electricity technologies [2, 5, 6].

In this paper, a foundation is laid for an engineering systems approach to studying the nexus, through the presentation of *first-pass* engineering models of the various salient exchanges of matter and energy in and between the electricity and water systems. The paper builds upon qualitative models previously provided in [9].

In Section 2, the models are presented alongside system activity diagrams for an integrated view of the electricity and water engineering systems. Section 3 presents an illustrative example in which the developed models are used to analyze the exchanges of energy and water in a given geographical region. Section 4 offers some insights that can be acquired with the aid of the discussed approach. Finally, Section 5 concludes the work and presents directions for future work.

2 Modeling

This section builds upon the previously developed qualitative model of the energy-water nexus[9] and proceeds in three parts: (i) delineation of the system boundary, (ii) presentation of SysML activity diagrams of the engineered electricity and water systems, and (iii) presentation of models of the functions identified in the activity diagrams . With the exception of the grid equations for both networks, which are modelled by means of edge-node incidence matrices, the models of the identified functions are developed using the bond graph methodology which readily facilitates the inter-domain modeling necessitated by the heterogenous nature of the energy-water nexus and which makes clear which variables are dependent and independent. The interested reader is referred to [8] for an introduction to bond graph modelling.

2.1 System Boundary and Context

Figure 1 chooses the system boundary around the three engineering systems of electricity, water and wastewater. It also depicts the high level flows of matter and energy between them and the natural environment. The labels A through J represent key flows of interest and are discussed in Section 4.

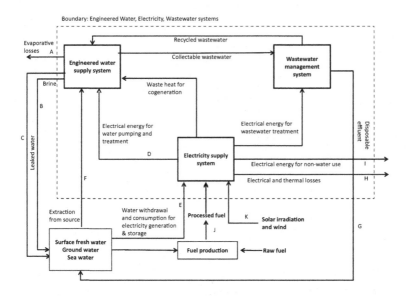

Fig. 1 System Context Diagram for Combined Electricity, Water & Wastewater Systems

Electricity, potable water, and wastewater are all primarily stationary within a region's infrastructure; in contrast, the traditional fuels of natural gas, oil, and coal are open to trade. Consequently, the fuel processing function, though it has a significant water footprint, is left outside of the system boundary. An advantage of this choice of system boundary is that the three engineering systems all fall under the purview of grid operators; and in some nations, such as the United Arab Emirates, all three grid operations are united within a single semi-private organization. The system context diagram shown in Figure 1 also makes it possible to relate a region's energy consumption to the required water withdrawals in a complex input-output model.

The following conventions are adopted in the bond graphs found in the following sections:

Engineered Water System:

- $F_{W_j} \in \mathbf{F_W}$ is water flowing through a pipe j
- $E_{W_i} \in \mathbf{E_W}$ is the pressure at a node i.
- $F_{Wnet_i} \in \mathbf{F_{Wnet}}$ is the water injected into the network at a node i.

Electric Power System:

- $F_{P_j} \in \mathbf{F_P}$ is current flowing through a transmission line j.
- $E_{P_i} \in \mathbf{E_P}$ is the voltage at a node i.
- $F_{Pnet_i} \in \mathbf{F_{Pnet}}$ is the current injected into the network at a node i.

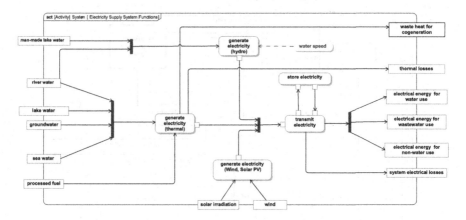

Fig. 2 Activity Diagram of Electricity System Functions

2.2 Electricity System Functions

The four most prominent electricity generation technologies are shown in figure 2. These technologies have varying water withdrawal and consumption footprints. Thermal generation requires large volumes of water for cooling purposes and is one of the chief concerns associated with the energy-water nexus. Hydroelectric power incurs evaporative losses due to the increase in exposed surface area created by the dam reservoir. Solar and wind generation do not have any significant water requirements in operation.Bond graph models of solar, wind and hydroelectric generation are presented below. Development of a thermodynamic bond graph model of thermal power generation is the subject of future work.

2.2.1 Generate Electricity-Hydro

Figure 3 shows a bond graph model of a hydroelectric power generation station in which a water drop H drives a turbine, represented here by an ideal gyrator with modulus β. The turbine, in turn, drives a generator, represented also by an ideal

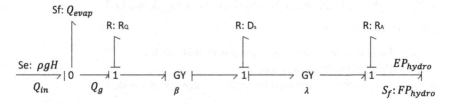

Fig. 3 Bond graph representation of hydroelectric power generation

gyrator with modulus λ. From the model, the voltage generated by the power station $E_{P_{hydro}}$ and the water withdrawn for generation Q_g can be determined as:

$$E_{P_{hydro}} = -\left[\frac{\lambda R_Q(\lambda^2 + D_S R_A) + \beta^3 R_A}{\beta^2 \lambda}\right] F_{Pnet_{hydro}} + \left[\frac{\rho g(\lambda^2 + D_S R_A)}{\beta \lambda}\right] H \quad (1)$$

$$Q_g = \frac{\lambda}{\beta} F_{Pnet_{hydro}}$$

where R_Q, D_S and R_A represent resistances and dampings of the penstock, turbine shaft and generator respectively. Of interest is the rate of water withdrawal Q_{in} from the source water body to support a given level of power generation as this requirement can impose a constraint on power generation. This withdrawal is given by: $Q_{in} = Q_{evap} + Q_g$; where Q_{evap} is the rate of water evaporation from the dam reservoir, a flow dependent on reservoir surface area, and ambient temperature and humidity conditions.

2.2.2 Generate Electricity – Wind

A bond graph model of a wind turbine is provided in [1]. Figure 4 reproduces the model with the simplifying assumptions that the wind turbine is operating in the steady state, that there is no stiffness between components and that all dampings are equivalent.

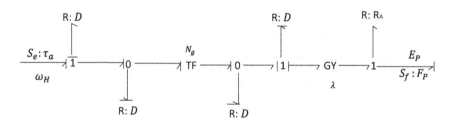

Fig. 4 Bond graph representation of wind turbine

From the model, the voltage, $E_{P_{wind}}$, imposed on the electrical distribution network by n identical, co-located wind turbines connected in series can be determined as:

$$E_{P_{wind}} = -\left[nR_A - \frac{n\lambda^2(3D - N_g^2)}{2D(N_g^2 + DN_g^2 - 3D)}\right] F_{P_{wind}} + \left[\frac{n\lambda DN_g(N_g^2 - 1)}{2D(N_g^2 + DN_g^2 - 3D)}\right] \tau_a \quad (2)$$

where

- N_g is the gearbox ratio of the wind turbine gearbox
- D is the damping of the various mechanical components and interfaces in the wind turbine

- τ_a is the aerodynamic torque exerted on each wind turbine, a function of the wind speed V, air density ρ_a, blade radius R and a dimensionless torque coefficient C_T given by [1]:

$$\tau_a = \frac{1}{2}\rho_a\pi R^3 C_T \times V^2$$

2.2.3 Generate Electricity – Solar Photovoltaic

Figure 5 is a simple bond graph model of solar photovoltaic installation approximated as a current source, a linear resistance, and a gyrator. In this model, the gyrator represents both the inverter switching circuit and the DC-to-DC conversion circuit. The function of the former is approximated by the change of causality from flow source to effort source effected by the gyrator while that of the latter is effected by the magnitude of the gyrator modulus. The resistance represents electrical losses in both processes.

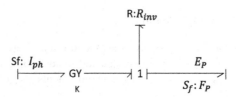

Fig. 5 Bond graph representation of solar PV installation

From the model, the voltage $E_{P_{solar}}$ imposed on the electrical distribution network by n such co-located installations connected in series can be determined as:

$$E_{P_{solar}} = -nR_{inv}F_{P_{solar}} + nKI_{ph} \tag{3}$$

where

- R_{inv} represents energy losses in the inverter
- K is the gyrator modulus
- I_{ph} is the photogenerated current induced in each solar PV installation. The photogenerated current is dependent on various material properties and the incident photon flux which is related to the solar irradiance and solar cell surface area[10].

2.2.4 Transmit Electricity

The transmission of electricity through an electricity distribution network can be easily described with the aid of an edge-node incidence matrix A_P, with the following equation:

$$\mathbf{F_{Pnet}} = A_P^T C_P A_P \mathbf{E_P} \tag{4}$$

where:

- C_P is a diagonal conductance matrix with each diagonal entry $C_{P_{jj}}$ being the conductance of transmission line j.
- $A_{P_{ji}} = 1$ if current in transmission line j leaves node i, $A_{P_{ji}} = -1$ if current in transmission line j enters node i, $A_{P_{ji}} = 0$ otherwise.

In typical power flow analysis, Equation 4 is multiplied by a diagonal matrix of bus voltages to yield the power flow equations.

2.3 Water System Functions

An activity diagram for the engineered water supply system is shown in Figure 6. All water grid functions are dependent on electrical or thermal energy input. Pumping, either for extraction or distribution, is responsible for the bulk of the energy consumed by the water system. Models of three of the indicated water supply options are presented below. A model of thermal desalination is to be developed in future work. Electricity consumption for municipal water use is not modelled as it represents a plethora of different processes that are typically not under the purview of grid operators.

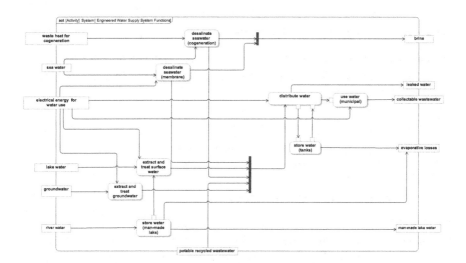

Fig. 6 Activity Diagram of Water System Functions

2.3.1 Extract and Treat Ground Water

The extraction and treatment of ground water can be effectively modelled as a pump as pumping consumes more than 98 percent of power at a ground water treatment plant [5].

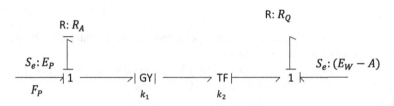

Fig. 7 Bond graph representation of pump

A bond graph model of a displacement pump that imposes a fixed pressure increment $(E_W - A)$ on a fluid is shown in Figure 7. The gyrator and transformer are representations of a motor and pump respectively. R_A and R_Q represent electrical and fluidic resistances in the motor and pump. From this bond graph representation, the current drawn by a groundwater treatment plant can be determined as:

$$F_{Pnet_g} = \frac{\rho g K (E_{W_g} - A_g) + R_Q E_{P_g}}{K^2 + R_A R_Q} \tag{5}$$

where A_g is elevation of the aquifer from which the groundwater is drawn above the selected piezometric datum, $K = k_1/k_2$ and E_{W_g} is the pressure to which water is pumped for distribution—a design parameter of the water distribution network.

2.3.2 Extract and Treat Surface Water

Similar to a groundwater treatment plant, a surface water treatment plant can be modelled as a pump. The current drawn by a surface water treatment plant can, therefore, be determined as:

$$F_{Pnet_s} = \frac{\rho g K (E_{W_s} - A_s) + R_Q E_{P_s}}{K^2 + R_A R_Q} \tag{6}$$

where A_s is elevation of the surface water body above the piezometric datum.

2.3.3 Desalinate Seawater (Membrane)

Reverse Osmosis, the dominant membrane desalination technology [7], is a pressure-driven process. A pressure-differential ΔP_{diff} that exceeds the natural osmotic pressure ΔP_{osm} must be developed across the membrane. The rate of water flow Q, across the membrane is then given by [19]:

$$Q = \frac{\kappa S (\Delta P_{diff} - \Delta P_{osm})}{d}$$

where κ is the membrane permeability coefficient for water, S is the membrane area and d is the membrane thickness. From an energy perspective, the reverse osmosis process can, therefore, be approximated by a pump that increases the pressure of water by ΔP_{diff} to sustain the flow $Q = F_{Wnet_{ro}}$. Further pumping is required to increase the piezometric head of the desalinated sea water from sea level A_{ro} to a design pressure $E_{W_{ro}}$ for distribution. A reverse osmosis plant can, therefore, be described with the following equations:

$$F_{Wnet_{ro}} = \frac{\kappa S}{d} \times (\Delta P_{diff} - \Delta P_{osm})$$

$$F_{Pnet_{ro}} = \frac{\rho g K (E_{W_{ro}} + \Delta P_{diff} - A_{ro}) + R_Q E_{P_{ro}}}{K^2 + R_A R_Q} \tag{7}$$

ΔP_{osm} is dependent on the dissolved solute concentration and is given by the van't Hoff equation [14] $\Delta P_{osm} = iMR$, where i is the van't Hoff factor, M is the molarity of the dissolved salts and R is the universal gas constant. The brine output can be determined from the average recovery ratio of the process. The recovery ratio RR is defined [3] as the ratio of the permeate (filtrate) to the feed seawater. The brine output can, therefore, be expressed as:

$$\text{Brine output} = \frac{F_{Wnet_{ro}}}{RR}(1 - RR) \tag{8}$$

The recovery ratio of a single reverse osmosis element is typically between 10 and 15% [3].

2.3.4 Distribute Water

The distribution of water through a water distribution network can be described with the aid of an edge-node incidence matrix A_W, with the following equation:

$$\mathbf{F_{Wnet}} = A_W^T C_W A_W \mathbf{E_W} \tag{9}$$

where:

- C_W is a diagonal conductance matrix with each diagonal entry $C_{W_{jj}}$ being the conductance of pipe j.
- $A_{W_{ji}} = 1$ if water in pipe j leaves node i, $A_{W_{ji}} = -1$ if water in pipe j enters node i, $A_{W_{ji}} = 0$ otherwise.

3 Illustrative Example

In this section, the models developed are used to solve for the various exchanges of matter and energy between the engineering systems. In particular, the inputs and outputs of interest, previously labelled in Figure 1 are found subject to the demands for water and electricity and the the the respective topologies. Figure 8 presents a conceptual illustration of a geographical region served by a number of different water and power sources for demonstration of the quantitative engineering systems model. The water distribution system is modelled as consisting of three water sources, as indicated, equidistant from aggregated demand node of $3m^3/s$ or approximately 70 million gallons a day. The required hydraulic pressure for the distribution is provided entirely by pumping to the clearwell at the treatment plants; in other words, there are no additional pumps in the distribution network.

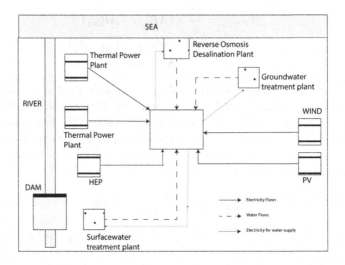

Fig. 8 Illustration inspired by Egypt (HEP- Hydroelectric generation plant, PV-Solar Photovoltaic plant, WIND- Wind power plant)

The electricity distribution network for this example is modelled with the standard IEEE14 bus system [11] with the following modifications:

1. The 40MW generator at bus 2 is replaced with a 100MW generator representing a hydroelectric power plant
2. The synchrononous compensators at buses 3,6 and 8 are replaced with 10MW generators representing a solar PV plant, a wind power power plant and a thermal power plant respectively. The generator at slack bus 1 is taken as a thermal power plant.
3. The reverse osmosis plant, groundwater treatment plant, and surface water treatment plant are attached to the network at buses 4,5 and 14

As machine constants, available inputs from the environment, and demands for both water and electricity are not determined by the effort and flow variables in either system, they are included in the model as parameters. A summary of the parameters and variables associated with the various modeled functions is given in Table 1 below. Subscript i in Table 1 refers to source nodes, while subscript j refers to demand nodes. The values of $\mathbf{E_{P_i}}$ are endogenously determined by the *'Generate electricity'* functions.

Table 1 Variables and Parameters

Function	Parameters	Variables
Generate electricity - hydro	$H, Q_{evap}, R_A, R_Q, D_S, \lambda, \beta$	$E_{P_{hydro}}, F_{Pnet_{hydro}}$
Generate electricity - wind	$V, C_T, R, N_g, D, \lambda$	$E_{P_{wind}}, F_{Pnet_{wind}}, \tau_a$
Generate electricity - solar PV	K, R_{inv}	$E_{P_{solar}}, F_{Pnet_{solar}}$
Transmit electricity	$C_P, A_P, \mathbf{F_{Pnet_j}}$	$\mathbf{E_{P_j}}, \mathbf{F_{Pnet_i}}$
Groundwater treatment	$R_A, R_Q, K, A_g, E_{W_g}$	F_{Pnet_g}, F_{Wnet_g}
Surface water treatment	$R_A, R_Q, K, A_s, E_{W_s}$	F_{Pnet_s}, F_{Wnet_s}
Reverse osmosis	$R_A, R_Q, K, A_{ro}, E_{W_{ro}}, \Delta P_{osm}, \kappa, S, d, RR$	$F_{Pnet_{ro}}, F_{Wnet_{ro}}$
Distribute water	$C_W, A_W, \mathbf{F_{Wnet_j}}, \mathbf{E_{W_i}}$	$\mathbf{E_{W_j}}, \mathbf{F_{Wnet_i}}$

4 Discussion

The model in Section 3 was implemented and solved in the General Algebraic Modelling System (GAMS). The values of the various inputs, outputs and exchanges, previoulsy labelled with the letters A through J in Figure 1, are shown in Table 2. The matter and energy flows not labelled with letters in Figure 1 are the subject of ongoing work and were not included in the model. Water withdrawal (E) and consumption (E^*) for thermal generation were modelled by means of empirically determined water withdrawal and consumption rates of $0.9 m^3/MW$ and $0.7 m^3/MW$ respectively for natural gas combined cycle plants with closed loop cooling systems [17]. Fuel energy consumption, J ($MWth$), was modelled by means of a typical heat rate of $0.0072MJ/MWh$ for a natural gas combined cycle plant[4]. As the wastewater system has not been modelled it is assumed that all water delivered to the aggregated demand node is returned to natural water bodies(G) in an untreated state. Electrical losses (H) are determined from the power flow analysis.

Table 2 Model Output

A	B	C	D	E	E^*
$1 m^3/s$	$6.3 m^3/s$	$0.3 m^3/s$	$29MWe$	$170 m^3/s$	$132 m^3/s$
F	G	H	I	J	K
$3.3 m^3/s$	$3 m^3/s$	$8MWe$	$230MWe$	$386MWth$	$20MWe$

Various measures of aspects of the nexus, that can be used to inform planning decisions, can be expressed in terms of the determined energy and matter flows:

- Ratio of useful electrical energy to priced energy resource consumption given by $(D+I)/J = 67\%$
- Ratio of energy wasted by electrical power system to priced energy resource consumption given by $H/J = 2\%$

- A measure of the degree of coupling between the electricity and water systems given by $D/(D+H+I) = 10.8\%$
- Water supply required to sustain the two systems given by $E+F = 173.3m^3/s$
- Ratio of water displaced from its original source to total water withdrawn for water and electricity systems given by $(A+C+E^*)/(E+F) = 77\%$
- Proportion of water withdrawn that is returned with significantly altered quality —a measure of environmental impact —given by $(B+G)/(E+F) = 5\%$

5 Conclusions and Future Work

This work has presented bond graph models of various elements of the engineered electricity and water systems and demonstrated their utility in a coupled engineering system model of these two connected systems. In future work, thermodynamic and chemical bond graph models will be developed to model the thermal generation, electricity and water cogeneration and wastewater treatment functions. In addition a numerical method will be developed to solve for the variables of interest in place of GAMS.

References

[1] Bakka, T., Karimi, H.: Wind turbine modeling using the bond graph. In: IEEE International Symposium on Computer-Aided Control System Design, pp. 1208–1213 (2011)
[2] Barcel, J., Codina, E., Casas, J., Ferrer, J.L., Garca, D.: A Review of Operational Water Consumption and Withdrawal Factors for Electricity Generating Technologies. Tech. Rep. NREL/TP-6A20-5090, Golden, CO (2011)
[3] Cipollina, A., Micale, G., Rizzuti, L.: Seawater desalination: conventional and renewable energy processes. Springer, Berlin (2009)
[4] Delgado, A.: Water Footprint of Electric Power Generation: Modeling its use and analyzing options for a water-scarce future. Master's thesis (2012)
[5] Goldstein, R., Smith, W.: Water & Sustainability (Volume 3): U.S. Water Consumption for Power Production - The Next Half Century. Tech. Rep. vol. 3. Electric Power Research Institute, Palo Alto, CA, USA (2002a)
[6] Goldstein, R., Smith, W.: Water & Sustainability (Volume 4): U.S. Electricity Consumption for Water Supply & Treatment - The Next Half Century. Tech. Rep. vol. 4. Electric Power Research Institute, Palo Alto, CA, USA (2002b)
[7] Isaka, M.: Water Desalination Using Renewable Energy. Tech. Rep. International Renewable Energy Agency (March 2012)
[8] Karnopp, D., Margolis, D.L., Rosenberg, R.C.: System dynamics: a unified approach, 2nd edn. Wiley, New York (1990)
[9] Lubega, W.N., Farid, A.M.: A Meta-System Architecture for the Energy-Water Nexus. In: IEEE SOSE Proceedings, Maui, Hawaii (to appear, June 2013)
[10] Markvart, T., Castaner, L.: Practical Handbook of Photovoltaics: Fundamentals and Applications. Elsevier, Amsterdam (2003)
[11] Milano, F.: Power System Modelling and Scripting. Springer (2010)
[12] Olsson, G.: Water and Energy: Threats and Opportunities. IWA Publishing, London (2012)

[13] Park, L., Croyle, K.: Californias Water-Energy Nexus: Pathways to Implementation. Tech. rep. GEI Consultants, Inc. (2012)
[14] Rao, S.M.: Reverse osmosis. Resonance 12(5), 37–40 (2007), doi:10.1007/s12045-007-0048-8
[15] Siddiqi, A., Anadon, L.D.: The water energy nexus in Middle East and North Africa. Energy Policy 39(8), 4529–4540 (2011)
[16] Stillwell, A.S., King, C.W., Webber, M.E., Duncan, I.J., Hardberger, A.: The Energy-Water Nexus in Texas. Ecology and Society 16(1), 2 (2011)
[17] United Nations, Managing Water under Uncertainty and Risk. Tech. rep. (2012)
[18] US DoE, Energy Demands on Water Resources. Tech. rep. (2006)
[19] Wilf, M.: The Guidebook to Membrane Desalination Technology. Desalination Publications, L'Aquila (2007)
[20] World Economic Forum, Energy Vision Update, Thirsty Energy: Water and Energy in the 21st Century. Tech. rep., Geneva, Switzerland (2009)

A Visual Logic for the Description of Highway Traffic Scenarios[*]

Stephanie Kemper and Christoph Etzien

Abstract. In this paper, we present the syntax and semantics of the visual logic (VL) used to specify (sequences of) traffic situations on the highway. VL was developed in the context of driver assistance system development, and is intended to bridge the (terminology) gap between system engineers and traffic psychologists developing driver assistance systems, and scientists modelling, analysing and verifying such systems. To achieve this goal, the logic is intuitive and simple, thus easy to understand and apply, yet it has a formal, automaton-based semantics which allows to use well-established tools and formalisms for further analysis of the system.

We show how VL can be used to specify scenarios on the highway involving interaction of driver and assistance system, and how it can be used in the context of observer-based verification to analyse and verify the assistance system with respect to these scenarios.

Keywords: Systems of systems, Visual Logic, Highway Traffic Scenarios, Automaton-based Semantics, Observer-based Verification.

1 Introduction

In this paper, we present syntax and semantics of a visual logic (VL) tailored to specify (sequences of) traffic situations on a highway. VL was developed in the

Stephanie Kemper · Christoph Etzien
OFFIS Institute for Information Technology, Escherweg 2, 26121 Oldenburg, Germany
e-mail: {stephanie.kemper,christoph.etzien}@offis.de

[*] This work was partly supported by Ministry for Science and Culture of Lower Saxony as part of the interdisciplinary research project "Integrated Modelling for Safe Transportation II" (IMoST2), and by European Commission funding the Large-scale integrating project (IP) proposal under ICT Call 7 (FP7-ICT-2011-7) Designing for Adaptability and evolutioN in System of systems Engineering (DANSE) (No. 287716).

M. Aiguier et al. (eds.), *Complex Systems Design & Management 2013*,   233
DOI: 10.1007/978-3-319-02812-5_17, © Springer International Publishing Switzerland 2014

context of the IMoST2 project.[1] Its purpose is twofold: on the one hand, it should be intuitive so that experts of different domains can use it to easily specify (sequences of) traffic situations from different points of view. On the other hand, the formal semantics allows to verify this abstract model against trajectories of real world systems. Such sequences of traffic situations are a subclass of the domain of Systems of Systems (SoS) and were in the focus of the IMoST2 project. Typically those systems are SoS as defined by Maier [19]. Especially in the traffic domain there might be detailed models for the participating vehicles and quite abstract models for traffic flows but there is still a need to fill the gap between those models to cover the interactions of vehicle, driver, road infrastructure and traffic rules. The purpose of modelling and analysing such an already running SoS is e.g. to adapt the existing Constituent Systems (CSs) to new features or to update existing CSs with new ones.

We take the five criteria from [19] to distinguish between complex systems and SoS. *Operational independence* respectively *managerial independence* characterise the CSs of an SoS to operate independently, the criteria are independently managed by different drivers/owners in our scenario. In our setting, the *evolutionary development of the CSs* would be the change of types of vehicles, but this is not (directly) addressed in this paper. *Emergent behaviour* characterises that the interaction of the CSs creates an emergent behaviour which non of the individual systems can show by itself; traffic jams are one famous example in the traffic domain. Within the scope of this paper, we will not take the *geographical distribution of the CSs* into account because we focus on segments of the highway and describe situations locally rather than globally. One of the most important challenges in the domain of SoS is to specify a model (abstraction) of the SoS which preserves the properties of interest.

Using VL, the state space of the model is only implicitly defined by the combination of all atoms. Each such atom characterises a potentially infinite set of precise traffic situations. This approach allows to check the specified model (1) against a real world model in case a behaviour of that model has been observed, or, as in the case of the IMoST project, (2) against trajectories of simulation runs. For a representative set of trajectories, e.g. simulation runs, the atoms of VL shall all be observable in this set of trajectories. Sequences of atoms can be verified in the same way but in this case not only the atom but also their order must be contained in the trajectories.

State of the Art and Related Work. A similar approach was followed in [4] by abstracting similar interaction relations into clusters—called Partner Abstraction—but the starting point in that work is the entire SoS and there is the need to verify that the transitions between abstracted clusters compares to the precise interactions of the CSs. We start in our approach by defining so-called atoms which represent structure similar to the partner clusters but in addition define the communication behaviour within such an atom. In [11] an approach for the specification of changing interaction relations between CSs is presented but only from the requirement perspective—not checking if the specification compares to the real world SoS. In that approach graph rewriting rules are annotated amongst others with timing constraints. Moreover, the formal semantics allows to check if the specification

---

[1] imost.informatik.uni-oldenburg.de

is contradictory in itself. In [10, 9] graph rules are used to describe the changes of structure—not only communication relations—of an SoS but including capabilities and requests of services which invoke the changes. Reconfiguration of systems to adapt to environmental changes is also a research topic in the domain of real-time systems. Verification of the real-time behaviour including the reconfiguration is supported by CHARON [15], Masaccio [12], and MECHATRONIC UML (e.g. [7, 14]). There are some approaches for modelling the structural aspects of adaptive systems [13, 20, 21] or the behavioural aspects [24, 2, 18, 8] but none of them considers both aspects [5]. This was first considered by the approach presented in [14].

Structure of This Paper. In section 2 we define the formal syntax and semantics of VL. After the general definition in Section 2.1, we define the syntax of atoms—the most basic constituents of VL—in Section 2.2. Section 2.3 defines how to compose atoms, and Section 2.4 defines the semantics of VL. The application of VL in observer-based verification is demonstrated in Section 3. Section 4 concludes the paper and gives some directions of future work.

2 The Visual Logic

In this Section, we describe syntax and semantics of the visual logic VL. We start with the syntax description in a top-down way.

2.1 Specifications

A *specification* $\mathscr{S}$ in VL, describing a traffic scenario on the highway, is a tuple $\mathscr{S} = (\mathscr{V}, \mathscr{L}, \mathscr{A})$, with $\mathscr{V} = \{v_1, ..., v_m\}$, $m \geq 1$, a set of *vehicles*, $\mathscr{L} = \{l_1, ..., l_n\}$, $n \geq 2$, a set of *lane separators*, and $\mathscr{A} = \{a_1, ..., a_j\}$, $j \geq 1$, a sequence of *(VL) atoms*.

An atom consists of two parts, one part graphically describing the spatial relations among vehicles on the highway, and the second part describing the communications and synchronisations among vehicles (or single vehicle components) sketched in the graphical part. For the first part, we have developed a new, tailored formalism called *Traffic View*, for the latter, we rely on (a subset of) the well-known formalism of *Live Sequence Charts* [17].

2.2 Syntax of Atoms

As mentioned above, an atom consists of two parts; formally, an atom $a \in \mathscr{A}$ is a pair $a = (tv, com)$, with *tv* a *traffic view* and *com* a *communication description*.

2.2.1 Traffic View

The idea of a traffic view *tv* is to describe a traffic situation on the highway, more precisely the spatial relations of vehicles in $\mathscr{V}$. A conceptual view of a traffic view is given in Figure 1, and a concrete example is given in Figure 2. Since we model traffic on the highway, we assume that all traffic proceeds in the same direction, in the pictures modelled from left to right. In accordance with this, we call the vertical dimension of the picture the *lateral* dimension of the traffic view, the horizontal dimension of the picture the *longitudinal* dimension of the traffic view (cf. Figure 1), and all other orientational notions (e.g. ahead, behind, left, right) change accordingly. We denote the set of vehicles in the traffic view *tv* of an atom a by $\mathcal{V}_a$. These vehicles are said to be *observable* in the current traffic view. We allow $\mathcal{V}_a \subset \mathcal{V}$, i.e. not all vehicles of the specification $\mathcal{S}$ need to be observable in the current traffic view.

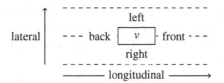

Fig. 1 Traffic View, Conceptual View

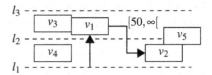

Fig. 2 Traffic View Example

The set of lane separators $\mathcal{L}$ spans the "canvas" of the traffic view, starting with l_1 at the bottom to l_n at the top, in this way defining $n-1$ lanes.[2] Vehicles are represented by equal sized rectangles carrying the vehicle identifier as defined in $\mathcal{V}$. Textually, we refer to the sides of a vehicle v as $v.l$, $v.r$, $v.b$ and $v.f$, cf. Figure 1. We write $\mathcal{V}|_l \stackrel{\text{def}}{=} \{v.l \mid v \in \mathcal{V}\}$ to denote the set of all left sides of vehicles in $\mathcal{V}$, equivalently for the other sides. The position of a vehicle is defined via the positions of its sides, we define the position separately for the two dimensions (conceptually, we project vehicle sides and lane separators onto the lateral respectively longitudinal axis). The relative positions of vehicles are defined as follows:

Longitudinal: each ordered pair of vehicles $(v_i, v_j) \in \mathcal{V}$ enjoys exactly one of the following longitudinal relations (we believe the intended meaning becomes clear from the examples; all examples refer to Figure 2):[3] v_i *before*$_\rightarrow$ v_j (v_1 and v_2), v_i *meets*$_\rightarrow$ v_j (v_3 and v_1), v_i *overlaps*$_\rightarrow$ v_j (v_2 and v_5), v_i *equals*$_\rightarrow$ v_j (v_3 and v_4), or the inverse of the former three (v_j *after*$_\rightarrow$ v_i, v_j *is met by*$_\rightarrow$ v_i, v_j *is overlapped by*$_\rightarrow$ v_i). These relations induce a total longitudinal order $\leq_\rightarrow$ on the set $Anc_{long} \stackrel{\text{def}}{=} \mathcal{V}|_b \cup \mathcal{V}|_f$ of *longitudinal anchors*, and we require $(v.b \leq_\rightarrow v.f)$ for all $v \in V$.

Lateral: similar to the longitudinal dimension, each ordered pair of vehicles $(v_i, v_j) \in \mathcal{V}$ enjoys exactly one of the following lateral relations: v_i *before*$_\updownarrow$ v_j (v_4 and v_3), v_i *meets*$_\updownarrow$ v_j (v_2 and v_5), v_i *overlaps*$_\updownarrow$ v_j (v_1 and v_3), v_i *equals*$_\updownarrow$ v_j (v_4 and v_2), or the inverse of the former three (v_j *after*$_\updownarrow$ v_i, v_j *is met by*$_\updownarrow$ v_i, v_j *is overlapped by*$_\updownarrow$ v_i). In addition, we distinguish whether a vehicle is placed in between two neighbouring lane separators l_i and l_{i+1} ("on a lane"), denoted $_{l_i}$-v-$_{l_{i+1}}$ (v_1), or

---

[2] Using lane separators instead of lanes allows to express lateral positions more easily.

[3] This definition is highly inspired by Allen's work on the possible relations of intervals [1].

on an *inner* (neither l_1 nor l_n) lane separator l_i ("in between two lanes"), denoted v_{-l_i} (v_5). These relations induce a total lateral order $\leq_\uparrow$ on the set $Anc_{lat} \stackrel{\text{def}}{=} \mathcal{V}|_r \cup \mathcal{V}|_r \cup \mathcal{L}$ of *lateral anchors*, and we require $(v.r \leq_\uparrow v.l)$ for all $v \in V$.

For two elements $x, y \in Anc_{long}$ (Anc_{lat}), we write $x \leq_\rightarrow y$ ($\leq_\uparrow$) if x is less than or equal to y under $\leq_\rightarrow$ ($\leq_\uparrow$), and consider $<_\rightarrow, =_\rightarrow, \geq_\rightarrow, >_\rightarrow$ ($<_\uparrow, =_\uparrow, >_\uparrow, \geq_\uparrow$) as abbreviations of conjunctions of such constraints in the usual way. We may omit indices $_\rightarrow$ and $_\uparrow$ if they are clear from the context. With this, the relative constraints can be expressed formally in a straightforward way. For example, $l_i\text{-}v_{-l_{i+1}}$ translates into $(v.l \leq l_{i+1}) \wedge (v.r \geq l_i)$, and $(v_i \text{ meets}_\rightarrow v_j)$ translates into $(v_i.f = v_j.b)$.

Note that for some constraints, there exists more than one straightforward (intuitive) way of expressing the constraint. For example, $(v_i \text{ equals}_\rightarrow v_j)$ can be expressed as $(v_i.f = v_j.f)$ or as $(v_i.b = v_j.b)$, which both have the same information content. We do not further constrain how to express relative constraints but leave the precise definition of the exact translation to the respective use cases.

So far, we have only considered the positions of anchors relative to each other. To be able to express constraints on absolute distances of anchors, we use *distance arrows*, which restrict the absolute distance of their end points. Distance arrows are labelled with intervals (open, half-open or closed) over $\mathbb{R}_{\geq 0}$, excluding intervals that represent the empty set. The default label, which can be omitted in the pictures, is $[0, \infty[$. *Lateral distance arrows* point upwards, from lane separators or left vehicle sides to right vehicle sides or lane separators, cf. the distance arrow $l_1 \longrightarrow v_1.r$ in Figure 2. *Longitudinal distance arrows* point right, from front vehicle sides to back vehicle sides, cf. the distance arrow $v_1.f \longrightarrow_{[50,\infty]} v_2.b$ in Figure 2.

The intended meaning of a distance arrow is that the absolute distance of its endpoints lies in the interval given in the label. Distance arrows can be expressed formally in a straightforward way, similar to relative constraints. For example, the longitudinal distance arrow $v_1 \longrightarrow_{[50,\infty]} v_2$ translates into $(50 \leq (v_2.b - v_1.f))$.

To handle absolute positions (which however is outside the scope of this paper), we assume the existence of a *lateral position function* $Pos_{lat} : Anc_{lat} \rightarrow \mathbb{R}_{\geq 0}$ and a *longitudinal position function* $Pos_{long} : Anc_{long} \rightarrow \mathbb{R}_{\geq 0}$. The intended idea of these functions is to assign values to lateral respectively longitudinal anchors (their position) while preserving the lateral respectively longitudinal order and the constraints obtained from the distance arrows. For example, we allow $(Pos_{long}(v_i.f) = Pos_{long}(v_j.b))$ only if $(v_i.f =_\rightarrow v_j.b)$.

The constraints of the lateral respectively longitudinal order are consistent by definition. In order to ensure consistency with the constraints obtained from distance arrows, and to guarantee the existence of Pos_{lat} and Pos_{long}, we forbid to add distance arrows that conflict with $\leq_\uparrow$, $\leq_\rightarrow$ or other distance arrows. For example, in Figure 2, additional longitudinal distance arrows $v_4.f \longrightarrow_{[10,20]} v_1.b$, or $v_4.f \longrightarrow_{[0,45]} v_2.b$, are not allowed, since they conflict with the fact that v_4 meets v_1 and—in the latter case—the other longitudinal distance arrow.

We may add superscript tv to all identifiers defined in this section (Pos_{lat}^{tv}, $\leq_\rightarrow^{tv}$, $equals_\uparrow^{tv}$, ...) to indicate that they reason about the constituents of traffic view tv.

2.2.2 Communication Description

Communication descriptions specify the communications and synchronisations that take place during the situation sketched in the traffic view. We use a feature subset of Live Sequence Charts (LSC) [17] for the specification of the communication descriptions, we also adopt the terminology related to LSCs. Apart from being a well-known and well-established formalism, this has the advantage that we get a translation into an automaton representation for free [17], cf. Section 2.4. Due to space limitations, we can only give a brief overview of the LSC features we are using, and refer to [17] for a complete and detailed description of syntax and semantics. An example of a communication description is shown in Figure 3.

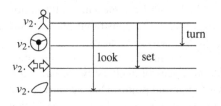

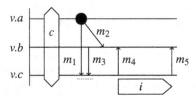

Fig. 3 Communication Description, Example

Fig. 4 Communication Description, Concepts

Simply speaking, a communication description is an LSC rotated counterclockwise by 90 degrees. It contains a number of communicating *instances*, depicted by horizontal lines. Each of the instance belongs to exactly one of the vehicles in $\mathcal{V}$. Every vehicle part that interacts with other parts in some way can be referred to as an instance. In Figure 3, the instances all belong to vehicle v_2 (prefix v_2.), and refer to (from top to bottom) driver, steering wheel, indicator lever and left side mirror. Communications are totally ordered (from left to right) per instance line, unless relaxed by coregions (see below). The communications depicted in Figure 3 could for example belong to (the beginning of) a lane change manoeuvre: the driver looks into the side mirror, sets the indicator and turns the steering wheel. Currently, we support the following LSC concepts in the communication descriptions, cf. Figure 4:

Synchronous messages: depicted by vertical arrows between instance lines, cf. m_1, m_3, m_4 and m_5. Messages are sent in the direction of the arrow; the arrow tail on the instance line is called *(message) send event*, the arrow head *(message) receive event*. Messages with empty labels are used for synchronisation.
Asynchronous messages: depicted by slanted arrows, cf. m_2.
Conditions: depicted by elongated hexagons orthogonal to the instance lines (cf. condition with constraint c). Conditions may involve one or more instances, instance lines of uninvolved instances are drawn through (cf. $v.b$). The intersection of a condition with an instance line is called a *condition event*.
Invariants: depicted by condition symbols parallel to the instance lines (cf. invariant with constraint i). Invariants always need reference points (send or receive events). A flattened end of the invariant includes its reference points (cf. left end

of the invariant), i.e. the reference point occurs simultaneously with the beginning/end of the invariant; a pointed end excludes its reference point (cf. right end of the invariant), i.e. the reference point occurs strictly before/after the beginning/end of the invariant. Invariants do not belong to a particular instance.

Coregions: depicted by a dotted line next to the instance line they belong to. In coregions, the sequential order of events is relaxed, i.e. they can occur in any order (cf. message receive events of m_1 and m_3).

Simultaneous regions: depicted by a black circle on the instance line. All events in a simultaneous region occur simultaneously (cf. send events of m_1 and m_2).

In particular, we do not (yet) support addition and removal of instance lines, or the concepts of temperatures (cf. [17]). The constraints in conditions and invariants are arbitrary, they only need to be evaluable in (i.e. reason about variables in the scope of) the current context.

Well-Formedness of Atoms (Vertical Composition). The idea of an atom is to describe two different aspects of the same scenario: spatial relations of vehicles in the traffic view, and communication behaviour of vehicle parts in the communication description. For an atom to be *well-formed*, the two constituents need to meet the following criterion to ensure that they describe the same scenario.

For each atom $a = (tv, com)$, only instances belonging to vehicles in $\mathcal{V}_a$ of tv are allowed to communicate in com, i.e., instances belonging to vehicles in $\mathcal{V} \setminus \mathcal{V}_a$ must not contain any events in com (com may contain such additional instances, which is required for horizontal composition, cf. Section 2.3 below).

2.3 Syntactical (Horizontal) Composition of Atoms

The *admissible successor* relation defines if two atoms $a_1 = (tv_1, com_1), a_2 = (tv_2, com_2)$ are composable (to be able to specify longer scenarios or manoeuvres).

Communication Description. Communication description com_1 is an admissible successor of communication description com_2 if it contains the same instances.

Traffic View. For tv_1 to be an admissible successor of tv_2, we need to ensure that position changes of vehicles (including appearance and disappearance) adhere to "physically reasonable" behaviour. [4] The following constraints need to hold:

- Appearance: new vehicles may appear only $before^{tv_2}_{\rightarrow}$ or $after^{tv_2}_{\rightarrow}$ all other vehicles (accelerating respectively slowing down into tv_2), several vehicles may appear at once. Formally, for $v \in V|_{a_2} \setminus V|_{a_1}$, we require that

  - $\exists \mathcal{V}^* \subset \mathcal{V}|_{a_2} : \forall v^* \in \mathcal{V}^*, \forall v' \in V|_{a_2} \setminus (\mathcal{V}^* \cup \{v\}) : (v \, equals^{tv_2}_{\rightarrow} v^*) \wedge (v \, before^{tv_2}_{\rightarrow} v')$, or
  - $\exists \mathcal{V}^* \subset \mathcal{V}|_{a_2} : \forall v^* \in \mathcal{V}^*, \forall v' \in V|_{a_2} \setminus (\mathcal{V}^* \cup \{v\}) : (v \, equals^{tv_2}_{\rightarrow} v^*) \wedge (v' \, before^{tv_2}_{\rightarrow} v)$

  If $\mathcal{V}^*$ is non-empty, then all vehicles in $\mathcal{V}^*$ appear at the same position as v.

---

[4] Another constraint is that the set of lane separators is identical for tv_1 and tv_2, but this is already guaranteed by definition (since tv_1 and tv_2 belong to the same specification, and the set of lane separators is defined on specification level, cf. the beginning of Section 2.1).

- Disappearance: vehicles may disappear only if they were before$^{tv_1}_{\rightarrow}$ or after$^{tv_1}_{\rightarrow}$ all other vehicles (slowing down respectively accelerating away from tv_1), several vehicles may disappear at once. Formally, for $v \in V|_{a_1} \setminus V|_{a_2}$, we require that

  - $\exists \mathscr{V}^* \subset \mathscr{V}|_{a_1} : \forall v^* \in \mathscr{V}^*, \forall v' \in V|_{a_1} \setminus (\mathscr{V}^* \cup \{v\}) : (v \, equals^{tv_1}_{\rightarrow} v^*) \wedge (v \, before^{tv_1}_{\rightarrow} v')$, or
  - $\exists \mathscr{V}^* \subset \mathscr{V}|_{a_1} : \forall v^* \in \mathscr{V}^*, \forall v' \in V|_{a_1} \setminus (\mathscr{V}^* \cup \{v\}) : (v \, equals^{tv_1}_{\rightarrow} v^*) \wedge (v' \, before^{tv_1}_{\rightarrow} v)$

  If $\mathscr{V}^*$ is non-empty, then all vehicles in $\mathscr{V}^*$ disappear from the same position as v.

- Relative (longitudinal or lateral) position change: in each dimension, a pair of vehicles may change its relative position only such that the new relation in tv_2 physically succeeds the relation in tv_1. Formally, for a pair of vehicles (v_i, v_j), $v_i, v_j \in V|_{a_1} \cap V|_{a_2}$, we require that

  - $(v_i \, before^{tv_2}_{\rightarrow} v_j)$ iff $(v_i \, before^{tv_1}_{\rightarrow} v_j)$ or $(v_i \, meets^{tv_1}_{\rightarrow} v_j)$
  - $(v_i \, meets^{tv_2}_{\rightarrow} v_j)$ iff $(v_i \, before^{tv_1}_{\rightarrow} v_j)$, $(v_i \, meets^{tv_1}_{\rightarrow} v_j)$ or $(v_i \, overlaps^{tv_1}_{\rightarrow} v_j)$
  - $(v_i \, overlaps^{tv_2}_{\rightarrow} v_j)$ iff $(v_i \, meets^{tv_1}_{\rightarrow} v_j)$, $(v_i \, overlaps^{tv_1}_{\rightarrow} v_j)$ or $(v_i \, equals^{tv_1}_{\rightarrow} v_j)$
  - $(v_i \, equals^{tv_2}_{\rightarrow} v_j)$ iff $(v_i \, overlaps^{tv_1}_{\rightarrow} v_j)$, $(v_i \, equals^{tv_1}_{\rightarrow} v_j)$ or $(v_j \, overlaps^{tv_1}_{\rightarrow} v_i)$
  - $(v_i \, is \, overlapped \, by^{tv_2}_{\rightarrow} v_j)$ iff $(v_i \, equals^{tv_1}_{\rightarrow} v_j)$, $(v_i \, is \, overlapped \, by^{tv_1}_{\rightarrow} v_j)$ or $(v_j \, is \, met \, by^{tv_1}_{\rightarrow} v_i)$
  - $(v_i \, is \, met \, by^{tv_2}_{\rightarrow} v_j)$ iff $(v_i \, is \, overlapped \, by^{tv_1}_{\rightarrow} v_j)$, $(v_i \, is \, met \, by^{tv_1}_{\rightarrow} v_j)$ or $(v_j \, after^{tv_1}_{\rightarrow} v_i)$
  - $(v_i \, after^{tv_2}_{\rightarrow} v_j)$ iff $(v_i \, is \, met \, by^{tv_1}_{\rightarrow} v_j)$ or $(v_i \, after^{tv_1}_{\rightarrow} v_j)$

  and the corresponding set of constraints for the lateral dimension.

- Lateral position change: in the lateral dimension, we have the additional constraint that single vehicles may change their position only such that they move to adjacent lanes/lane separators. Formally, for $v \in V|_{a_1} \cap V|_{a_2}$, we require that

  - $l_i \textrm{-} v^{tv_2}_{-l_{i+1}}$ iff $v^{tv_1}_{-l_i}$, $l_i \textrm{-} v^{tv_1}_{-l_{i+1}}$, or $v^{tv_1}_{-l_{i+1}}$, and
  - $v^{tv_2}_{-l_i}$ iff $l_{i-1} \textrm{-} v^{tv_1}_{-l_i}$, $v^{tv_1}_{-l_i}$, or $l_i \textrm{-} v^{tv_1}_{-l_{i+1}}$

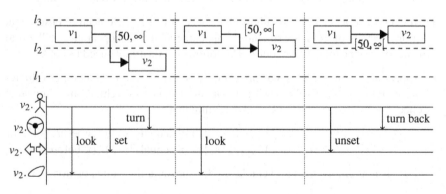

Fig. 5 Composition of Atoms, Example

An example of a composition of three atoms is shown in Figure 5. Subsequent atoms are separated by vertical dotted lines. This example is an extension of the example introduced in Section 2.2.2, it shows one possible specification of a lane change manoeuvre from the right to the left lane. In the first atom, vehicle v_2 is on the right lane, its driver prepares the lane change by looking into the side mirror, setting the indicator and turning the steering wheel. In the second atom, vehicle v_2 is in between the two lanes, the driver looks into the side mirror again, for example to check that the distance to v_1 is still large enough to continue. Finally, in the third atom, v_2 has arrived on the same lane as v_1 and has thus completed the lane change, its driver unsets the indicator and turns back the steering wheel. Throughout the whole manoeuvre, the vehicles keep a (safety) distance of at least 50. We leave it to the reader to verify that the above constraints are satisfied.

2.4 Semantics of a Visual Logic Specification

We define an automaton-based semantics for a VL specification. This allows us to use existing tools (e.g. UPPAAL [22], KRONOS [23]) and formalism (e.g. translation into a formula representation [16]) for further analysis of a VL specification.

Since communication descriptions are a subset LSCs (cf. Section 2.2.2), we can rely on the translation presented in [17] for this part. The translation presented there yields a *timed symbolic automaton*, which is essentially a *timed automaton* (TA) [3] that allows Boolean combinations of alphabet symbols as transition guards instead of single alphabet symbols. Since nowadays, most formalisms and tools around TA can simulate (UPPAAL) or directly accommodate ([16], KRONOS) this, in the remainder of this paper we shall refer to the resulting automaton as a TA.

Due to space limitations, we do not repeat the formal translation of communication descriptions into TA here, but refer the reader to [17] instead (a more concise yet less detailed description of the translation is given in [6]).

It remains to explain how the traffic view is integrated into this translation: for every traffic view, we define a single *spatial constraint sc*, which is the conjunction of the constraints obtained from lateral and longitudinal order, and lateral and longitudinal distance arrows. This spatial constraint is added to the communication description as a condition over all instances at the beginning of the communication description, and as an invariant having as reference points the first and last message send/receive event of the communication description (if there are no messages in the communication description, then the invariant is omitted). Figure 6 shows the resulting communication description for the first atom of the sequence in Figure 5.

At this point, we conceptually understand the communication descriptions of the atoms in a sequence as one single communication description, as already indicated by the continuous instance lines in Figure 5. We then apply the translation of [17] to the resulting (single) communication description, such that the semantics of the atom sequence is given by the semantics (i.e., the set of runs) of the resulting TA. Note that conditions and invariants are already handled by the translation, so we do not need to adapt the translation itself. Figure 7 shows the resulting TA when applying the entire translation to the atom sequence in Figure 5 (we refer to [3] for a detailed definition of the syntax and semantics of TA), where *sc1*, *sc2* and *sc3* denote the spatial constraints of the first, second and third atom, respectively.

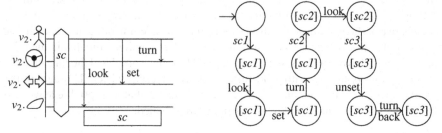

Fig. 6 Spatial Constraints added to the   **Fig. 7** TA for Atom Sequence in Figure 5
Communication Description, Example

3   Application: Observer-Based Verification

In the preceding Section, we have presented syntax and semantics of VL. In this Section, we show how VL specifications and the TA generated from them can be used as observers in an observer-based verification setting.

The idea of observer-based verification is to analyse the behaviour of a system by using observers. These observers observe the evolution of the system under consideration, and check whether a set of (desired) properties is satisfied, or whether a a certain (undesired) behaviour occurs. The observation can be performed in a simulation of the system as well as during normal operation.

As mentioned in the introduction, VL was developed in the context of the IMoST2 project. The main focus of this project was to develop methods, techniques and tools for model- and simulation-based driver assistance system development. Amongst others, this included the development of a provisional driver assistance system (to be used for further analysis of the newly developed features) and of a driver model (to simulate a driver driving with the assistance system). VL specifications were used to describe the (intended/desired) behaviour of the driver assistance system, i.e., how it aids and interacts with the driver, during typical manoeuvres and scenarios on the highway (for example a lane change manoeuvre, or keeping the safety distance). The system under consideration was a set of sample trajectories generated by human test drivers in a driving simulator combined with a prototypical implementation of the driver assistance system.

After a suitable translation of the simulator output values (e.g., the vehicle locations had to be translated from absolute (xy) coordinates into relative positions satisfying $\leq_\uparrow$ and $\leq_\rightarrow$), the generated TA could be used as observers on the simulator trajectories. This allowed to analyse whether for a certain manoeuvre, the behaviour described in the VL specifications complied with the behaviour of the test drivers. If this was not the case, the specification had to be adapted and tested again before the corresponding behaviour could be implemented in the final assistance system.

4 Conclusion

In this paper, we have presented syntax and semantics of a visual logic VL used to specify highway traffic scenarios with communication. The major advantage of the logic is the fact that scenarios and manoeuvres can be specified on a purely graphical level ("drawing pictures"), so the logic is intuitive and easy to understand; yet it has a formal automaton-based semantics, which allows to use well-established methods and tools for further analysis. The logic serves as an interface between system engineers and traffic psychologists developing driver assistance systems, and scientists modelling, analysing and verifying the behaviour of such systems. We have shown how the logic can be applied in the context of observer-based verification, to for example analyse properties of driver assistance systems.

At the moment, VL has no means to express whether the scenario expressed by a specification describes a desired or forbidden behaviour, whether (in the former case) the behaviour should be observed once or repeatedly, what (in the latter case) should happen if a forbidden behaviour is observed (e.g. major system failure, or just manoeuvre abortion) or what should happen if only a prefix of the specified behaviour has been observed (consider this a non-occurrence of the specified behaviour, or as a major failure?). We plan to extend the communication descriptions to the full set of LSC features (cf. [17] for a full list and detailed discussion), such that we will be able to precisely define and clarify these ambiguities.

We further plan to extend VL with (1) vehicle-specific properties like speed and acceleration, which can then be used in constraints and invariants in the communication descriptions, (2) more powerful distance arrow labels like variables (in addition to constants) and functions over these variables and the vehicle-specific properties (safety distance, time-to-collision), and (3) appearance/disappearance of lanes.

Finally, we plan to make specifications more concise: at the moment, vehicles can only change to a new position that physically succeeds the previous one (cf. Section 2.3). We plan to define how and under which conditions a sequence of position changes can be combined, such that it is for example possible to define a change to a neighbouring lane in one step rather than in two (cf. Figure 5). This will significantly decrease the size of a VL specification, and thus also of the resulting TA describing the semantics.

Acknowledgements. We would like to thank Ernst-Rüdiger Olderog, Hardi Hungar, Sven Linker, Werner Damm, Thomas Peikenkamp and the members of the SAV group at OFFIS for fruitful discussions about syntax and semantics of VL.

References

1. Allen, J.: Maintaining knowledge about temporal intervals. CACM 26(11), 832–843 (1983)
2. Allen, R., Douence, R., Garlan, D.: Specifying and analyzing dynamic software architectures. In: Astesiano, E. (ed.) ETAPS 1998 and FASE 1998. LNCS, vol. 1382, pp. 21–37. Springer, Heidelberg (1998)

3. Alur, R.: Timed automata. In: Halbwachs, N., Peled, D.A. (eds.) CAV 1999. LNCS, vol. 1633, pp. 8–22. Springer, Heidelberg (1999)
4. Bauer, J., Wilhelm, R.: Static analysis of dynamic communication systems by partner abstraction. In: Riis Nielson, H., Filé, G. (eds.) SAS 2007. LNCS, vol. 4634, pp. 249–264. Springer, Heidelberg (2007)
5. Bradbury, J., Cordy, J., Dingel, J., Wermelinger, M.: A survey of self-management in dynamic software architecture specifications. In: Garlan, D., Kramer, J., Wolf, A. (eds.) WOSS, pp. 28–33. ACM (2004)
6. Brill, M., Damm, W., Klose, J., Westphal, B., Wittke, H.: Live Sequence Charts: An Introduction to Lines, Arrows, and Strange Boxes in the Context of Formal Verification. In: Ehrig, H., Damm, W., Desel, J., Große-Rhode, M., Reif, W., Schnieder, E., Westkämper, E. (eds.) INT 2004. LNCS, vol. 3147, pp. 374–399. Springer, Heidelberg (2004)
7. Burmester, S., Giese, H., Tichy, M.: Model-driven development of reconfigurable mechatronic systems with mechatronic UML. In: Aßmann, U., Akşit, M., Rensink, A. (eds.) MDAFA 2003. LNCS, vol. 3599, pp. 47–61. Springer, Heidelberg (2005)
8. Canal, C., Pimentel, E., Troya, J.: Specification and refinement of dynamic software architectures. In: Donohoe, P. (ed.) WICSA, IFIP Conference Proceedings, vol. 140, pp. 107–126. Kluwer (1999)
9. Etzien, C., Gezgin, T., Fröschle, S., Henkler, S., Rettberg, A.: Contracts for evolving systems. In: 4th IEEE Workshop on Self-Organizing Real-Time Systems (to appear, 2013)
10. Gezgin, T., Etzien, C., Henkler, S., Rettberg, A.: Towards a rigorous modeling formalism for systems of systems. In: ISORC Workshops, pp. 204–211. IEEE (2012)
11. Henkler, S., Hirsch, M., Priesterjahn, C., Schäfer, W.: Modeling and verifying dynamic communication structures based on graph transformations. In: Engels, G., Luckey, M., Schäfer, W. (eds.) Software Engineering. LNI, vol. 159, pp. 153–164. GI (2010)
12. Henzinger, T.A.: Masaccio: A formal model for embedded components. In: van Leeuwen, J., Watanabe, O., Hagiya, M., Mosses, P.D., Ito, T. (eds.) TCS 2000. LNCS, vol. 1872, pp. 549–563. Springer, Heidelberg (2000)
13. Hirsch, D., Inverardi, P., Montanari, U.: Graph grammars and constraint solving for software architecture styles. In: ISAW, pp. 69–72. ACM (1998)
14. Hirsch, M., Henkler, S., Giese, H.: Modeling collaborations with dynamic structural adaptation in mechatronic UML. In: SEAMS, pp. 33–40. ACM (2008)
15. Ivancic, F.: Modeling and analysis of hybrid systems. Ph.D. thesis, Univ. of Pennsylvania (2003)
16. Kemper, S.: SAT-based Verification for Timed Component Connectors. Sciene of Computer Programming 77(7-8), 779–798 (2012), http://www.sciencedirect.com/science/article/pii/S0167642311000499
17. Klose, J.: Live sequence charts: A graphical formalism for the specification of communication behaviour. Ph.D. thesis, Universität Oldenburg (2003)
18. Kramer, J., Magee, J.: Analysing dynamic change in software architectures: a case study. In: CDS, pp. 91–100. IEEE (1998)
19. Maier, M.: Architecting principles for systems-of-systems. Syst. Eng. 1(4), 267–284 (1998)
20. Oreizy, P., Medvidovic, N., Taylor, R.: Architecture-based runtime software evolution. In: Torii, K., Futatsugi, K., Kemmerer, R. (eds.) ICSE, pp. 177–186. IEEE CS (1998)

21. Taentzer, G., Goedicke, M., Meyer, T.: Dynamic change management by distributed graph transformation: towards configurable distributed systems. In: Ehrig, H., Engels, G., Kreowski, H.-J., Rozenberg, G. (eds.) TAGT 1998. LNCS, vol. 1764, pp. 179–193. Springer, Heidelberg (2000)
22. UPPAAL: modeling, simulation and verification of real-time systems, http://www.uppaal.com/
23. Yovine, S.: Kronos: A verification tool for real-time systems. STTT 1(1-2), 123–133 (1997)
24. Zhang, J., Cheng, B.: Model-based development of dynamically adaptive software. In: Osterweil, L., Rombach, H., Soffa, M. (eds.) ICSE, pp. 371–380. ACM (2006)

Early Stage Verification and Validation of Cyber-Physical Systems through Requirements Driven Probabilistic Certificate of Correctness Metric

Alex Van der Velden, David Fox, and Jeff Haan

Abstract. Cyber Physical Systems feature a tight coupling of computational and physical elements. The behavior of geometry (deformations, kinematics), physics and controls need to be certified over a wide operational range. However, current system verification and validation practices focus on comparing virtual prototype behavior with physical prototype behavior, which is not available early in the design process. This significantly increases the risk that errors are found late in the development process, which is the leading cause of program cost overruns.

To address this issue, a virtual prototype metric called the Probabilistic Certificate of Correctness (PCC) was developed that was applied to a multi-tier design process. This metric computes the probability that the actual physical prototype will meet its benchmark acceptance tests, based on virtual prototype behavior simulations with known confidence and verified model assumptions. To rigorously and efficiently compute PCC it is important to account for all sources of uncertainty in a scalable manner, including model verification and behavior simulation accuracy and precision, manufacturing tolerances, context uncertainty, human factors and confidence in the stochastic sampling itself. This process is illustrated through the example of safety certification of an Unmanned Aerial Vehicle (UAV).

1 Introduction

The complexity of engineered systems continues to increase, and conventional approaches to Verification & Validation [MIL-STD-499A , 1969] do not scale to

Alex Van der Velden · David Fox · Jeff Haan
SIMULIA Dassault Systèmes Americas, Corp, USA
e-mail: {alex.vandervelden,david.fox,jeff.haan}@3ds.com

M. Aiguier et al. (eds.), *Complex Systems Design & Management 2013*, 247
DOI: 10.1007/978-3-319-02812-5_18, © Springer International Publishing Switzerland 2014

the large, complex, adaptable, smart (cyber-physical) systems of today. A next-generation approach to systems engineering is needed that enables a requirements-driven engineering process for the verification and validation of complex cyber-physical systems where the veracity of requirements is measured, quantified, and traded off against cost, mission targets, and other performance objectives.

A good example of verifiable requirements for ground combat vehicles is the NATO Reference Mobility Model [R. B. Ahlvin, and P. W. Haley, 1992] as applied by the Canadian Department of National Defense [2020] which for example includes:

Mission essential functions, such as engine start without external assistance in temperatures down to -46 deg. Celsius, fording a hard bottom water obstacle to a depth of 600mm, etc.;

The probability of successfully completing a mission (STANAG 4158) must be 96.2%;

Essential durability requirement: 15 year lifespan, assuming annual usage of 7000 km.

The Dassault Systèmes 3DEXperience Platform delivers a virtual testbed to verify the optimal design of complex, heterogeneous, and tightly coupled cyber-physical systems in order to achieve dramatic reductions in development costs and schedules compared to traditional iterative physical prototyping approaches, as well as dramatic increases in system understanding and management of complexity. This engineering approach uses RFLA hierarchical decomposition shown in Fig. 1 to enable scalability for complex systems:

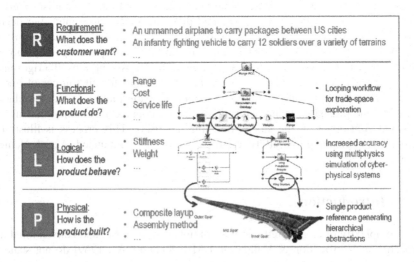

Fig. 1 Scalability via Hierarchical Decomposition

2   The Probabilistic Certificate of Correctness

This hierarchical decomposition enables the calculation of a Probabilistic Certificate of Correctness (PCC) for system requirements. PCC is a metric used to capture risk in the engineered systems development processes, especially the risk of not meeting performance requirements. For maximum leverage, it is important that a requirements verification process with a PCC metric is implemented in a consistent way for all requirements. Our definition of PCC includes the following key elements:

Probability of satisfying requirements (Ps), which gives the expected behavior as an estimated probability for a given confidence level (as a project risk parameter).

Confidence in the probability of satisfying requirements, which gives a statistical confidence in the estimated probability. Confidence relates directly to the number of samples we need to use.

Determining measurement or simulation accuracy: Deploying measurements or simulations with known low accuracy will generally result in faster cycles. Low accuracy typically results in increases in power, size, and weight of the product.

Modeling the variation manufacturing & operational variables, such a wind, temperature etc. Higher variation will lead to increases in power, size and weight.

Systematically modeled and verified set of assumptions, which builds assumptions into the simulation model and facilitates a comprehensive approach to model verification, error reduction, and inconsistency identification (such as inconsistent units).

Our approach for PCC uses sensitivity analysis to identify which model errors have the largest impact on the variability of performance, affecting the ability to meet requirements. Identifying which model errors drive this variability allows a focused effort whereby higher fidelity component and subsystem simulation models are used to improve model accuracy and achieve better probabilistic predictions, and hence higher probability of satisfying requirements.

PCC can be implemented as a scalable engineering practice for certifying complex system behavior at every milestone in the product lifecycle. This is achieved by: capturing methods by process flow automation; creating virtual prototypes at different levels of model fidelity for efficient simulation and integrating these models into a simulation process flow; verifying requirements in parallel by deploying virtual prototypes across large organizations; reducing cycle time proportional to additional computational resources and trading off in real time system sizing; modeling accuracy, technology selection and manufacturing tolerances against requirements and cost.

3   Example: Safety Certification of an Unmanned Aerial Vehicle

In order to demonstrate the systems engineering approach described above, we use a virtual model of a UAV designed for point-to-point same day package delivery in regional US markets [Van der Velden 2012]. In this example we will focus on two safety requirements:

Safe Range – to ensure the ability cross certain geographical regions, to a high probability under all weather conditions, the UAV range must be greater than 600 km to a probability level of 99.999%.

The aircraft should be free of flutter up to speeds of 120 m/s.[EASA 2009]

4   Requirements and Functional, Logical and Physical Models

Figure 2 shows the Tier 1 functional process, and the Tier 2 logical and physical processes for calculation of PCC of an UAV in IsightTM [2012].

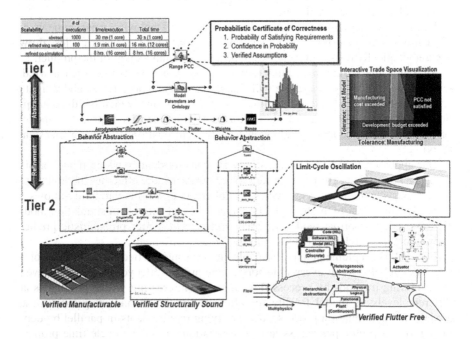

Fig. 2 Example of multi-tier Cyber-Physical System RFLP process

It includes subsystem or sub-discipline calculations for aerodynamics, ultimate load, wing weight, overall weights, and the Breguet range equation. These subsystems are collected and, grouped into a task for this UAV's range requirements where model parameters and ontology are characterized. This includes constants, geometric parameters, cruise related parameters, and the verification assumptions necessary for probabilistic certification (PCC).

All parameters include initial/default values and units. The entire range task is encapsulated within a probabilistic analysis task, which facilitates the automated application of sampling algorithms and estimation of the probability of satisfying the requirements.

5   Tier 1 and 2 Structural Wing Design

Following detailed investigation we find that the Safe Range requirement   was hard to satisfy. At a reasonable economic payload of 70 kg (replacing the pilot weight), the probability of satisfying the constraint with 97.5% confidence was no greater than 99.7%, implying a fair number of "unplanned" landings for every 1000 flights. This is acceptable for real motor gliders, but not for commercial systems.

Given the results of the sensitivity analysis, the decision is made to improve the accuracy of modeling such that we would meet the safe range for a payload as close as possible to 70 kg.

In order to make the decision process as transparent and objective as possible, we conducted a complete investigation of the design space using a Design of Experiments approach and a collaborative visual tradeoff of the modeling options. We concluded that improvements that impact wing weight have the most "bang for the buck." The nominal UAV carries little fuel compared to the wing weight, so improving the accuracy of the wing weight calculation has a significant impact on safe range.   10 kg of wing weight represents a 25% increase in the nominal fuel.

Accordingly, we target a more accurate wing structure shown in Fig. 3:

Define a parametric manufacturable wing (spars, ribs, skin)

Size this wing structure using high fidelity FEM so that it will pass the ultimate stress test milestone reliably, while including the effect of manufacturing tolerances and not changing the assumptions of the Tier 1 simulation (same wing span, airfoil thickness, area)

Create a high fidelity approximate wing weight mode from a DOE study as a function of load and manufacturing tolerance with a tolerance of 5% which can be deployed in the Tier 1 analysis that approximates the FEM model results.

According to the structural engineer, he was able to achieve a 7% accuracy (lower than the 5% target), but at a 23 kg lower mean (within the confidence interval of the preliminary value).

Because this more refined weight model includes the effect of carbon fiber selection, we are now able to trade off ply thickness versus max material cost for the wing skin (2% total weight) and achieve significant cost savings per aircraft while keeping remaining modeling cost below 10,000 hours and achieving a safe range of 600 km.

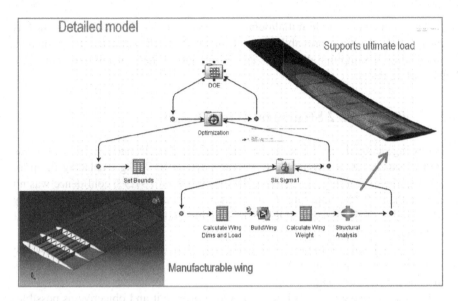

Fig. 3 High Accuracy FEA Wing Model

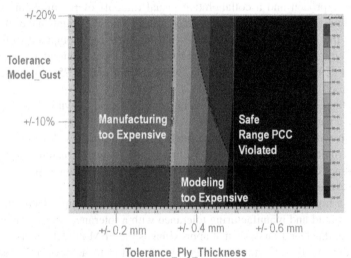

Fig. 4 Effect of Manufacturing Tolerance versus Modeling on PCC

6 Tier 2 Flutter Control System Design

As part of the wing structural design we need to prove the implied assumption that meeting the flutter speed of 120 m/s can be done with an active flutter control system without additional structural weight.   The flutter control system (FCS)

has a trailing edge flap deflected by a servo actuator controlled by an electronic control unit.

To simulate our wing with active flutter control we co-simulate our system in various modeling domains, as illustrated in Fig 5, which reflect both the physical nature of subsystem components and our choice of simulation fidelity: In the time domain the control software is modeled using a discrete time model of computation while the remaining physical components are modeled with continuous time responses. In the spatial domain lower fidelity representations are made in the form of ordinary differential equations (ODEs) while the higher fidelity descriptions of the structure and fluid are made in the form of partial differential equations (PDEs) solved using the finite element method.

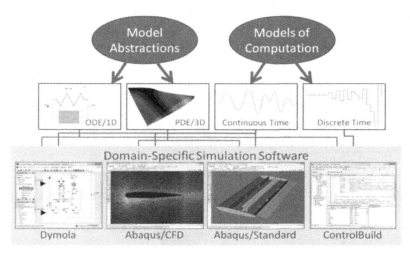

Fig. 5 Domain Specific Simulation Software for the FCS

	Dymola (Monolithic Solution)	Low Fidelity (Lumped)	Mixed Fidelity test geometry	Mixed Fidelity final geometry	High Fidelity (FSI)
structure	ODE	Point mass/rotary inertia (low-fi)	Coarse mesh (medium-fi)	Finer mesh (high-fi)	Finer mesh (high-fi)
aeroloads	ODE	Equations (low-fi)	Nodal traction (Medium-fi)	Nodal traction (Medium-fi)	CFD (high-fi)
controller	ODE	Discrete (high-fi)	Discrete (high-fi)	Discrete (high-fi)	Discrete (high-fi)
actuator	K, K+1PID, 2PIDs	2PIDs (high-fi)	2PIDs (high-fi)	2PIDs (high-fi)	2PIDs (high-fi)

Fig. 6 Multi-Tier co-simulation models for the FCS

Figure 6 illustrates the multi-(sub) tier component modeling treatment choices available in order to trade off computational effort against model fidelity. At the lower fidelity levels most components are modeled using the Functional Mockup Interface standard [Modelisar, 2010] A prototype co-simulation engine was used to provide the N-code coupling. We employ various tools for creation of these component models: ControlBuildTM [2012] to generate the electronic control unit software, DymolaTM [2012], to compute the response of the servo actuator, AbaqusTM [2012]Standard to compute structural deflections, and AbaqusTM/CFD to compute the aerodynamic loads. We confirmed the accuracy of the flutter onset speed predicted by our co-simulation we compared our multi-fidelity co-simulation results with results from a known experiment [Strganac, 2000]

Just like we did with the static wing structure, we execute a DOE study varying flow stream velocity versus wing velocity, wing stiffness and control system technology. Figure 7 shows the tradeoff between velocity and sensor sampling period.

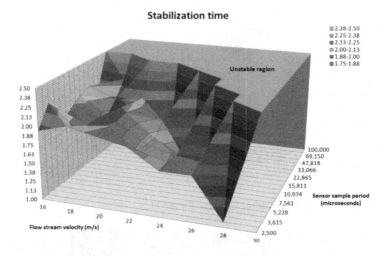

Fig. 7 Flutter onset speed as a function of sensor sampling speed and velocity for the experimental model of Strganac [Strganac, 2000]

We now create an approximation of the flutter onset speed as a function of system technology for our wing and repeat our Tier 1 trade study. Higher bandwidth sensors and actuators will be more expensive. And the sensor sampling period is used to throttle the overall system bandwidth.

Fig. 8 shows a different slice of the same space where we trade off bandwidth against aspect ratio for flutter speed. The minimum flutter speed of 120 m/s is about 20% above the flutter speed of the structure alone. A minimum bandwidth around 20 Hz is required to achieve the higher flutter speed. As the aspect ratio is increased a somewhat higher bandwidth is required. Higher bandwidths allow for still higher flutter speeds. This contour plot also shows that we cannot afford an aspect ratio beyond 26 in order to meet the safe range requirement.

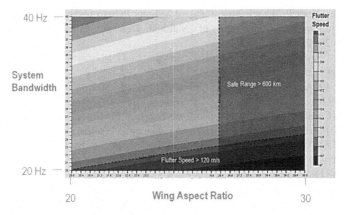

Fig. 8 Effect of bandwidth versus aspect ratio for flutter speed

7 Summary

The process presented is an improvement over the traditional systems engineering V -cycle because verification and validation happens at every process stage from concept to manufacturing using multi fidelity virtual prototypes. This approach identifies problems early and reduces rework significantly in the more expensive implementation and physical certification phase.

Acknowledgement. This work was sponsored by the Defense Advanced Research Projects Agency, Tactical Technology Office, META-II program issued under contract by DARPA/CMO HR0011-11-C-0049. The views and conclusions contained in this document are those of the authors and should not be interpreted as official policies, either expressly or implied, of the Defense Advanced Research Projects Agency or the US Government. "Approved for Public Release, Distribution Unlimited".

References

AbaqusTM, computer software (2012), http://www.3ds.com
Ahlvin, R.B., Haley, P.W.: NATO Reference Mobility Model, Edition II, NRMM II User's Guide, Technical report GL-92-19, (AD 170301) US Army Engineer Waterways Experiment Station, Vicksburg, MS (1992)
ControlBuildTM 2012, computer software (2012), http://www.3ds.com
DymolaTM 2012, computer software (2012), http://www.3ds.com
EASA: European Aviation Safety Agency. Certification specifications for Very Light Aeroplanes. In: CS-VLA 2009 (2009)
IsightTM 5.7, computer software (2012), http://www.3ds.com

Light Utility Vehicle Wheeled the Standard Military Pattern Fleet, Statement of Operational Requirements, Department of National Defense, Canada, Project File 2349 (March 2000),
http://www.vcds-vcemd.forces.gc.ca/boi-cde/jvm-emv/doc/luvwsor-ebouvir.pdf

MIL-STD-499A Engineering Management, United States Department of Defense (May 1974)

Modelisar Consortium "Functional Mock-up Interface for Co-simulation" version 1 (October 2010), http://www.functional-mockup-interface.org

NATO STANAG 4158: Guidelines for Classifying Incidents for Reliability Estimation of Tracked and Wheeled Vehicles (November 1982)

Shaping the Future of the Canadian Forces: A Strategy for 2020,
http://www.cds.forces.gc.ca/str/index-eng.asp

Strganac, T.W., Ko, J., Thompson, D.E., Kurdila, A.J.: Identification and control cycle oscillations in aeroelastic system. Journal of Guidance, Control and Dynamics 23, 1127–1133 (2000)

Van der Velden, A., Koch, P., Devanathan, S., Haan, J., Naehring, D., Fox, D.: Probabilistic Certificate of Correctness for Cyber Physical Systems. In: ASME 2012 International Design Engineering Technical Conferences & Computers and Information in Engineering Conference, Chicago, IL, USA, August 12-15, DETC2012-70135 (2012)

Re-using SysML System Architectures

Andreas Korff

Abstract. In the development of complex systems, Model-based Systems Engineering (MBSE) has been introduced successfully into many projects. The OMG SysML as architectural modeling language is part of that success, because it offers the right perspectives to design complex systems. However, even when using MBSE methods and the SysML, the risk of re-inventing the wheel has to be overcome as well. This implies that System Engineers have to know in advance, what is already available as re-usable system asset and fits to the needs of the on-going project.

After introducing a suitable standard with OMG RAS, this paper will show different use cases for component-based design expressed in SysML using asset definition and propagation. This includes top-down exchange of system component specifications as well as bottom-up construction of new systems based on existing components. Asset reuse is based on communication. We will present the necessary communication means for propagating SysML-based system assets, which allows efficient collaboration in teams designing complex systems.

1 Setting the Scene: What Is an Asset?

In times when systems get more and more complex, because their functional and non-functional requirements increase in numbers and interconnectness, in parallel the system design team is dispersed into many sites and organizations, the time available until the system is ready and proven to work is expected to be more and more limited. Currently, there are specific trends visible in a lot of domains, which can help in this situation:

Andreas Korff
Atego Systems GmbH
Major-Hirst-Str. 11
38442 Wolfsburg
Germany
e-mail: Andreas.korff@atego.com

M. Aiguier et al. (eds.), *Complex Systems Design & Management 2013*,
DOI: 10.1007/978-3-319-02812-5_19, © Springer International Publishing Switzerland 2014

First, the on-going appreciation of system engineering as a necessary and important role in the organizations. Systems engineers are able to focus on a proper solution design, before going into details too fast. They are bridging between the other domains, project management; hardware, software and mechanical engineering; the regulatory and certification authorities of concern; and last, but definitively not least, customers and project sponsors.

Second, the inclusion of modeling, growing from pure implementation concept into design and specification. Model-based systems engineering is the essential driver for the success of a systems engineer, as he is able to balance all the different perspectives like system specifications, requirements documents, interface control documents, test specifications and so on, which he would have to keep consistent otherwise purely manually.

Third, the growing acceptance of the OMG SysML as to primary standard for a language expressing system models. Of course there are many other, also graphical languages out, but the SysML is capable to integrate them as specific notations for detailing single perspectives. Given a proper tooling support, a SysML model is the central element, the cap stone in the arch between pure textual requirements and implementation artifacts, enabling the bi-directional traceability required by not less than all current development standards like ISO 26262, or DO-178C.

Another aspect to speed up time-to-market time is of course to avoid the famous "Reinventing the wheel". Although very obvious for everyone, the reality in design projects shows that re-using existing artifacts is harder than it should be. Beside the fact that a lot of times artifact designs are not made to be re-usable – and this is applicable for mechanical, hardware and software artifacts – the biggest problem is to transfer and manage the knowledge what is actually available for re-use.

As a consequence, there is a standard needed defining a meta-model supporting re-usability. In fact, there is a standard available already since 2005: The Object Management Group (OMG), which is responsible for well-known standards like the Unified Modeling Language (UML) or the Systems Modeling Language (SysML), offers in its portfolio of publicly available standard definitions the Reusable Asset Specification, OMG RAS. Originally focusing purely on software artifacts, the definitions in OMG RAS can be interpreted and extended to describe generic assets. As Figure 1 shows, there are profiles foreseen in the RAS meta-model, for applying the asset definitions to different domains.

For component-based design, which is feasible for both software and systems engineering, there are specific elements necessary for a system asset: First of all, we need ways to store assets according their nature and dependent on the projects they are used in. Therefore Asset Libraries, for the general context, and containers or catalogs have to be available. Containers within libraries allow defining access rights and groups of interest for the team communicating the assets. In addition, optionally projects can be used to group assets together independently from their storage in different containers.

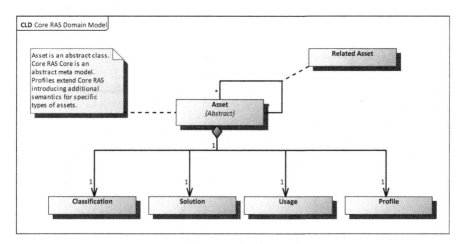

Fig. 1 Major sections in the Core RAS Domain Model

2   Use Case 1: Bottom-Up Construction

The asset itself can contain one or more solutions, in which sets of artifacts, component specifications or interface specifications are stored. The UML class diagram in Figure 2 shows these elements, including their main attributes and their associations. Using the full set of meta-elements defined in the extended RAS meta-model, all the basic principles of re-use can be supported. For instance, Service-oriented Architectures (SOA), using operations as services of assets, can be easily mapped onto RAS. The same applies to Component-based Design (CBD), as not only interfaces describing the services are relevant for a reusable asset, but also other public component attributes, as we will see later. The System-of-Systems (SoS) paradigm can be seen as an extension to the CBD, and thus this is well supported as well. If the context of a system changes, it becomes a component or even sub-component in the bigger and more complex composite system above it. The anti-locking system in a car is quite a complex system, including mechanical, hardware and software elements. On the other hand, it is just a component in the chassis subsystem within the car. But also the car can be seen as a sub-component of the traffic infrastructure.

In heterogeneous software designs, there are means to integrate different software components using another OMG standard, called Interface description language (IDL).

Also the UML itself is capable to describe re-usable assets, using the class model, for instance. The same applies for the SysML for system assets, and we will have a closer look on them later. Using SysML, it is possible to describe a bottom-up construction of a system by re-using pre-defined component assets.

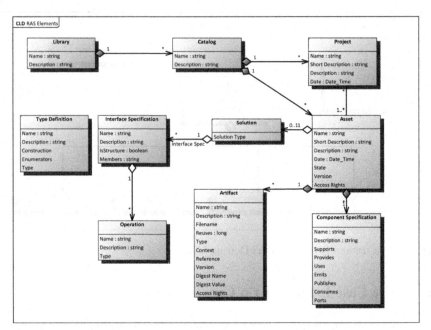

Fig. 2 Extended RAS elements

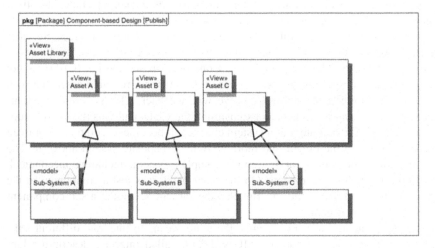

Fig. 3 Publishing building blocks

Figure 3 shows the publication of assets as a SysML package diagram. Subsystems A, B, and C exist as solutions is assets. They might be using IDL, UML, or SysML as descriptive language, and they are published as Assets A, B, and C in an Asset Library, where the asset views abstract the real information within the different models. Only elements described as public as propagated, and dependent on the mapping between the meta-model used in the Subsystem asset and the extended RAS, the information describing the asset are transformed in the representing view of the asset.

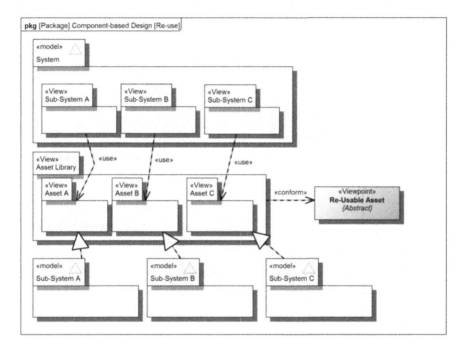

Fig. 4 Re-Use of Building Blocks

The full picture of the bottom-up use case can be seen in Figure 4. The System as a model is composed by the subsystems A, B, and C, which are using the assets form the asset library. This re-use is based on another transformation of the information in the asset library into the system model. If e.g. the asset B describes a software asset, the view of this subsystem consists of interfaces owning the operations which can be called when using this asset. This is not a "clone and own" propagation of subsystems, as the matters of concern are acknowledged. The user of an asset does not need to see all the internal details, but only the functionality which is offered.

3   Use Case 2: Design a System When You Need New Components

Very often the available components do not fulfill all the needs to be used in the overall system. The success of requirements-based engineering shows that it is already understood that in this case it is essential to communicate the requirements of the future component or system in the right way. Model-based specifications are the ultimate way to reduce the misunderstandings or pure errors in textual requirement specifications. However, even a textual specification can be seen as a re-usable asset, but of course a SysML-based specification of a subsystem is much more precise.

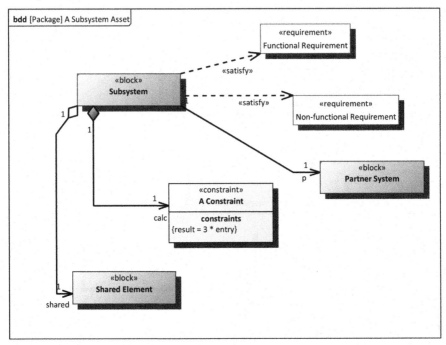

Fig. 5 Subsystem Specification

In Figure 5, the needed subsystem is described using a SysML block Definition Diagram. Contained and shared parts, valid constraints and requirements, which the subsystem shall satisfy, are all necessary parts of the future subsystem. In addition, the ports of the subsystem can be further on modeled in an internal block diagram, like shown in Figure 6. The subsystem itself contains subsystems, and ports of different types.

This specification of the subsystem can be propagated into an Asset Library, in which it can be found and identified by someone, who wants to reuse a specification, i.e. who wants to implement this subsystem. Searching for elements within the Asset library can be based on names, descriptions, projects, or solutions.

As a starting point, the SysML-originated asset information can be imported as a SysML reused asset into a suitable SysML modeling tool, e.g. Atego Artisan Studio.

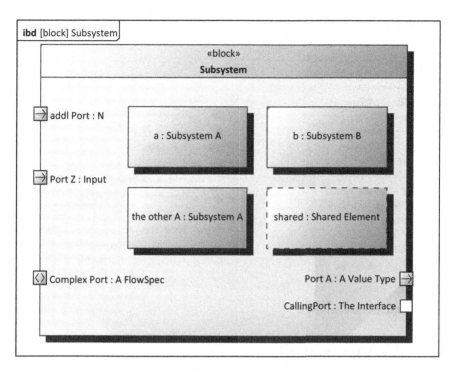

Fig. 6 Ports and Parts of a Subsystem Specification

The result of the asset reuse can be seen as shown in Figure 7, which is the browser content for the subsystem. Under the asset name, the subsystem is available as SysML block, containing all public parts and ports. To complete the subsystem specification, the requirements within the asset (contained or related to the subsystem), the constraints and all types of parts and ports are part of the reusable asset as well.

The subsystem can be further on developed and a design solution proposal can be published as an available building block according the bottom-up construction in use case 2. The specification and the system design are related, but not the same, of course. SysML offers suitable traceability links, like for instance allocations.

The subsystem asset propagations offers a model-in-model approach, because the system model only contains the black-box information of the subsystem, whereas the subsystem model contains all white-box information to fully understand all aspects of the subsystem as a system model in SysML. This enables team collaboration, which takes intellectual property considerations automatically into account. Internal implementation details or buy-or-build calculations are kept in the right model, only exposing the relevant asset information.

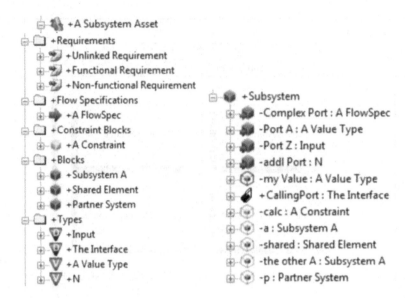

Fig. 7 SysML-based asset reuse

4   Asset Communication Means

Both basic asset library use cases make it obvious that reuse is more than making information somehow available. Communication means are essential for an efficient reuse for all types of assets. The implementer of the reused subsystem specification needs a way to identify the specification as something he can provide. The systems engineer reusing building blocks needs a way to find suitable subsystems. In general there is the requirement to control the access of the asset information, and to allow the versioning of assets. Assets must be accessible without any specific technical pre-requisite, as they need to be communicated within and in between possibly dispersed organizations.

Therefore the enabling technology for asset libraries has to implement the basic use cases as shown in Figure 8. We propose a web-based technology, with a browser front-end, access control, and a powerful storing technology as a server backbone. A good example of this asset library implementation is the Atego Asset Library.

Users accessing the Asset Libraries are equipped with the adequate rights to search, publish, and reuse assets in defined libraries and catalogs. As assets can come from UML or SysML models, their publication and reuse needs to be linked with the relevant modeling tool, possibly integrated in the tool UI. In order to store the relevant asset data, UML and SysML models can be annotated with an Asset Profile, so the necessary communication data, like asset source, can be stored directly in the model.

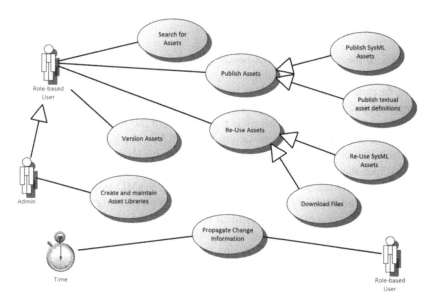

Fig. 8 Basic Use Cases for an Asset Library

Assets can change over time, of course, and these changes have to be carefully handled. It is a big difference if a subsystem model is versioned due to internal changes or if interface or requirement changes have to be published. The reusing partner has to be informed about that change, and he needs to define a suitable time when updating the reused asset on his side. In our implementation of the Atego Asset Library, there is the option to flag interest for a single asset, a project, or a container. If something "interesting" has been updated, the user is informed by mail, including a URL to the asset which was changed.

5   Difference to Typical CM Support

Reuse is not something new, of course. Every developer in reasonably sized projects is already using configuration management (CM) or application lifecycle management (ALM) tools to fully control changes, reasons for changes and relationships to other artifacts.

However, there are big differences to asset libraries, as CM or ALM tools are built to control the development of artifacts, but not to propagate reuse. The version information can contain internals, which are valuable for the developer, but not for possible re-users. Usually, the access to CM or ALM tools is strictly internal, and external communication has to manually abstract the internal information.

Therefore, asset libraries are complementary to any form of internal development management systems. They enable proper reuse and shall motivate the development of reusable assets. Integrating reuse metrics, the time and cost savings can be made visible, influencing the management to re-think about the concentration on short-term results, building artifacts which can be used only once.

6 Summary

In order to meet time-to-market needs, reuse of existing components is key. This requires a new form of communicating reusable assets, as the traditional ways of reuse by pure clone and own does not work efficiently especially for system component assets. In addition, the proper reuse requires a standard for reusable assets, so the generic definition of assets is possible. With the Reusable Asset Specification of the Object Management Group (OMG RAS), there is an existing standard available, which can be extended to suit also systems engineering asset. If an asset library like the Atego Asset Library implements this standard, the exchange of text-based, software interface-related, UML, and SysML assets is possible. This allows the contemporary system development, implementing a design-time Supplier-Consumer process, which is ideal for System-of-Systems design. Dependent on the levels of concern, this breaks up complex SysML models into smaller parts, thus simplifying the different systems and subsystems perspectives. In addition, efficient intra- and inter-organization communication about reusable system assets is therefore possible.

References

[OMG01] Reusable Asset Specification, Version 2.2, formal/05-11-02, Object Management Group (November 2005)

[OMG02] OMG SysML 1.3, formal/2012-06-01, Object Management Group (June 2012)

[OMG03] OMG SysML 1.2, formal/2010-06-01, Object Management Group (June 2010)

[OMG04] UML 2.4.1 Superstructure, formal/2011-08-06, Object Management Group (August 2011)

[OMG05] UML 2.4.1 Infrastructure, formal/2011-08-05, Object Management Group (August 2011)

[OMG06] OMG IDL 3.5, in-process version of IDL, Beta 1, Object Management Group (February 2013)

[ATEGO07] Atego Asset Library, Product Datasheet, Atego (2012),
http://www.atego.com/products/atego-asset-library/

Water Saving in a Complex Industrial System – Evaluation of the Sustainability of Options with System Dynamics

Katharina M. Tarnacki, Thomas Melin, and Sabina Jeschke

Abstract. Great amounts of water are used in the steel industry for cooling, cleaning and as transport medium. In this study the water management of an integrated steelworks is analysed in order to identify bottlenecks, suggest measures and evaluate their water saving potential. For the assessment of the status quo and the identification of bottlenecks, the water balance methodology is applied. Based on the results of the water balance a list of possible measures for enhanced water management in the plant is suggested. The implementation of the suggested options is modelled with system dynamics in order to assess the water saving potential. The specific focus in this study is set on the simulation of the complex water system with the implemented options under specific weather conditions. The information about the behaviour of the system with changing weather scenarios delivers conclusions on the sustainability of the options concerning the local water resources. The approach delivers the support in the decision-making process for water management related investments in the plant.

1 Introduction

In this study water balance tools and system dynamics are applied in order to improve the complex water management in an industrial system. In a previous study

Katharina M. Tarnacki · Sabina Jeschke
IMA/ZLW & IfU, RWTH Aachen University, Aachen, Germany
e-mail: katharina.tarnacki@ifu.rwth-aachen.de,
    sabina.jeschke@ima-zlw-ifu.rwth-aache

Katharina M. Tarnacki · Thomas Melin
AVT – Aachener Verfahrenstechnik, Chemical Process Engineering,
RWTH Aachen University, Aachen, Germany
e-mail: melin@avt.rwth-aachen.de

M. Aiguier et al. (eds.), *Complex Systems Design & Management 2013*,   267
DOI: 10.1007/978-3-319-02812-5_20, © Springer International Publishing Switzerland 2014

the water balance has been applied in order to analyse the water management in the plant, identify the bottlenecks in the water management and suggest efficient water management options (Dimova et al. 2009; Dimova et al., 2007). Furthermore, the overall water resources management of the water system has been modelled for different weather scenarios (Vamvakeridou-Lyroudia et al., 2007). The objective of this study is the evaluation of the suggested improvement options in terms of their water saving potential and the consideration of their sustainability under specific weather scenarios.

1.1 Water in Industry

Water is one of the major resources in many industrial activities and depending on the process has different functions. It is used as part of the product, for cleaning and cooling purposes as well as transport medium (EC, 2012; Lens et al., 2002). The European industrial water cycles are mostly not closed, however the recycling rates are steadily increasing, various water saving measures and technologies are often applied (, WSSTP, 2009).

The steel production is a very water and resources intensive activity. Great amounts of water are used for the cleaning and cooling of products and gases. Depending on the location of the production site and the technology applied around 10-50m³ of freshwater are used for the production of 1t of a sub-product in the steelworks. The processed water is treated mechanically, cooled in cooling towers and recycled in the process (EC, 2012).

1.2 Investigated Water System

The investigated steelworks, initially constructed in 1963, was designed for a complete metallurgical cycle. The main technological stages include the sinter plant, blast furnaces, top blown converters and electric arc furnaces, as well as hot and cold rolling mills. The products were mainly cast iron and steel which were exported to the European Union, Turkey, some of the former Yugoslav countries, USA and China.

The plant is supplied with two water types: drinking water and industrial water. Drinking water is applied for drinking and high quality needs. Industrial water has lower than drinking water quality in particular in respect to suspended solids and soluble heavy metals..

Drinking water consumption is around 5,300,000 m³/a. Most of the technological cycles use completely recycled water. Water losses during the manufacturing process are supplemented through supply of fresh industrial water from different water sources. The annual industrial fresh water use amounts to over 55,000,000 m³ and the amount of recycled water is over 580,000,000 m³/a.

The industrial water supply scheme consists of both fresh water and reused water sources. Three reservoirs, groundwater sources, as well as river water abstractions provide process water to the plant. Process water is used in two types of

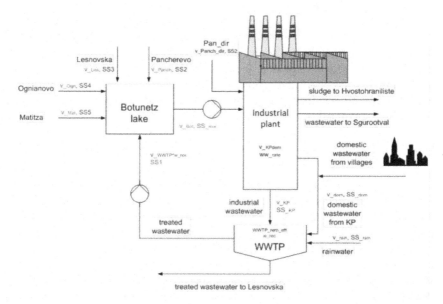

Fig. 1 Investigated water system

turnover cycles: clean cycle for cooling of machines and installations and dirty cycle - for cleaning or cooling of gasses and hot metals. The process and accidental losses are compensated either with water from the clean cycle or with fresh water (Dimova, 2007).

In the plant there are three different industrial wastewater/sludge discharging points to the Lesnovska river which can affect the environment directly: treated wastewater from the wastewater treatment plant for industrial and rain waters (WWTP IRW), from Hvostohraniliste (sludge pond) and from Sgurootval (waste disposal pond). All those wastewater streams underlie controls and different discharge norms. The norms concentrate on the main contaminants from steelworks (heavy metals, organics: among them phenols, suspended solids). Actually, the water stream from WWTP IRW should only be discharged in emergency cases; it should be recycled through the buffer reservoir Botunetz as much as possible. The present recycling rates from WWTP amount to 62% due to water quality problems in this water stream. The discharge norms correspond with the general European limit values for discharges of steelworks.

In the region the climatic conditions are responsible for alternating droughts and floods during the year. Especially the flooding periods pose a problem to the plant due to the fact that the risk of pollutant release from the ponds rises. Especially the waste disposal pond is under risk: solid waste containing heavy metals is disposed there and its release would also contaminate the Lesnovska river and the surrounding area.

In dry periods the water supply for the plant is limited because the freshwater reservoirs (Pancherevo and Ognianovo) are also used by other users (agricultural and domestic) and priority is given to public water supply. In those emergency cases the permitted water abstractions from all available natural freshwater resources (rivers, groundwater sources, reservoirs) are also reduced, partly to less than 50% of the usual abstraction. In this case, water supply cannot fully cover the total water demand of all production units in the plant. For those cases, the industrial plant has a crisis management plan in which priorities in closing of production plants are defined (Dimova et al., 2009).

1.3 System Dynamics

System dynamics (SD) modelling has been introduced by Forrester (1961) for modelling and simulation of long-term decision-making in dynamic industrial problems. It has also been applied for complex environmental and water systems (Ford, 1999; Mulligan and Wainwright, 2004; Simonovic, 2003; Chung et al 2008). It can be applied for complex feedback systems when analytical models do not exist but system simulation can be performed for existing feedback mechanisms (Vamvakeridou-Lyroudia et al., 2007). SD modelling can be applied to determine the relevance of specific options for the overall water system and reveal potential of further water saving in the system.

2   Methodology

2.1   Modelling

Modelling with SD requires systems with feedback loops in order to analyse the system behaviour. Simulations can be run for defined time steps (in this study days) to determine the change in the system. Here the considered system elements were the two reservoir capacities: Pancherevo and Ognianovo. The sustainable use of these reservoirs has been evaluated. In the model different elements occur: stocks (reservoirs, water users; rectangular symbols), converters (water flows; arrows) and influencing parameters (wastewater reuse rate, rainwater amount etc.; small circles) (Figure 2). SIMILE has been applied as SD modelling tool. For the optimisation of the Botunetz reservoir an additional MATLAB simulation has been performed and linked to the SD model.

2.2   Scenarios and Assumptions

For the evaluation of the suggested options, different scenarios are modelled. They include different suggested options and different weather scenarios.

For the operation of Botunetz which is the central node of the recycling system (Figure 1) the following assumptions are relevant for all scenarios:

- ideal mixing of different water streams in the Botunetz lake
- negligible sedimentation of suspended solids, due to small particles and the values from the monitoring programme which confirm this assumption
- optimised lowest and maximal water flows from all reservoirs according to maximum allowed releases
- suspended solids concentration of the resulting water flow to the plant should not exceed 25 mg/L
- incoming wastewater flow to the wastewater treatment plant (WWTP) is a mixture of wastewater from the plant, domestic wastewater from the plant and neighbouring villages and rainwater collected on the territory of the industrial plant

Water flows are calculated with the following equations (Eq. 1-3):

$$\dot{v}_{Bot} = \dot{v}_{Les} + \dot{v}_{Panch} + \dot{v}_{Mat} + \dot{v}_{Ogn} + \dot{v}_{WWTP} * w_{rec} \tag{1}$$

$$\dot{v}_{WWTP} = \dot{v}_{rain} + \dot{v}_{dom} + \dot{v}_{KP} \tag{2}$$

$$\dot{v}_{KP} = \dot{v}_{KP_{dem}} * ww_{rate} \tag{3}$$

Suspended solids concentrations of the mixtures are calculated with the following equations (Eq. 4-5):

SS in Botunetz:

$$SS_{mix} = \frac{SS_1 * \dot{v}_{WWTP} * w_{rec} + SS_2 * \dot{v}_{Panch} + SS_3 * \dot{v}_{Les} + SS_4 * \dot{v}_{Ogn} + SS_5 * \dot{v}_{Mat}}{\dot{v}_{WWTP} * w_{rec} + \dot{v}_{Panch} + \dot{v}_{Les} + \dot{v}_{Ogn} + \dot{v}_{Mat}} \tag{4}$$

with
SS from treated wastewater from WWTP:

$$SS_1 = \frac{\dot{v}_{rain} * SS_{rain} + \dot{v}_{dom} * SS_{dom} + \dot{v}_{KP} * SS_{KP}}{\dot{v}_{rain} + \dot{v}_{dom} + \dot{v}_{KP}} * (1 - WWTP_{rem\_eff}) \tag{5}$$

$\dot{v} = water\ flow\ rate$

$SS = suspended solids concentraion$

$w_{rec} = recycling\ rate\ from\ WWTP$

$ww_{rate} = wastewate\ rate\ from\ the\ plant$

$WWTP_{rem\_eff} = removal\ rate\ in\ the\ wastewater treatment\ plant$

Table 1 and Table 2 summarise the applied modelling parameters. Table 1 shows the quantity and quality of flows and Table 2 reservoir capacities and withdrawals.

Table 1 Applied modelling parameter values - reference scenario 0.1

PARAMETER	DESCRIPTION	UNIT	VALUE
KP_dem	Process water demand for Kremikovtzi plant (KP)	10^3m³/month	4373
ww_rate	waste water rate from KP (in % of total water inflow to KP)	%	51
ww_Botu_rate	Treated wastewater rate to Botunetz	%	62 (normal condition)
rain	Rainwater	10^3m³/month	419
KP_dom	domestic wastewater	10^3m³/month	190
Panch_dir	Pancherevo direct	10^3m³/month	600
$\dot{v}_{Panch}$	Pancherevo to Botunetz	10^3m³/month	899
$\dot{v}_{Les}$	Lesnovska	10^3m³/month	391
$\dot{v}_{Ogn}$	Ognianovo	10^3m³/month	629
$\dot{v}_{Mat}$	Matitza	10^3m³/month	87
WWTP_rem_eff	WWTP SS- removal efficiency	%	56
SS2	SS concentration from Pancherevo	mg/L	18.2
SS3	SS from Lesnovska river	mg/L	34.2
SS4	SS from Ognianovo	mg/L	10.6
SS1_1	SS rainwater	mg/L	5
SS1_2	SS domestic wastewater	mg/L	200
SS1_3	SS wastewater from KP	mg/L	55

Table 2 Reservoirs capacities and withdrawals in different scenarios (water availabilities)

	TOTAL CAPACITY (m³)	AVERAGE WATER WITHDRAWAL (x10³m³/month)		
		Normal year (0.1)	Dry year (0.2)	Very dry year (0.3)
Ognianovo	31.6x10⁶m³ (oper. volume 20x10⁶ m³, min 11x10⁶m³)	629	315	157
Pancherevo	64.7x10⁶m³ (minimum 16.16x10⁶m³)	through Botunetz: 899	As low as possible	As low as possible
		direct: 600	450	300
Botunetz (buffer reservoir)	1.1x10⁶m³	As much as possible	As much as possible	As much as possible
Matitza	-	87	-	-
Rudnik (domestic water)	-	340	340	340
Lesnovska river	-	< 500	< 500	< 200

2.3 Conceptual Model of the System

Figure 2 presents the system dynamics model of the whole investigated water system which is the basis for the simulation. The major system elements are: stocks, converters and variables. The Botunetz lake is the central node of the system. In this reservoir all incoming flows are mixed.

The major influencing variable is the monthly average rainwater which defines the weather scenarios (dry, wet, normal year). This variable has direct impact on the following system components:

- Total water demand of the industrial plant
- Rainwater inflow to the WWTP
- Max allowed release from Matitza river
- Max allowed release from Ognianovo
- Inflow to Ognianovo
- Max allowed released from Pancherevo
- Inflow to Pancherevo
- Max allowed release from Lesnovska

Each time step (one day) delivers for all stocks in the system (reservoirs and water demands) new values according to the parameter values.

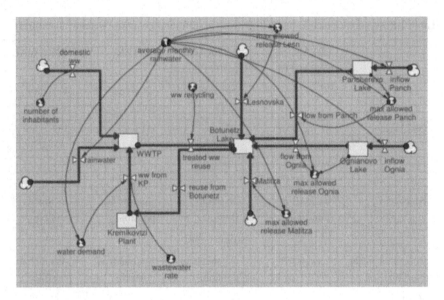

Fig. 2 System Dynamics model of the water system

2.4   Weather Conditions Scenarios (Scenarios 0.x)

The region where the plant and its water system are located is specially affected in dry and very dry time periods. Use restrictions for the external water resources (Pancherevo and Ognianovo reservoirs) are introduced to avoid overexploitation of these valuable water sources which are also an important water source for other local users (domestic and agriculture). However, also water abstractions from rivers (Matitza, Lesnovska) are restricted in dry periods to ensure minimal river flow.

To overcome the water deficit due to use restrictions of external water sources own water sources (treated wastewater from wastewater treatment plant, supplied through the buffer reservoir Botunetz) are exploited.

Scenarios are defined for normal (scenario 0.1), dry (scenario 0.2) and very dry years (scenario 0.3). For those three scenarios following parameters are affected:

- precipitation (normal year: average monthly rainfall given at the WMO website www.worldweather.org for Sofia region, dry year: 50% of normal precipitation, very dry year: 25% of normal precipitation, wet year: +40% of normal precipitation)
- abstraction restrictions from external water sources (reservoirs and rivers supplying also other users, see Table 2)
- KP water demand (with shut down of different priority production plants)

2.5   Scenarios for Single Options (1.x-5.x)

Based on the results of the water balance a couple of improvement options have been suggested. The implementation of those options has specific impact on the water management and has therefore different water saving potentials which are not easy to be calculated without applying any simulation, in this case system dynamics.

The following options have been modelled:

- Optimisation of Botunetz operation (scenarios 1.x):
- Advanced coke wastewater treatment (scenarios 2.x)
- Renovations of cooling towers in the blast furnace (scenarios 3.x)
- Hot rolling mill (chemical treatment of dirty cycle water, adjustment of water flow and energy consumption in high pressure rotary pumps) (scenarios 4.x)
- Improvement of WWTP efficiency (scenarios 5.x)

2.6   Option Implementation and Long-Term Weather Scenarios (Scenarios 7.x)

These scenarios model how sustainable in terms of reservoirs capacity the implementation of specific options will be under specific weather scenarios, which means that time periods with normal, wet, dry and very dry weather conditions are simulated and the impact of the implemented options on the water levels in the reservoirs is shown and discussed.

- Scenario 7.1: very dry year, normal year, very dry year, normal year
- Scenario 7.2: very dry year, dry year, 2 x normal year
- Scenario 7.3: normal year, very dry year, 2 x wet year

3   Results and Discussion

For complex water systems with feedback loops like the investigated steelworks, system dynamics modelling proves to be a suitable tool to predict impacts of option implementation. The impact of the suggested options has been modelled Especially, the optimised distribution of water flows from available freshwater sources, water savings and resulting water qualities can be simulated.

Due to the complexity of the water system and due to the fact that also water qualities (in this study only suspended solids concentrations have been considered) need to be taken into account it is difficult to calculate the impacts without any previous simulation. Suspended solids concentration of water supplied to KP should not exceed 25 mg/L (set as boundary condition of the modelling). The quality of the water flow entering the central wastewater treatment plant influences the treatment performance so that the SS concentration should be kept as low as possible.

The investigated scenarios with considered boundary conditions and assumptions the models deliver following optimised minimal flows from the reservoirs (Figure 3). In case of normal year conditions (scenarios x.1) for Ognianovo significantly reduced flows compared to reference scenario 0.1 can be especially observed for scenario 1.1 (optimisation of Botunetz operation) and scenario 3.1 (measures for blast furnace). For Pancherevo especially in scenario 1.1 the flows are significantly reduced, slight decrease is also noticed for scenario 4.1. In all other scenarios the flows are almost the same as in scenarios 0.x.

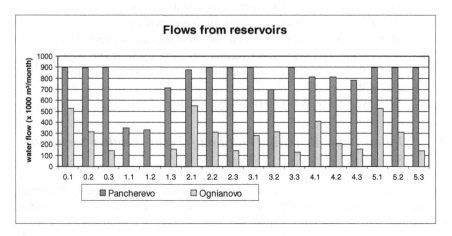

Fig. 3 Reservoir flows for all scenarios

Sustainable reservoir operation is very much linked to occurring weather conditions. The defined weather conditions considered as scenarios x.1, x.2 and x.3 are related to average values for normal, dry and very dry years. However, the unsustainable reservoir operation (overexploitation) is specially occurring when long dry or very dry periods occur and the following normal or wet conditions are too short to ensure the refilling of the reservoirs. For the assessment the consideration of longer time periods, such as several years, is necessary.

The modelled scenarios are based on the implementation of an option combination (option 1 and 5) in order to consider also limitations of the new improved water management.

For the Ognianovo reservoir for the performed weather scenarios it can be concluded, that two normal years can almost compensate the water used during one dry and one very dry year (scenario 7.2). For two very dry years, the refilling during two normal or two wet years (almost the same refilling) is not sufficient (scenario 7.1). The sustainable operation for the Ognianovo capacity is for one very dry year the refilling during two normal years. The considered dry and very dry year scenarios are modelled under the condition that the plant has a reduced water demand in times of water shortages. However, it has to be taken into account that reduced water demands are a consequence of shutting down low priority production plants which cannot be done for a long time period. So in longer dry and very dry time periods the exploitation of freshwater sources (mostly the reservoirs) would be even higher than in the modelled scenarios.

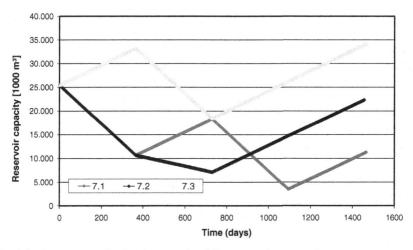

Fig. 4 Ognianovo capacity development for different weather scenarios

For the Pancherevo reservoir the dry and very dry years have almost the same impact on the reservoir capacity (scenario 7.2 – first year is very dry, second is dry) and the refilling during the normal year compensates almost completely the losses during very dry years (scenario 7.3 – first year normal, second very dry). The wet years have a more positive impact on the Pancherevo water level than for Ognianovo (scenario 7.3 third and fourth years).

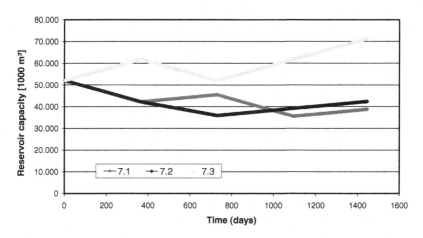

Fig. 5 Pancherevo capacity development for weather scenarios

4 Conclusions

System Dynamics turned out to be a powerful tool in order to assess water saving potential of the suggested options and resulting reservoirs capacity development in a complex industrial water system. SD supported to identify the interrelations between the system components. Especially, the impact of water saving in the plant, the interactions with the water reuse cycle and the resulting recycling rates were modelled and ensured the optimal reservoir Botunetz operation which was the central recycling node. It was able to foresee the impact on other water sources and choose the best operation strategy. The additional weather scenarios proved that only option combination can deliver a long-term sustainable water management of the plant.

For the full implementation of the options a more detailed analysis of the water quality is necessary. In this study only the loads of the critical water quality parameter suspended solids have been simulated. In the steelworks also heavy metal concentration can be also a limiting value, especially ferrous.

This evaluation is only based on the water quantity and water quality considerations. Other criteria such as costs, maintenance and construction efforts are not modelled in this study. The whole decision-taking process requires as well a multi-criteria analysis for the defined criteria.

References

AQUASTRESS. Final project report – Aquastress Case Studies (2009),
    http://environ.chemeng.ntua.gr/aquastress
Chung, G., Lansey, K., Blowers, P., Brooks, P., Ela, W., Stewart, S., Wilson, P.: A general water supply planning model: Evaluation of decentralised treatment. Environmental Modelling & Software 23, 893–905 (2008)
Dimova, G., Tarnacki, K., Melin, T., Ribarova, I., Vamvakeridou-Lyroudia, L., Savov, N., Wintgens: The water balance as a tool for improving the industrial water management in the metallurgical industry – Case study Kremikovtzi Ltd., Bulgaria. In: Proc. 6th IWA Specialist Conf. on Wastewater Reclamation and Reuse for Sustainability, October 9-12. Antwerp (2007)
EC, 2012 Europ. Com. – Integrated Pollution Prevention and Control (IPPC) – Ref. Doc. on Best Available Techniques on the Iron and Steel Production (2012)
Ford, A.: Modelling the Environment: An Introduction to System Dynamics Modelling of Environmental Systems. Island Press, Washington DC (1999)
Forrester, J.: Industrial Dynamics. Pegasus Communications, Waltham (1961)
Kojiri, T., Hori, T., Nakatsuka, J., Chong, T.-S.: World continental modelling for water resources using system dynamics. Phys. Chem. Earth 33, 304–311 (2008)
Lens, P., Hulshoff Pol, L., Wilderer, P., Asano, T.: Water Recycling and Resource Recovery in Industry. IWA Publishing (2002) ISBN 1 84339 005 1
Mazzoleni, S., Rego, F., Giannino, F., Legg, C.: Eco-system modelling: vegetation and disturbance. In: Wainwright, J., Mulligan, M. (eds.) Environmental Modelling: Finding Simplicity in Complexity, pp. 171–186. John Wiley & Sons Ltd., West Sussex (2004)
Mulligan, M., Wainwright, J.: Modelling and Model Building. In: Wainright, J., Mulligan, M. (eds.) Environmental Modelling: Finding Simplicity in Complexity, pp. 7–73. John Wiley & Sons Ltd., West Sussex (2004)

Rehan, R., Knight, M.A., Haas, C.T., Unger, A.: Application of system dynamics for developing financially self-sustaining management policies for water and wastewater systems. Water Res. 45, 4737–4750 (2011)

Ribarova, I., Assimacopoulos, D., Jeffrey, P., Daniell, K., Inman, D., Vamvakeridou-Lyroudia, L.S., Melin, T., Kalinkov, P., Ferrand, N., Tarnacki, K.: Research-supported participatory planning for water stress mitigation. Journal of Environmental Planning & Management 54, 283–300 (2011)

Simonovic, S.: Assessment of Water Resources Through System Dynamics Simulation: From Global Issues to Regional Solutions. In: Proc. 36th Hawaii International Conf. on System Sciences IEEE (HICSS 2003), Hawaii, USA (2003)

Sušnik, J., Vamvakeridou-Lyroudia, L.S., Savić, D.A., Kapelan, Z.: Integrated System Dynamics Modelling for water scarcity assessment: Case study of the Kairouan region. Science of the Total Environment 440, 290–306 (2012)

Tarnacki, K.: Evaluating Industrial Water Saving and Water Management Options in Order to Mitigate Water Stress. Thesis (PhD), RWTH Aachen (in press, 2013)

Vamvakeridou-Lyroudia, L.S., Savic, D., Tarnacki, K., Wintgens, T., Dimova, G., Ribarova, I.: Conceptual/System Dynamics Modelling Applied for the Simulation of Complex Water Systems. In: Ulanicki, B., Vairavamoorthy, K., Butler, D., Bounds, P.L.M., Memon, F.A. (eds.) Water Management Challenges in Global Change, Proceedings of the International Conference CCWI 2007 and SUWM 2007, Leicester UK, September 3-5, pp. 159–167. Taylor & Francis Group, London (2007)

Wintgens, T., Dimova, G., Ribarova, I., Druzynska, E., Caruk, M., Vamvakeridou-Lyroudia, L.S., Melin, T., Tarnacki, K.: Industrial water management as a water stress mitigation option. In: Koundouri, P. (ed.) The Use of Economic Valuation in Environmental Policy, pp. 125–195. Routledge, Oxon (2009) 978-0-415-45323-3; Winz, I., Brierley, G.: The Use of System Dynamics Simulation in Integrated Water Resources Management. In: Proc. of the 27th International Conference of the System Dynamics Society, Albuquerque, New Mexico (2009)

WSSTP: Water Supply and Sanitation technology Platform (WSSTP) Thematic Working Group 3: Water in Industry. Final Draft Document (July 2009)

Handling Complexity in System of Systems Projects – Lessons Learned from MBSE Efforts in Border Security Projects

Emrah Asan, Oliver Albrecht, and Semih Bilgen

Abstract. Well-established systems engineering approaches are becoming more inadequate as today's systems are becoming more complex, more global, more COTS/re-use based and more evolving. Increased level of outsourcing, significant amount of subcontractors, more integration than development, reduced project cycles, ecosystem like collaborative developments, software product lines and global development are some of the changes in the project life cycle approaches. By structuring data into views with a common language and format, architecture frameworks can be utilized as tools for managing system complexity. In EADS, both architecture frameworks and MBSE are considered as major areas in the future of contemporary systems engineering practices. In this paper, we share our experiences towards adopting MBSE in border security projects where the role of EADS is the system of systems integrator. From the practitioner's point of view,

Emrah Asan
EADS Deutschland GmbH,
Landshuter Str. 26
D-85716 Unterschleissheim, Germany
e-mail: emrah.asan@gmail.com

Oliver Albrecht
BMW
Knorrstrasse 147
D-80788 Munich, Germany
e-mail: Oliver.albrecht@bmw.de

Semih Bilgen
Department of Electrical & Electronics Engineering,
Middle East Technical University
06800 Ankara, Turkey
e-mail: bilgen@metu.edu.tr

M. Aiguier et al. (eds.), *Complex Systems Design & Management 2013*,
DOI: 10.1007/978-3-319-02812-5_21, © Springer International Publishing Switzerland 2014

lessons learned presented in this paper can help companies to increase the speed of shift from traditional systems engineering approaches to architecture driven, knowledge focused and model-based systems engineering practices.

1   Introduction

A lot of things have changed since F.P. Brooks made the famous statement that there was no silver bullet for the software problem back in 1987 [1]. Changeability and complexity, essential properties of software, are still there and no silver bullets have been produced to deal with them. The only thing that is changing about this fact is the increasing size and complexity of the software. Today we are talking about large scale, complex software intensive systems of systems (SoS) in almost all sectors.

In this new world even for a simple daily business process, a large number of systems, users and services are interacting. We can say that the Software Intensive Systems of Systems (SISOS) concept was not created but evolved as a natural technological response to the customer demands and needs [2].

It is true that the driving force behind today's complex systems of systems is our demands and needs. However, today, it is open to discussion which one is driving the other. SISOS has also changed the concept of time and its value. In the past, the army which has more information sources has the superiority in the field of war. However, today the ability of using the acquired information in rapid decision making is the determining factor of the success. It can sometimes turn into a disadvantage if the huge amount of information you have slows down your decision making process.

Size and complexity of projects/systems, increased use of COTS and subcontractors, global team structure, issues of interoperability with legacy systems, re-use considerations and large number of stakeholders with conflicting interests constitute only a subset of the problems that result from the changes in the systems engineering (SE) environment [3-8].

In response to these trends we have somehow changed the way we conduct the system projects. In the past a complex standalone system was being developed by a single company. A small part of the system was being outsourced and only a limited amount of COTS usage was preferred. In traditional SE processes, systems engineers were spending most of their time in top-down design. In such a traditional approach requirements analysis, design and implementation processes were the main SE processes [9] [10].

Today, SISOS are developed by collaborating companies using significant amounts of COTS/re-use. One prime contractor which is called the lead systems integrator or system of systems integrator has the overall system responsibility. However, this lead systems integrator outsources much of the system's capability. Projects turn into integration projects rather than pure top-down development projects [11].

In the past systems engineers preferred COTS/re-use/outsourcing as a risk mitigation approach. Today, this is no more a preference but the dominant way of developing complex systems. In such an environment, systems engineers of the lead systems integrator company spend most of their time in capturing the architectural knowledge that spans operational objectives, major physical components and their interfaces, end-to-end functional behavior of the integrated system, etc.

A lead systems integrator does not have the full control over the collection of systems. However, it is possible to set some constraints on their development, structure or operation in order to achieve overall objectives. Under such a scenario, the level of detail of the information captured in terms of architectural knowledge is much less than is relevant to the detailed development of each individual system in the SoS [12]. As a result, the SE activities in this environment are more of a knowledge management kind than top-down design activities.

This is one of the reasons that architecture frameworks are becoming more popular in the large scale SoS development environments. By structuring data into views with a common language and format, architecture frameworks can be utilized as tools for managing system complexity. They serve as a communication tool to stakeholder communities with different views of the system.

Model-Based Systems Engineering (MBSE) is one of the major areas that are highlighted in The International Council on Systems Engineering (INCOSE) 2020 vision statement [13]. MBSE is a transition from document-centric SE to model-centric practices. MBSE promises improvements in SE by providing a holistic view of the system using integrated models, ensuring design integrity, easing requirements traceability and enhancing knowledge management across the project team and stakeholders.

Due to large operating environment and diverse missions, border security projects are mostly SoS projects and exhibit many complexities along various dimensions (Fig. 1).

EADS is a worldwide SoS integrator providing large scale solutions to civil and military customers around the globe. Air systems (aircraft and unmanned aerial systems), land, naval and joint systems, intelligence and surveillance, cyber security, secure communications, test systems, missiles, services and support solutions are among the solution portfolio of EADS.

Border security projects constitute the top strategic objectives of EADS. Although it has been more than 10 years since EADS started to invest in this area, we are still facing challanges in managing border security projects as programs or as a business line.

EADS manages the border security business line by projects. There are 4 ongoing border security projects in EADS. In all of these projects, EADS acts as the lead system integrator for large scale software intensive border security systems. The subsystems are standalone systems developed for different domains such as command and control, communication, intelligence, surveillance, etc.

Operational — Diverse missions: patrol, intercept, SAR, traffic control, disaster management, fishery control.

Functional — Wide range of application domains: Surveillance, command & control, intelligence, communications, reporting, etc.

Physical — Many subsystems, software integration, hardware integration, distributed systems, diverse environments, several hundred sites

Technical — Interoperability, performance, adaptability, robustness, obsolescence, harsh environments, civil works, electrical engineering

Social — Many stakeholders. Buyer is not the user. Several organizational units and departments are involved. Diverse and contradicting expectations.

Fig. 1 Border Security Projects Exhibit Many Complexities in Various Dimensions

All these subsystems are available in the market as almost COTS systems. Minor customizations on the hardware and software are done according to project specific needs. Obviously the end system is a SoS.

In EADS, both architecture frameworks and MBSE are considered as major areas in the future of contemporary SE practices. For an organization, adoption of new approaches such as MBSE requires careful consideration of the introduction mechanisms. In this paper, we share our experiences towards adopting MBSE in border security projects where EADS performs the role of lead systems integrator. From the practitioner's point of view, lessons learned presented in this paper can help organizations to increase the speed of shift from traditional SE approaches to architecture-driven, knowledge-focused and model-based SE practices.

2 Handling "Complexity" with Decomposition Approaches

Decomposition is a widely used approach in divide-and-conquer type of strategies to deal with complexity. A wide variety of strategies are in common use for accomplishing decomposition such as decomposition by structures, behaviors, goals and etc.

"Classic" SE methods apply the dominant decomposition along physical units (Fig. 2a). It is the natural match when complexity and cost is driven by physical units or their integration. That is the reason that traditional SE literature advises to build the WBS over the PBS [14].

Contemporary software engineering methods apply the dominant decomposition along operational and functional units (Fig. 2b). It is the natural match when the driver of complexity are use cases/features and their development.

Many other disciplines (e.g. rollout, logistics, and training) apply the dominant decomposition along topological units (Fig. 2c). It is the natural match when complexity and cost is driven by topographic units.

System developing companies or the companies who act as the system of systems integrators often apply the dominant decomposition along social units (Fig. 2d). It is the natural match when complexity and cost is driven by governance and by the need to manage a large amount of people coming from different departments.

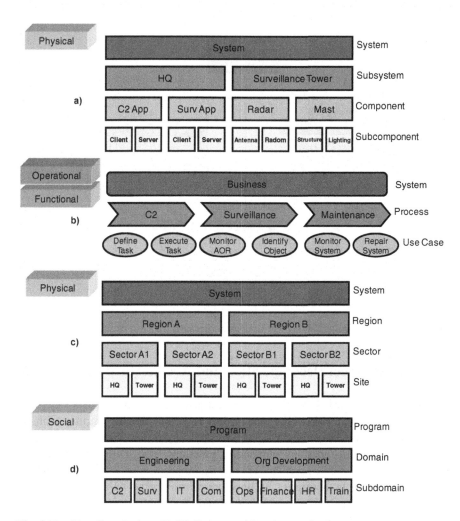

Fig. 2 Handling Complexity with 2D Decomposition Approaches

Each dimension of complexity is addressed with its own decomposition based on the primary stakeholders dealing with that dimension. As a result, in a single project, several decompositions are reflected as "separate" branches in the WBS (Fig. 3).

In large scale, socio technical, software intensive system of systems (SISOS) projects, such as border security projects, the classic dominant decompositions are no longer sufficient. The solution has many complex aspects. Therefore, no particular decomposition is dominant. Trying to coerce all aspects into one decomposition scheme is misleading and usually leads to interpretation misalignments.

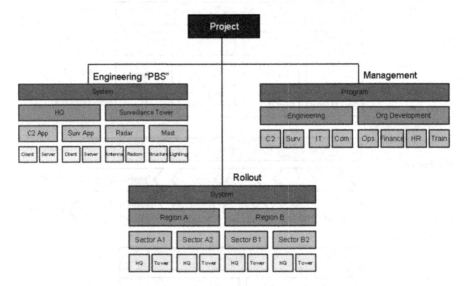

Fig. 3 Project Effort Reflects Different Decompositions

3   Multi-aspect Decomposition and Architecture Frameworks

An alternative to two dimensional classical decomposition schemes can be a multi-aspect decomposition. For one aspect (e.g. functional), starting with an abstract level, one can increase the level of detail by decomposition (e.g. system functions, subsystem functions, component functions, etc.). Different aspects can be introduced as semantic layers to separate decomposition in different domains as shown in Fig. 4. In enterprise architecture frameworks (AF) such a layered approach can be observed in terms of views and viewpoints.

Layered decomposition is a result of the top-down nature of traditional SE. In such a layered approach, dependency between elements of different layers

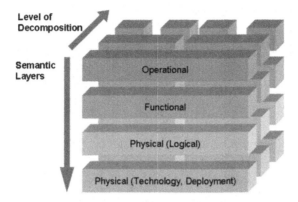

Fig. 4 Layered Decomposition Approach to Handle Complexity

represents a "cause-effect" or "problem-solution" relationship. Each layer depends on the layer above it (Fig. 5). Each element in a layer needs to be justified and validated against the layer above it.

This is a way of performing analysis and there is no problem with this approach if one prefers to view the problem in different layers or from different aspects. However, this approach becomes problematic if one tries to perform the SE tasks in the same manner that he/she analyzes the problem. This results in a waterfall like process and prevents systems engineers from focusing on value adding activities and prevents an engineering organization from effectively reacting to changing needs and constraints in each of -but mostly the top- design layers.

As shown in Fig. 6, in such a classic approach phases are linearly planned. In every phase a layer is completed and the results are captured in documents. For the next phase, the knowledge of this layer that is captured in the documents is an input for the next layer. Gray areas in Fig. 6 represent the unknown areas and colored areas represent documented knowledge.

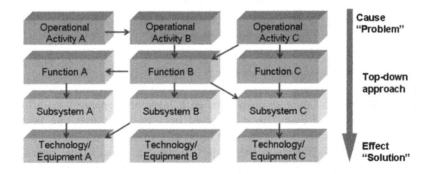

Fig. 5 Cause-Effect Relationship between Layers

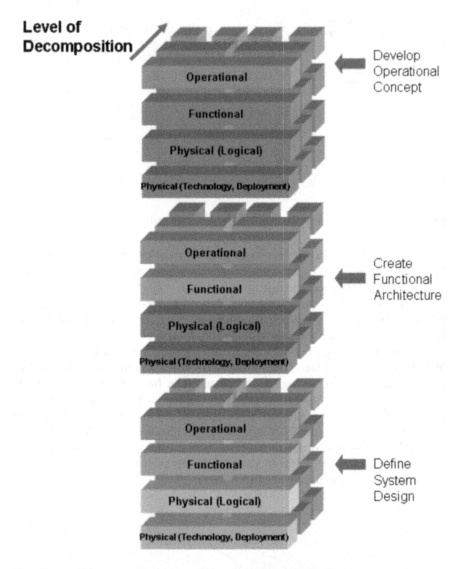

Fig. 6 Layered Decomposition Approach Results in Waterfall-like Process

The engineers that are responsible for each layer are separated in different teams. The WBS and plans are constructed in a way that first one layer is completed and then activities for the next layer start. Completion is determined by the release of the documents at the project milestones. Most of the inter-team interaction occurs at the milestone reviews where document hand over occurs. In non-MBSE environments this is usually accomplished by transforming one design model (e.g. the operational design) into textual requirements for the next design model, a process that by its manual nature is error-prone and costly.

Such a classical approach implicitly ignores the existing knowledge at the beginning of the project. Operational knowledge is assumed to be mostly captured in the contract and customer requirements. However, existing knowledge of the system functions, subsystems and technologies are ignored. Knowledge becomes available when it is captured in a formally released documentation. This problem is validated by investigating the SE tasks that are defined in the WBS of the 4 border security projects. SE tasks do not give you a clue about what is known by the SE team and what is not known. This is really a problem in planning and monitoring the SE activities.

A more realistic representation of the initial system knowledge is depicted in Fig. 7. Our discussions with the systems engineers revealed that required knowledge to develop the solution is not completely unknown at the beginning of the project. For various reasons such as customer preference, buying strategy of the company or legacy systems there were significant constraints (lower layer known knowns) on each of the layer of the system design at the very beginning of the project. Although Fig. 7 depicts a more realistic situation, it still does not emphasize the importance of the interaction/relationships between different knowledge domains (i.e. semantic layers).

3.1 Develop the Architecture like Solving a Rubik's Cube

System functions together with the operational activities constitute the interface between the operational domain and the technical domain. In other words, they bridge the gap between the problem space and the solution space. In SE, interface definition activities cannot be done by a single team but requires collaboration between the teams sitting on two sides of the interface.

Our approach to capture the operational activities and system functions is like solving a Rubik's cube. The correct strategy to solve a Rubik's cube is to focus on fixing all the faces at once. Beginners try to solve the faces one by one. Once they fix one face they think that this is a step forward. In reality it is almost impossible to fix other faces without breaking the already fixed one.

Architecture knowledge is like a Rubik's cube. It is not possible to justify knowledge in one domain (e.g. operational domain) in isolation from the other knowledge domains (e.g. functional and physical domains). Consider each face of the Rubik's cube as representing a knowledge domain (or a view/viewpoint from an enterprise architecture framework perspective).

In a traditional approach, different aspects of the solution are generally derived in a linear and sequential manner. This is also observable in the definition of the major milestone reviews like stakeholder requirements review, system requirements review, system design review, critical design review, etc. This is like trying to solve a Rubik's cube one face at a time.

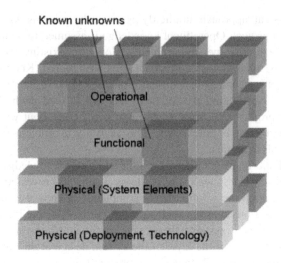

Fig. 7 Initial Knowledge at the Beginning of a Project is Spotty

For example, stakeholder requirements analysis phase ends when the operational knowledge is captured in the CONOPS document and in operational requirements document. Once these documents are completed (i.e. this operational domain face is fixed), system requirements phase starts and uses these documents as input.

However, according to the situation depicted in Fig. 7, we already have significant knowledge (i.e. known knowns) about existing system/subsystem functions, equipment and technology during the stakeholder requirements analysis phase. Traditional, top-down driven approach has a tendency to ignore this. As a result, operational architecture is developed with little understanding of the system and the system level decisions are made with little understanding of the subsystems (i.e. ignoring the existing knowledge in the next layer).

When the next layer's knowledge becomes available in the process, it becomes necessary to change the decisions made in the upper level. This is similar to the need to break the already fixed face of the Rubik's cube while trying to fix the next face.

The problem with this situation is, since the earlier decisions are generally the most critical ones, and changes in those decisions require significant rework, projects are reluctant to make changes in those decisions. In the best case scenarios where stakeholders are ready to make changes such changes bring significant costs as in large scale system projects this breaking and fixing faces cycles are long in duration due to the linear nature of the project phases. In other words you can never fix all faces together.

A variant we encountered in our case study occurred when – in an attempt to fix more than one side in a one-sided approach – high-level requirements were written using undocumented assumptions about implementation constraints, leaving subsequent readers with an incomplete and confusing picture. This led to conflict situations when during detailed design better design options were found, which would have achieved the system's overall purpose but which were are in conflict with the previously written requirements.

3.2 Organizational Problems in Knowledge Management

There is also an organizational aspect of this problem. System knowledge is bound to individuals and no individual team member has the complete picture of the solution. As demonstrated in Fig. 7, majority of the system knowledge space consists of the known knowns. This means that the required knowledge to design and describe the solution exists in the engineering organization. However, clusters of known knowns are distributed among the individuals and across the teams (Fig. 8).

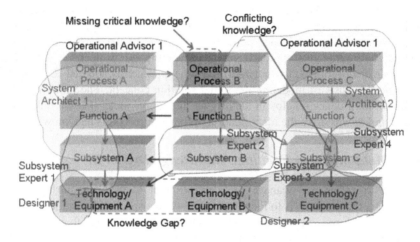

Fig. 8 Existing system knowledge is bound to individuals

It is observed that traditional team structures are not helping in this kind of situations. Such organizations are set up to work in fixed predetermined knowledge areas and not to address the particular knowledge gaps of a project. In static engineering organizations issue oriented collaboration is also hampered by each organizational branch choosing its own tools and methods. Instead, collaborative modeling environments should be preferred, which allow engineers to share knowledge and work on issues, such as they occur. Establishing short term knowledge circles should be preferred to erecting static organizations to bring necessary knowledge and skill resources to deal with engineering issues as they arise.

Success Factor: It is vital to determine the nature of a project by charting available and missing knowledge and understanding which available resources are needed to fill the knowledge gaps. Instead of fitting the problem(s) to pre-existing organizations the most efficient use of knowledge owners is to arrange them according to the problem.

4   Towards a Knowledge Driven Approach

In order to deal with the problems mentioned above, a knowledge driven process is defined as follows:

- Identify known knowns and known unknowns,
- Capture existing knowledge (i.e. known knowns),
- Acquire missing knowledge

This process can be applied at any level of the system development and in any project lifecycle. The critical point is to focus on the knowledge that prevents the teams from making timely decisions. However, one of the most critical lessons learned is to adopt this approach very early during the project. As a result, in border security projects, two phased life-cycle approach is adopted (Fig. 9). The first phase consists of modeling the initial solution concept. The initial modeling of the solution concept helps in identifying the "gray" areas (known unknowns). Known unknowns drive the identification of the required skills and knowledge set.

4.1   Meta-model to Facilitate Knowledge Driven SE

In order to define and manage SE activities that focus on acquiring missing knowledge, the first step is to identify the knowledge space of the solution (i.e. the socio-technical system) to be developed. This is the knowledge that we need for the definition and evaluation of the system. Part of the required knowledge is already known by us (i.e. KKs) and part of it is missing (i.e. KUs). As a result, the solution knowledge space represents the union of known knowns and known unknowns. As the project team has no control over the unknown unknowns, it is not necessary to include them in the knowledge model.

The question is how to describe your system. What are the alphabet, words and sentences that you will use in describing your solution architecture? There must be some knowledge elements as building blocks. As soon as these knowledge elements are defined, one needs to define the attributes of these elements and the critical relationships among these knowledge elements. That is how we can capture the knowledge in a consistent manner.

Case study showed us that meta-models can provide great support to SE teams at this point. Engineers from different backgrounds and organizations do misunderstand each other for a significant amount of time, because they associate different concepts with deceptively simple words (such as event). In [15] the necessity of establishing a ubiquitous language is elaborated in the context of customer-implementer relationships. We observed that the joint elaboration of a meta-model serves exactly the same purpose in the work relationship of engineers. It helps engineers to establish a common understanding about the key concepts and knowledge items in their shared problem.

A meta-model contains definition of architectural elements and all allowable relationships between those elements. An oversimplified meta-model is shown in Fig. 10. This meta-model has three architectural elements and two relationships defined among them. In a knowledge focused SE approach, such a meta-model can guide the systems engineers to develop complete and consistent knowledge on an operational activity. For example, according to Fig. 10, in order to define an operational activity, one needs to know which operational user performs the operational activity and which system functions are supporting the realization of that operational activity.

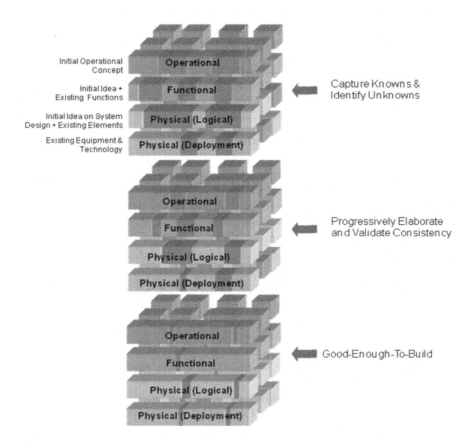

Fig. 9 Two Phased Life-Cycle Model Adopted in Border Security Projects

During the case study we have observed that the teams first try to start with an existing Architecture Framework (AF) but soon spend significant time on adopting the associated meta-models. This results in complex meta-models which are not used efficiently. AFs like DODAF, which created to ease strategic portfolio management in the defense domain, are beyond the scope of SE. This is very critical to understand. Only part of the AFs is within the responsibility of the SE teams, other parts are merely distracting. However, like the other concepts, understanding the main idea behind the meta-model and defining the expectations precisely helps solving the problems of systems engineers.

We also observed that formalism on the side of the meta-model enables automation and verification of a complex system model. As an example one project had a self-written tool which verified that all demanded relationships between model elements were satisfied. In the example above e.g. that each defined function support at least one activity.

Success Factor: A jointly elaborated and evolving formal meta-model is the foundation for creating complex knowledge models. Formalism enables automation and verification of a model.

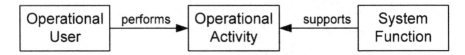

Fig. 10 Example of an Oversimplified Meta-model

4.2   *Model-Based Systems Engineering Is "Systems Engineering"*

Models and text-based documents are two different ways of representing knowledge. In the document based approaches the knowledge is captured in terms of requirements, design explanations, etc. and gathered in documents. These documents become the sources of necessary knowledge when a decision is to be made.

Representing the knowledge in terms of graphical models has many advantages over text based representations. This is one of the benefits promised by MBSE. However, it is critical to understand that MBSE is not the art of drawing diagrams. MBSE is model based + systems engineering. First of all it is SE. MBSE is a shift from document-centric to model-centric. SE part (the SE process) remains essentially the same. In other words, requirements that constitute a good application of SE remain unchanged.

Case study pointed out that in most of the cases systems engineers do not know what to expect from MBSE. That is one of the reasons that they define their MBSE attempts as unsuccessful. In reality, the part they were unsuccessful was not the MBSE but the SE. Modeling or MBSE can help you in communicating the existing knowledge in an easier and faster way. However, MBSE cannot acquire the missing knowledge for you. MBSE cannot perform the trade off analysis for you. If some knowledge is missing, it can be relatively easier to identify missing knowledge in MBSE. However, if prototyping is required to elicit some knowledge or some operational analysis is required to identify the needs of the users MBSE is not the solution for this.

During the interviews, all systems engineers stated that their main problem was to understand the customer's operating environment and their problems in this environment such as customer's business processes, organizational structure, decision hierarchy, user types, their interaction, etc. Without understanding the real operational problem of the customer, it is not possible to design a system to solve that problem. As a result, whether you adopt a document-based approach or a model-based one, the problem here is the missing knowledge to justify the SE decisions. Decisions (i.e. the architecture and design of the system) can be captured in models or in documents. A MBSE approach cannot turn bad decisions into good ones or MBSE cannot make a bad architecture a good one.

Success Factor: Apply SE principles to your own problems. First learn your problems related to SE and understand the benefits promised by MBSE. Clearly identify what kind of benefits you expect from MBSE and measure your projects against those expectations. Models reflect your SE process. Do not blame the mirror for your ugly processes.

4.3   Requirements First Approach vs. Requirements Last Approach

Traditional SE approaches are also known as "document-driven" or "document-heavy" processes. We observed that producing a lot of documentation is not the main problem of the traditional SE processes. Every piece of information in the documents somehow reflects a decision. Systems engineers need to deliver the documents with the required decisions within a pre-determined timeframe ended at a project milestone. In this approach they are expected to make the decisions when the necessary information is missing. The rest of the design and development activities (and the associated documentation) are based on such ill-made decisions. Systems engineers state that most of their time is spent on changing the decisions when the required information is available and re-working on the documents in the later phases of the projects.

Findings also indicate that the existing processes are weak in measuring the quality and completeness of the documents. Systems engineers use most of their time to write further requirements on the problem domain that they are familiar with. In such an approach while the details of the well-known areas increase with a lot of redundant requirements, the unknown space remains to be unknown. Time is used to increase the thickness of the documents. However, knowledge creation and hence the value provided to the systems engineers is not directly proportional with the thickness of the documents.

In order to deal with such issues, a requirements last approach is adopted. This is in contrast to the traditional approach where every activity starts with a set of requirements from the layer above. In this completely model driven approach no requirements are written until the model is frozen. All the knowledge management activities are performed on the model. Existing knowledge is captured in terms of models that are supported by textual explanations when necessary. Until it is evaluated that the knowledge captured in the model repository is mature and stable enough no requirements are written. As the model is frozen, "shall" bases textual requirements are derived from the knowledge that is captured in the model. We have experienced that this approach reduced rework significantly and increase the agility of SE activities.

The more the various design models and semantic layers are combined into one holistic model of the system, the less need there actually is for classic requirements writing. In the classic approach textual requirements are merely an extract from one design model with the intent of allowing the generation of another design model. In an integrated design model this activity can become totally superfluous, as causality is expressed through model element relationships.

Success Factor: Convey the message that requirements (or any other text based design document) are not the ultimate goal. Documents are tools to communicate knowledge. Experience showed us that documenting stable knowledge does not take time. First stabilize the common and consistent knowledge through models and then generate the documents. This reduces the time lost in updating the documentation.

4.4   Definition of the Problem Is Part of Solution Development

Customers perform their daily missions based on their operational processes. All the processes are decomposed down to task level where every task can be allocated to an organizational unit (whether it is an individual or it is a team). They have difficulties in achieving these tasks. Therefore, they ask for technical support from the system developers. Their main interest is the efficiency of their organization. They are not interested in the outputs of the technical systems only but the output of the socio-technical system including (but not limited to) the processes, operational users and technical systems.

In the traditional SE understanding, customer is expected to perform all the necessary analysis activities before the project to identify their problem. The problem must be captured in the contractual requirements and maybe in the accompanying ConOps document. Systems engineers are expected to develop the solution based on this well-defined and well-scoped problem definition. On the other hand this is never the case and this deadly assumption is one of the main reasons for classical SE approaches to be unsuccessful.

This assumption is still widely accepted in defense industry projects. Following are some of the findings from a recent case study [16] that was performed with 59 senior systems engineers from 10 different defense companies (4 from Turkey, 3 from Europe, 2 from USA and 1 from Australia) and from 18 different projects:

- Only %24 of the systems engineers stated that they have a defined ConOps development (or a similar one) process in their organization.
- None of the projects/organizations has a defined process (or a guideline) for operational scenario development, prototype development or stakeholder analysis.
- None of the organizations has a business analyst/business process engineer (or a similar) role in the team.
- In any of the projects, there isn't a task defined in the WBS to capture and document the existing business/operational processes of the customer.

Success Factor: It is the responsibility of the system developer to understand the operational problem and justify the solution against the problem. Operational problem is never fully and consistently defined in the contract. Project-wide visibility of this fact can be achieved by explicitly defining work packages in the WBS for capturing the operational domain knowledge.

4.5   Systems Engineers Are Not Business Analysts

Including tasks in the WBS for eliciting and capturing the operational problem does not solve the problem. Teams and individuals are needed to be allocated to these tasks. Assigning these tasks to SE teams is of course one of the options. However, the case study revealed that it is not easy to perform these tasks effectively with traditional SE teams.

First of all, systems engineers do not have the necessary operational domain knowledge. Secondly, systems engineers are engineers. Engineers are not educated for identifying the problems but solving "well-defined" problems. In order to elicit the operational needs of the customers, one needs very good communication and question asking skills. The knowledge resides on the customer side and you will get what you ask for. Asking a question like "what is your problem" will not help you at all.

Success Factor: Operational domain must be handled by the business analysts. If these tasks will be performed by the SE team, develop business analysis skills in your SE competency framework. Define specific roles for analyzing the problem space.

4.6   Facilitate Transition to MBSE by Modeling Experts

Case study revealed that using modeling experts increase the acceptance of the MBSE approach by the systems engineers. Instead of asking the engineers to start modeling from the first day of the new approach, we used modeling experts to perform the modeling activities. Systems engineers are used as the knowledge sources and analysis experts. Modeling experts asked the questions to collect the necessary information from the systems engineers and developed the models based on the knowledge provided by the systems engineers. When the systems engineers do not feel the pressure to learn new modeling languages and new modeling tools they do not react the MBSE approach. In time, they develop skills to use the language and the tool themselves.

5   Conclusions

In EADS, both architecture frameworks and MBSE are considered as major areas in the future of contemporary SE practices. For an organization, adoption of new approaches such as MBSE requires careful consideration of the introduction mechanisms. In this paper, we shared our experiences towards adopting MBSE in border security projects where EADS performs the role of lead systems integrator. From the practitioner's point of view, lessons learned presented in this paper can help organizations to increase the speed of shift from traditional SE approaches to architecture-driven, knowledge-focused and model-based SE practices.

References

[1] Brooks, F.P.: No Silver Bullet. IEEE Computer 20(4), 10–19 (1987)
[2] Wade, J., Madni, A., Neill, C., Cloutier, R., Turner, R., Korfiatis, P., Carrigy, A., Boehm, B., Tarchalski, S.: Development of 3-Year Roadmap to Transform the Discipline of Systems Engineering. Final Technical Report-SERC-2009-TR-006, Systems Engineering Research Center (2010)
[3] Highsmith, J., Cockburn, A.: Agile software development: the business of innovation. IEEE 34(9), 120–127 (2001)
[4] Gilb, T.: Competitive Engineering. In: Competitive Engineering. Elsevier (2005)
[5] Boehm, B.: Some future trends and implications for systems and software engineering processes. Systems Engineering 9(1), 1–19 (2006)
[6] Turner, R.: Toward Agile Systems Engineering Processes. CrossTalk The Journal of Defense Software Engineering, 11–15 (April 2007)
[7] Kennedy, M.R., Umphress, D.A., Ph, D.: An Agile Systems Engineering Process The Missing Link? CrossTalk The Journal of Defense Software Engineering 4(3), 16–20 (2011)
[8] Bosch, J., Bosch-Sijtsema, P.: From integration to composition: On the impact of software product lines, global development and ecosystem. Journal of Systems and Software 83(1), 67–76 (2010)
[9] ISO and IEC (International Organisation for Standardisation and International Electrotechnical Commission). ISO/IEC 15288. Systems and software engineering-System life cycle processes (2008)
[10] Haskins, C. (ed.): Systems Engineering Handbook: A Guide for System Life Cycle Processes and Activities. Version 3.2. INCOSE, San Diego (2010), revised by M. Krueger, D. Walden, and R. D. Hamelin.
[11] Asan, E., Bilgen, S.: Agile collaborative systems engineering-motivation for a novel approach to systems engineering. INCOSE, Rome (2012)
[12] Maier, M., Emery, D., Hilliard, R.: ANSI/IEEE 1471 and Systems Engineering. Systems Engineering 7(3) (2004)

[13] INCOSE, Systems Engineering Vision 2020, INCOSE (2007)

[14] NASA, Systems Engineering Handbook (2007)

[15] Evans, E.J.: Domain-Driven Design: Tackling Complexity in the Heart of Software. Addison-Wesley Longman (2003)

[16] Asan, E., Bilgen, S.: Agility problems in traditional systems engineering – a case study. In: CSDM, Paris, pp. 53–70 (2012)

The Multidimensional Hierarchically Integrated Framework (MHIF) for Modeling Complex Engineering Systems

Ahmad Alabdulkareem, Anas Alfaris, Vivek Sakhrani,
Adnan Alsaati, and Olivier de Weck

Abstract. In recent years, the interest in modeling complex engineering systems has grown rapidly. Due to their highly nonlinear nature coupled with their critical importance, the task of modeling such systems becomes a high-risk, high-payoff endeavor. There are many complexities associated with understanding and modeling the dynamics of such systems, these include functional, spatial and temporal complexities. The main objective of this paper is to present a modeling framework that enables the planning and evaluation of complex engineering systems taking into account these multidimensional complexities. The framework uses a multidimensional hierarchical decomposition of complex engineering systems combined with modeling techniques such as System Dynamics and Agent Based Modeling. Decomposition and modeling are integrated to simulate possible future states of the system and to evaluate different plans and scenarios for system evolution using key performance indicators (KPIs). The paper also presents a case study demonstrating the use of the proposed framework for an integrated water model applied to the kingdom of Saudi Arabia.

1   Introduction

Complex Engineering Systems, such as water and electricity grids have a direct impact on the social, environmental, and economic health of society and are critical for the quality of life [2]. These systems are interconnected, large-scale, and complex requiring rigorous planning to maximize their sustainability. Such

Ahmad Alabdulkareem · Adnan Alsaati
King Abdulaziz City for Science and Technology, Saudi Arabia
e-mail: kareem@mit.edu, a.alsaati@cces-kacst-mit.org

Vivek Sakhrani · Anas Alfaris · Olivier de Weck
Massachusetts Institute of Technology, United States
e-mail: {Sakhrani,Anas,deweck}@mit.edu

M. Aiguier et al. (eds.), *Complex Systems Design & Management 2013*,
DOI: 10.1007/978-3-319-02812-5_22, © Springer International Publishing Switzerland 2014

complex systems evolve over time from a current state into uncertain future states as a consequence of different decisions and policies. During the planning phase, decision-makers are faced with the daunting task of choosing the 'right' decisions and evaluating their different impacts, gains, and repercussions in terms of sustainability. Models that can take proposed plans for a complex engineering system, simulate its future states, and provide metrics for evaluation can be very useful in accomplishing this task. Decision-makers can use these models to derive behaviors and knowledge from data and relationships for their planning processes [20], to develop more robust and resilient systems. However, modeling complex engineering systems is a complex process in itself.

The paper is structured as follows. The following 'Background' section discusses the motivation for developing the Multidimensional Hierarchically Integrated Framework (MHIF). The subsequent 'Framework' section describes the methodology and implementation process in detail. An application of the framework for studying a complex water infrastructure system then follows. Finally, the last section contains a summary of the work, identifies next steps, and areas for future research.

2   Background

Complex Engineering Systems are multi-dimensionally complex. One dimension of such complexity is the *functional complexity* denoting the linkages and interdependencies between different infrastructure networks [6]. For example, housing and urban structural layers depend on underlying water and electricity networks, which in turn are linked to each other. Information and telecommunications (ITC) networks are embedded within the housing layer as well as other networks (Figures 1 and 2). In previous work, it has been shown that these different systems can be modeled by representing their intra-layer and inter-layer linkages through Design Structure Matrices (DSM) and Domain Mapping Matrices (DMMs) [1]. Functional interdependencies could thus be classified as physical interdependencies (e.g. material or energy flows between electricity, water networks), cyber interdependencies (i.e. information, controls and instrumentation), geospatial interdependencies (i.e. co-location of infrastructure elements), or even logical interdependencies (i.e. economic, environmental, or social links) [13].

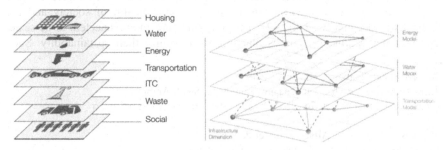

Fig. 1 Many Infrastructures   **Fig. 2** Inter-infrastructure complexity

Another dimension of complexity within complex engineering systems is *spatial complexity*. For example, large scale infrastructure networks could be distributed and disaggregated over different spatial segments, such as regions, cities, or even neighborhoods (Figure 3).

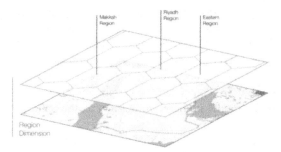

Fig. 3 Spatial complexity

Finally, another dimension of complexity that needs to be considered is in the *temporal dimension,* where each point in time for the system is determined by historical decisions and events. However, since time flow is unidirectional, most models simplify the problem by the Markov assumption that states: *"the future is independent of the past given the present"* [7]: $\left(S_{t+1}S_{(0:(t-1))}|S_t\right)$.

A number of frameworks have been developed to tackle different dimensions of system complexity. Several of these frameworks focus on some specific dimensions of the system [12,16,18]. For example, the Complex Large-scale Integrated Open System process ("CLIOS PROCESS") is a process for both analyzing and managing the complex sociotechnical systems, enabling users to go from the problem and goal identification to the implementation and adaptation of strategic alternatives [18]. Another framework is the DPSIR (Driving forces, Pressure, State, Impact, and Response) framework which is used to analyze systems and develop and categorize different system indicators [16]. However both frameworks do not develop a method for producing mathematical models to test out different scenarios and decision alternatives for supporting decision-making.

Our proposed framework takes a holistic approach to modeling complex engineering systems. The framework integrates several modeling techniques to accomplish the task at hand. These include System Dynamics (SD) [17] and Agent Based Modeling (ABM), which in recent years has been gaining traction, and has now sufficiently matured [5]. SD & ABM have been compared in the literature [5, 15] to highlight their similarities and differences, but have also been integrated in multiple ways because of their complementarities [14, 19].

3   The Proposed Framework

Within the MHIF, a system is represented by a series of *States $(S_{t0}, S_{t1} \dots S_{tn})$*. Each state S_x consists of four main components, which include: *Decision*

Variables (DVs), General Drivers (GDs), Model Variables (MVs) and Key Performance Indicators (KPIs). Inputs to the model are both the *DVs* and *GDs*, while KPI*s* are outputs, and MVs are internal model variables. To illustrate the different components, let us consider a simple model for planning a national water system.

DVs are within the control of the decision maker and define the plans that need to be evaluated; the 'what if scenario' to be tested. In our illustrative water model, this could be the need to test a proposed plan to increase water production capacity. GDs are parameters that affect the state of the system but are exogenous and out of the decision maker's control, e.g. population growth. MVs are internal model variables that are not observed by the user (except for their initial values which are inputs), but are calculated for certain equation needs of the model. A state at time *t* in the model is therefore defined as the values of DVs, GDs and MVs.

The *KPIs* are predefined indicators that are calculated within the model and are exposed to help evaluate the system at a particular state. In our illustrative example a KPI could be CO_2 emission levels. The KPIs can include different metrics (economic, environmental, social, or technical). This enables decision makers to evaluate and compare different simulation runs and choose the appropriate plan, or define new DVs. This is an iterative process, where different plans and scenarios are tested, evaluated, and compared leading to a more robust planning process. It is possible to generate new hybridized scenarios or split larger scenarios into sub-scenarios over multiple rounds of running the MHIF.

Consider the following mathematical formulation of how the different variables would interact within a model:

$$Inputs = \{DV, GD, MV_0\}$$

$$DV = [DV1 \quad \dots \quad DVh] = \begin{bmatrix} DV1_1 & \cdots & DVh_1 \\ \vdots & \ddots & \vdots \\ DV1_n & \cdots & DVh_n \end{bmatrix} = \begin{bmatrix} DV_1 \\ \vdots \\ DV_n \end{bmatrix}$$

$$GD = [GD1 \quad \dots \quad GDj] = \begin{bmatrix} GD1_1 & \cdots & GDj_1 \\ \vdots & \ddots & \vdots \\ GD1_n & \cdots & GDj_n \end{bmatrix} = \begin{bmatrix} GD_1 \\ \vdots \\ GD_n \end{bmatrix}$$

Internal Model Variables:

$$MV_t = [MV1_t \quad \dots \quad MVm_t]$$

$$MVx_{t>0} = f(DV_t, GD_t, MV_{t-1}, MVx_t^{-1}) \quad : f \ is \ a \ model \ function$$

$$MV = [MV1 \quad \dots \quad MVm] = \begin{bmatrix} MV1_1 & \cdots & MVm_1 \\ \vdots & \ddots & \vdots \\ MV1_n & \cdots & MVm_n \end{bmatrix} = \begin{bmatrix} MV_1 \\ \vdots \\ MV_n \end{bmatrix}$$

State at time t:

$$S_t = \{DV_t, GD_t, MV_t\}$$

KPIs and Outputs:

$$KPI = [KPI1 \quad ... \quad KPIk] = \begin{bmatrix} KPI1_1 & \cdots & KPIk_1 \\ \vdots & \ddots & \vdots \\ KPI1_n & \cdots & KPIk_n \end{bmatrix} = \begin{bmatrix} KPI_1 \\ \vdots \\ KPI_n \end{bmatrix} \in MV$$

$$Outputs = \{KPI\}$$

Where,

MV_0: Initial values of model variables (the only MV that is an input)

n: number of time steps

DV: Decision Variables' vector,   h: number of DV variables,

GD: General Drivers' vector,   j: number of GD variables,

$MVx_{t>0}$: The xth variable of MV at time t, while $t > 0$

x: Model variable index $1 \leq x \leq m$

MVx_t^{-1}: The model variables' vector at time t (MV_t) excluding MVx_t or any MVy_t that would lead to referencing MVx_t (preventing circular referencing)

MV: Model Variables' vector,   m: number of MV variables,

S_t: List of all variables at time t

KPI: Key Performance Indicators' vector

k: number of KPIs where $1 \leq k \leq m$

Each of the model vectors DV, GD, MV, and KPI is of size $n \times h$, $n \times j$, $n \times m$, $n \times k$ respectively. For each one of the model vectors V (*either DV, GD, MV, or KPI vector*), there are (h, j, m, or k respectively) variables Vx (columns), each Vx is a time series vector of the variable's value over time, and V_t (row) is the value of all of the V variables at time t. The KPIs are a subset of the variables within MV, but spanning the same time steps (from 1 to n).

In our illustrative water model, let us consider the production capacity (Cap) decision variable, while the general drivers are population Growth Rate (GR) and Consumption Per Capita (CPC), and the model variables are Demand (D), Supply (S), and Population (Pop). Finally, the key performance indicator of interest Shortage (Sh)[1]:

$$Inputs = \{DV, GD, MV_0\}$$

$$DV = [Cap] = \begin{bmatrix} Cap_1 \\ Cap_2 \\ Cap_3 \end{bmatrix} = \begin{bmatrix} DV_1 \\ \vdots \\ DV_n \end{bmatrix}$$

$$GD = [GR \quad CPC] = \begin{bmatrix} GR_1 & CPC_1 \\ GR_2 & CPC_2 \\ GR_3 & CPC_3 \end{bmatrix} = \begin{bmatrix} GD_1 \\ \vdots \\ GD_n \end{bmatrix}$$

---

[1] Generally the goal is to design an infrastructure system to drive shortage close to zero either from above or below (negative or positive capacity margins).

$$MV_0 = [Pop_0 \quad D_0 \quad S_0 \quad Sh_0] \text{ (Input)}$$

$$MV_t = [Pop_t \quad D_t \quad S_t \quad Sh_t]$$

$$Pop_{t>0} = f(Pop_{t-1}, GR_t) = Pop_{t-1} * \left(1 + \frac{GR_t}{100}\right)$$

$$D_{t>0} = f(Pop_t, CPC_t) = Pop_t * CPC_t$$

$$S_{t>0} = f(D_t, Cap_t) = min\,(D_t, Cap_t)$$

$$Sh_{t>0} = f(D_t, S_t) = max\,(0, D_t - S_t)$$

$$MV = [Pop \quad D \quad S \quad Sh] = \begin{bmatrix} Pop_1 & D_1 & S_1 & Sh_1 \\ Pop_2 & D_2 & S_2 & Sh_2 \\ Pop_3 & D_3 & S_3 & Sh_3 \end{bmatrix} = \begin{bmatrix} MV_1 \\ \vdots \\ MV_n \end{bmatrix}$$

$$KPI = [Sh] = \begin{bmatrix} Sh_1 \\ Sh_2 \\ Sh_3 \end{bmatrix} = \begin{bmatrix} KPI_1 \\ \vdots \\ KPI_n \end{bmatrix} \in MV$$

$$Outputs = \{KPI\}$$

In terms of the multi-dimensional complexity described previously, the functional complexity is handled by decomposing the different systems into different layers. In the spatial decomposition, the functional layers are distributed spatially into segments; such that some of the functional components in different spatial sectors could have different parameters (e.g. different spatial regions have different production capacities). The temporal complexity is captured in the different states of the system.

At this stage, within a specific system functional layer and spatial segment, the governance of resources and their flows need to be maintained. For example, in the illustrative model these would include supply or demand components that represent a plant and a city. Figure 4 illustrates that sub components can cross feed (i.e. supply from a specific spatial segment feeds demand from another segment), to allow for the network to be represented and simulated while still keeping the hierarchical decomposition intact. Figure 4 contains only spatial cross feeding (e.g. water being pumped from region A to region B), while functional cross feeding is possible as well (between functional layers, e.g. diesel fuel required to power irrigation pumps).

When addressing multidimensional complexities, models can have different scales, ranging from a micro-scale (accurate depictions of reality that have high

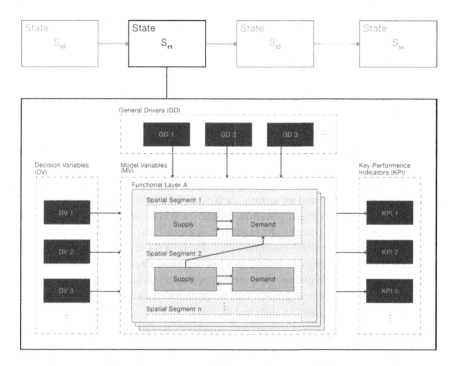

Fig. 4 An illustration of a system state in the conceptual MHIF model

granularity), to a macro-scale (where the high level dynamics are captured to produce useful system insights and evaluations that in most cases have a low granularity). A wide spectrum of granularities, also called meso-scale, exists between these two extremes. How coarse grained or how aggregated the different decompositions in the model are, depends on the objectives of the decision-makers. The proposed MHIF lends itself to the user's preference of which scale is desired.

After hierarchically decomposing and identifying the scale and the building blocks of the model, the next step is to implement them as autonomous integrated agents in an Agent Based Model (ABM) environment. This enables us to separately model each component, and define the rules by which it will interact with other homogeneous or heterogeneous agents. The cognitive structure of the agents can be modeled in different ways. In our framework System Dynamics (SD) models are chosen to represent the abstract high-level dynamics of the system component. The values of an agent's SD represent the agent's state, and they can depend on endogenous or exogenous agent variables which constitutes some of the links between interdependent agents. This enables us to reach different levels of granularity by the scale of these building blocks (Table 1).

Table 1 Characteristics of different scales

Model Scale	Techniques used	Focus	Spatial Scale	Temporal Scale (time horizon)
Macro	SD	high level dynamics	National	up to ~30 years
Meso	SD+ABM	mid-level interdependencies	Regional	up to ~15 years
Micro	SD+ABM	course grained / accurate depictions of reality	Urban	up to ~3 years

In our illustrative example, the functional elements of the modeled system (plant, city) are represented as agents, each with their own embedded SD models. The benefit of ABM is the replication capability within different spatial or functional segments. For example, different regions can have multiple 'city demand' agents replicated to simulate multiple cities. Each of these agents could have customized parameter settings to reflect local or regional peculiarities. In addition, it is possible to have different agents contain uncertain random variables to simulate uncertainty (e.g. population growth can be probabilistically distributed and have different random values within different agents).

Figure 5 shows two agent's cognitive structures modeled as an SD that correspond to water supply and demand components within the illustrative model. These SD models simply project demand following specific population growths and evaluate the effectiveness of the supply side expansion plan.

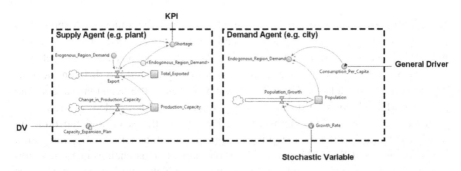

Fig. 5 SD implementation of the cognitive structure of the functional agent components

The agents shown in Figure 5 contain DVs (e.g. capacity expansion plan), GDs (e.g. 'stochastic' population growth, or consumption per capita), MVs (e.g. exogenous demand 'link from another agent'), and KPIs (e.g. 'shortage', i.e. the unmet demand for water). The inherent stochasticity in the sub-systems suggests that outputs will not be identical for every run of the model, even if the DVs (inputs) are held constant. Therefore, Monte Carlo simulations are carried out to show the different impacts and possibilities that a single input plan (DV) can have (2d-histogram Figure 6).

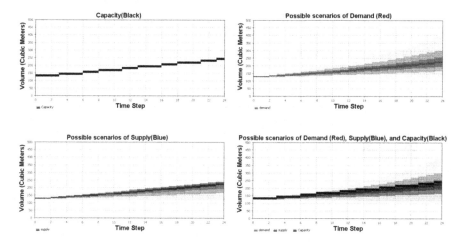

Fig. 6 2d-histogram of a Monte Carlo simulation for the illustrative model

As can be seen in Figure 6, most scenarios in the illustrative model have demand and supply below the supply capacity line (DV), but in some scenarios demand can go above that line (creating shortages). This means the decision maker would choose a level of certainty that is acceptable for that event.

4   The Integrated Water Model Case Study

To demonstrate the proposed framework, we applied it to investigate national and regional water capacity expansion as part of the larger Sustainable Infrastructure Planning System (SIPS) Project for the Kingdom of Saudi Arabia. The spatial resolution was decomposed into 13-regions that represent the political regions within the kingdom. Figure 7 shows a network representation of the final model implementation for only two regions with a sample of their special crossing links. The objective of the model is to investigate how balanced regional development can be achieved taking into consideration the supply side components: groundwater (g), wastewater treatment (wt), desalination (d) and demand side components: agricultural (a), municipal (m), and industrial (i) uses.

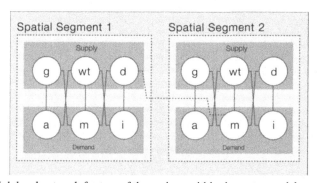

Fig. 7 The high level network for two of the regions within the water model

Figure 8 is a zoom-in of the supply-side components, where all the supply components are integrated into one SD structure for illustration. Demand is similarly structured, but is shown as a black box in the diagram.

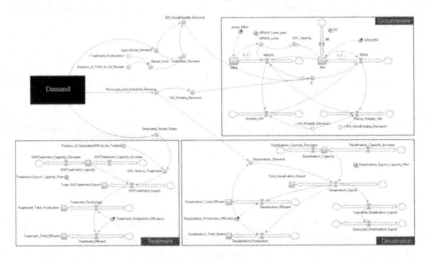

Fig. 8 Supply side components, with their SDs connected for illustration purposes

In its actual decomposed form, each component resides within a separate agent, which can be replicated to represent as many homogenous elements as needed, allowing for higher granularity and more realistic stochastic variability. For example, the agent representing groundwater (g) in a "region" of the model contains a groundwater SD structure, as shown in Figure 9. Agents are replicated as many times as necessary across regions and connected to other agents (e.g. water treatment, or agricultural demand) to reflect the different possibilities and physics-based constraints of the real world systems.

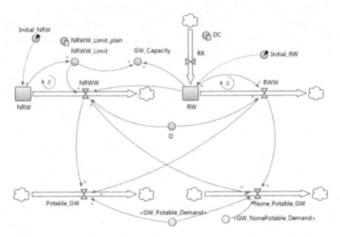

Fig. 9 SD of the groundwater (g) agent. Zoom-in of the upper right region in Figure 8.

Within an agent's SD model, the governing equations of the flows and auxiliary variables (MV) contain DVs, GDs, and links to other components (other agents' MVs). Consider this example for the state of the system at time step t:

$$RR_t = DC_t + normal(RFm_t, RFstd_t)$$
$$RWW_t = \min(D_t, RW_t)$$
$$RW_t = RW_{t-1} + RR_{t-1} - RWW_{t-1}$$
or (following the previously presented mathematical formulation):
$$MVa_t = DVb_t + normal(GDc_t, GDd_t)$$
$$MVe_t = \min(MVf_t, KPIg_t)$$
$$KPIg_t = KPIg_{t-1} + MVa_{t-1} - MVe_{t-1}$$

Where:

RR_t : Recharge Rate at time t (MV)
DC_t : Dam Capacity at time t (DV)
$normal(m, sd)$: Gaussian random number generator with mean = m, standard deviation = std (stochasticity)
RFm_t: Rain Fall mean at time t (GD)
$RFstd_t$: Rain Fall standard deviation at time t (GD)
RWW_t: Renewable Water Withdrawal at time t (MV)
D_t: Sum Demand on this groundwater agent at time t (MV)
RW_t: Renewable Water quantity at time t (KPI $\in$ MV)
a,b,c,d,e,f,g: variable indices

D_t, the total demand for groundwater at time t, is a variable that links this agent to other agents (its value is an aggregate of demands from agents using this agent's resources). An initial implementation of the Integrated Water Model was done in "Anylogic" which provided a platform through which ABM and SD could be seamlessly integrated. Based on the chosen capacity expansion schedule (DV) for desalination within one of the regions, the model produces a set of graphs representing the response of system to that scenario. Figure 10's plot (A) shows the interaction of a few model variables pertaining to desalination, and plot (B) shows the 'shortfall' variable (KPI) from a Monte Carlo simulation run, where: shortfall = max(0,demand-supply).

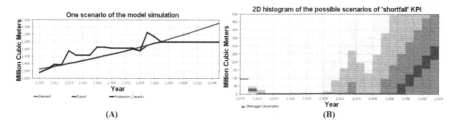

Fig. 10 (A) Capacity (Black), Export (Blue), and Demand (Red). (B) Shortfall.

We observe that the probability density of shortfalls is much higher further out in time. This is consistent with the intuition that capacity addition plans will need to be reevaluated for a longer time horizon, beyond the year 2026, based on new demand projections.

5   Conclusion

In this paper, the Multidimensional Hierarchically Integrated Framework (MHIF) was presented to concurrently address functional, spatial and temporal complexities that are inherent in Complex Engineering Systems. The framework argues for a multidimensional hierarchical decomposition. It also identifies taxonomy of the different variables within the system model, linking them together in a mathematical formulation and describing the information flow. The implementation of the framework is carried out in an Agent Based Model and System Dynamics integrated environment. The framework can be used to model complex engineering systems such large infrastructure systems, simulating the effects of decision makers' plans on future progression of those systems, and evaluating the results to allow for comparisons to be done between different plans and scenarios. The framework was then applied to an Integrated Water Model case study. Validation and verification of the models developed by the framework are still needed, yet the initial model development and testing phases show the potential for a powerful framework. Preliminary alpha-tests with several stakeholders elicited many comments and some suggestions for changes to the model. These changes included additions such as conveyance (pumping station components) and new Indicators (virtual water returns from the agricultural sector) to add to the overall usability and functionality of the system. The alpha-tests show that such models can provide a fast and relatively cheap method for evaluating different infrastructure plans, can assist in giving insights, and works well in demonstrating the high level dynamics of the system.

Acknowledgments. The authors would like to thank King Abdulaziz City for Science and Technology (KACST) for funding this work and the Center for Complex Engineering Systems (CCES) at KACST and MIT for their support. The authors would also like to thank Mohammed Hadhrawi and Waleed Gowharji for their technical assistance.

References

[1] Alfaris, A., et al.: Hierarchical decomposition and multi-domain formulation for the design of complex sustainable systems. Journal of Mechanical Design (2010)
[2] Aschauer, D.A.: Why Is Infrastructure Important (1989)
[3] Bar-Yam, Y.: Dynamics of complex systems (1997)
[4] Bartolomei, J.E., Hastings, D.E., de Neufville, R., Rhodes, D.H.: Engineering Systems Multiple-Domain Matrix: An organizing framework for modeling large-scale complex systems. Syst. Engin. (2012)

[5] Borshchev, A., et al.: From System Dynamics and Discrete Event to Practical Agent Based Modeling: Reasons, Techniques, Tools (2004)

[6] de-Weck, et al.: Engineering systems: Meeting human needs in a complex technological world. MIT Press, Cambridge (2011)

[7] Koller, D., et al.: Probabilistic Graphical Models: Principles and Techniques (2009)

[8] Macal, C.: To Agent-Based Simulation From System Dynamics. In: Proceedings of the 2010 Winter Simulation Conference (2010)

[9] Moldan, B.: How to understand and measure environmental sustainability: Indicators and targets. Ecological Indicators (2012)

[10] Moldan, B.: Structuring problems in sustainability science: The multi-level DPSIR framework. Geoforum (2012)

[11] Niemeijer, D., et al.: A conceptual framework for selecting environmental indicator sets. Ecological Indicators (2012)

[12] Pederson, P., et al.: Critical Infrastructure Interdependency Modeling: A Survey of Critical Infrastructure Interdependency Modeling (2006)

[13] Rinaldi, S.M.: Modeling and Simulating Critical Infrastructures and Their Interdependencies (2004)

[14] Schieritz, N.: Integrating System Dynamics and Agent-Based Modeling

[15] Schieritz, N., et al.: Modeling the Forest or Modeling the Trees A Comparison of System Dynamics and Agent-Based Simulation (2001)

[16] Smeets, E., et al.: Environmental indicators: Typology and overview. Prepared by: Project Managers

[17] Sterman, J.D.: System Dynamics Modeling: Tools for Learning in a Complex World (2001)

[18] Sussman, J.M., et al.: The " CLIOS PROCESS" (2009)

[19] Teose, M., et al.: Embedding System Dynamics in Agent Based Models for Complex Adaptive Systems (2007)

[20] Ullman, D.G.: Making Robust Decisions: Decision Management For Technical, Business, and Service Teams

Natural Systems Engineering

Derek Hitchins

Abstract. Natural systems have evolved on Earth over some 542MY, through trial and error, by natural selection, with such exemplary results as the human body, one of the most complex organisms on the planet.

The human body exhibits different 'levels of organization:' organization appears at each level. 'Nature's design' evidences a higher degree of coupling and mutual interdependence between structures at the same level of organization, for example between organ systems, than is the usual case for supposedly equivalent manmade systems. The human brain also exhibits levels of organization, but these do not find correspondence in manmade processing systems, while the operation of the brain's remarkable memory is still uncertain, with many clues, but conflicting ideas amongst neuroscientists.

Homeostasis in organisms can be singularly complex, with stabilizing 'mechanisms' that differ markedly from those employed by cyberneticists and engineers: could the latter learn to advantage from successful, natural systems?

Nature has also created extensive insect social systems, so-called 'super organisms.' Honeybees have existed for more than 100MY, ants over 120MY and termites over 150MY, suggesting they have a variety of successful survival strategies. Modern humans have been on the planet barely 2MY so far...and already threaten their own survival together with that of many other species.

Overall, Nature has an enviable 'track record' of evolving efficient, adaptable, effective, survivable organisms and super-organisms that exist in more-or-less mutual harmony with other organisms and super-organisms.

Biomimetics, similar to biologically inspired design, is the study of the structure and function of biological systems as models for the design and engineering of materials and machines. Together with ecology, biomimetics (Bar-Cohen, 2011) and bio-mimicry (Benyus, 1997) may also offer sophisticated models, processes and procedures for:

Derek Hitchins
144 Castle Street
Salisbury Wilts, SP1 3UA, UK
e-mail: prof@hitchins.net

M. Aiguier et al. (eds.), *Complex Systems Design & Management 2013*,   315
DOI: 10.1007/978-3-319-02812-5_23, © Springer International Publishing Switzerland 2014

- systems thinking,
- systems conception,
- systems design,
- systems architectures,
- systems lifecycles,
- system survival strategies

… and many more as yet largely untapped. Can such models be employed to advantage for manmade systems, human systems, societal systems, business and industry systems, systems engineering, even economic systems? Are they applicable, can they be trusted, if they are different, *why* are they different, is their efficacy amenable to proof? These are questions to ponder, while at the same time admiring the legacy and insight the models may provide.

1   Levels of Organization

Biologists and anatomists identify so-called 'Levels of Organization' within living things. Levels of Organization describe the way in which organisms are synthesized, starting with the cell, the smallest living entity. Cells are complex miniature factories creating proteins; they have emergent properties.

Cells of differing kinds may be organized into tissues, the next level of organization. Tissues are formed from the emergent properties of their constituent cells. Tissues are more complex than cells; they 'contain' the complexity of many interacting cells. Tissues may be: connective, muscle, nervous and epithelial.

Tissues of different kinds may come together to form organs. An organ is a collection of tissues joined in a structural unit to serve a common function. In the human body, the heart, liver, kidneys, etc., are organs, formed from the emergent properties of their constituent tissues. A kidney comprises nephrons, the principal functioning units of the kidney, filtering blood and producing urine; nephrons comprise complementary tissues made from many different cells, all set in a collagen framework or scaffold.

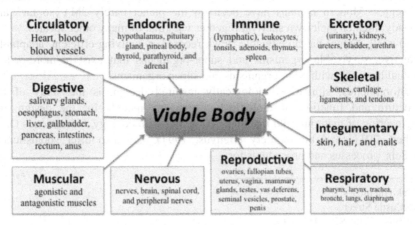

Fig. 1 Organ Systems in a Viable Body

The body is comprised of organs that work together as organ systems; there are some ten to twelve major organ systems in the body, depending upon how they are counted. See Figure 1, which shows the viable body's organ systems together with their constituent organs.

1.1   Learning from Natural Levels of Organization

There are discernable levels of organization within the human body:

5. Organism	4. Organ System	3.Organ	2. Tissue	1. Cell

Each level is formed from the emergent properties of assemblages and combinations of the level below. At each level there may be many of same- or similar- kind, mutually interacting entities.

Biologists identify some 210 different human cell types. Different cell types organize into different tissues. Different tissue types organize into organs. Different organs organize into organ systems. And, different organ systems organize into the human organism. There is, however, a formative sequence within the embryo.

A human embryo grows around an axis, which will become the spine and spinal cord. The heart, brain, spinal cord and gastrointestinal tract form initially, followed by limb buds and internal organs. Each of these follows (depends upon?) the prior existence in the embryo of heart, brain and nerves; they do not appear independently and then 'join up,' as it were… so there is no reduction, only synthesis and development, with nerves, blood vessels, etc., developing within, and at much the same time as, the tissues and organs. Remembering, of course, that the embryo is developing according to a well established "master plan" in the DNA.

1.2   Extending the "Levels of Organization" Paradigm

Figure 2 shows the cell-to-organism hierarchy diagrammatically from bottom to top: cell, tissue, organ, organ system, organism and then extended beyond to population and society (and, not shown, continuing above to ecosystem and biome, both of which include non-living 'things.') At right is a supposed correspondence with manmade systems, such that Organ System corresponds with System at Level 4, Organism with Platform, at Level 5, etc.

Correspondence at levels 1 and 2 is more questionable. While a cell is a living thing, a man-made component, an artefact, is not; neither is a composite of such inert components. However, components may be energised and may be active; for instance, a field-programmable gate array (FPGA) may be programmed to perform a variety of functions. While energized, such electronic devices/ components may not unreasonably correspond with cells. Similarly, tissues may

Biology/Anatomy   Man-made Systems

Community ◄ ► Company

Population ◄ ► Group

Organism 5 Platform

Organ System 4 System

Organ 3 Subsystem

Tissue 2 Composite

Cell 1 Component

* Population - all the organisms that belong to the same species, in the same geographical area
** Community - a group of interacting living organisms sharing a populated environment

Fig. 2 Levels of Organization – Organic and Manmade Systems

reasonably correspond with energized assemblages of such active devices, referred to as composites. Neither component nor composite, however, seems to quite merit the epithet of 'system,' since they are not 'organized coherent wholes.'

Levels 3 and 4 are more promising: an organ system corresponds sensibly with a system consisting of interacting subsystems that perform functions, and have 'organized whole' characteristics. Similarly, an organism corresponds sensibly with platform, in that a platform, like an organism, carries its organ systems around with it 'internally,' i.e., sensors, processors, etc.

Population is less clear. All workers in a hive of honeybees, for example, are sisters. Such cannot be said of humans forming a group, and the difference may be important: hive behaviour is highly motivated towards the survival of the Queen and the next generation, and this behaviour may be dependent upon their being closely related genetically, and therefore willing to supress their own need to reproduce as they are, in effect, furthering their own genes in caring for the offspring of the Queen.

The implication at community level is of different species interacting in the same populated environment, which might be less obvious for humans in interacting groups, since all humans are of the same species. Perhaps we may consider human groups with different cultures, disciplines, faiths or beliefs as corresponding conceptually with different species—at least in some degree.

One curious note presents itself; that a hierarchy of man-made systems bears such apparent similarity with that found in human biology/anatomy. Three explanations come to mind:

1. That the notion of hierarchy is simply a naturally occurring property of complex systems in that emergent properties and behaviour happen at various scales in complex systems.
2. That the hierarchy for man-made systems is innate to humans as designers and creators, and may have emerged from their subconscious
3. That the hierarchy for man-made systems has been consciously based upon the natural hierarchy found by biologists and anatomists

Then again, many of those engaged in early work on systems were biologists, one of whom was instrumental in developing General Systems Theory (Bertalanffy, 1968).

2 Levels of Organization and Layers of Systems Engineering

Levels of Organization indicate successive levels of complexity. Systems engineering seeks to manage complexity in the conception, design and synthesis of complex manmade systems. There is an evident correspondence between Levels of Organization, and various 'flavours' of systems engineering.

Figure 3 presents correspondence between three parallel hierarchies: levels of organization; manmade systems; and, layers of systems engineering. An 'anchor' point of correlation/correspondence is between Organ System: Level IV: System; and, Layer 2, Project Systems Engineering.

This sits comfortably above Organ: Level III: Subsystem; and, Layer 1 Product/Subsystems Engineering, where an organ is evidently a functional subsystem of an organ system, and product/subsystems engineering is evidently one level down on project systems engineering, since a project is likely to comprise a number of interacting functional products.

That leaves Business Systems Engineering addressing Company and Group; Industry Systems Engineering addressing Organizations; and, Socioeconomic /Societal Systems Engineering operating at national level.

LEVELS OF ORGANIZATION V. LAYERS OF SE

Biology/Anatomy		Man-made Systems	SE Layer
Nation	IX	Nation	5. Socioeconomic/societal SE
Region	VIII	Organization	4. Industry Systems Engineering
Community	VII	Company	3. Business Systems Engineering
Population	VI	Group	
Organism	V	Platform	2. Project Systems Engineering
Organ System	IV	System	
Organ	III	Subsystem	1. Product/Subsystem Engineering
Tissue	II	Composite	Artefact Engineering
Cell	I	Component	

* Population - all the organisms that belong to the same species, in the same geographical area
** Community - a group of interacting living organisms sharing a populated environment

Numbers refer to 5-layer SE Model:see Hitchins D.K. (2003) *Advanced Systems Thinking and Management*, Artech House, MA

Fig. 3 Levels of Organization and Layers of Systems Engineering

The figure suggests that 'ascending' layers of systems engineering address 'increasing' levels of organization, therefore increasing degrees of complexity. Further, the figure suggests that systems engineering, at any layer, is essentially the integration of parts from the level below. So, Industrial Systems Engineering is the integration of several/many interacting businesses into a coherent Industry. Business Systems Engineering is the integration of several/many projects into a coherent business. And so on.

The figure also indicates that Business/Enterprise, Industry and Socioeconomic Systems Engineering are primarily concerned with increasingly complex *people* systems, rather than technology. As such, systems engineering will be concerned, less with technology, more with decision-making and behaviour, operations, management, organization, resourcing, training, control, contribution, coordination and cooperation, belief systems, *Weltanschauungen*, group psychology, social anthropology, etc.

Meanwhile, at the lower end, Levels II and I, the figure suggests that there is 'less systems engineering, more conventional engineering,' since both Components and Composites are manmade artefacts, not systems in the sense of being whole, internally-organized entities performing functions.

The potential value of this structural relationship is for a cross fertilization between the three hierarchies at each and any level. It presents a different view of systems engineering as being about integrating subsystems where each is mutually interdependent on all the others…

3   The Triune Brain

The Triune Brain is a model of the evolution of the vertebrate forebrain and behaviour proposed by the American physician and neuroscientist, Paul D. McLean. The Triune ('three in one') Brain consists of the reptilian complex, the paleo-mammalian complex, and the neo-mammalian complex (neocortex), viewed as structures sequentially added to the forebrain in the course of evolution:

1. The reptilian brain: responsible for species-typical instinctual behaviours involved in aggression, dominance, territoriality, and ritual displays.
2. The paleo-mammalian brain, otherwise the Limbic System: responsible for the motivation and emotion involved in feeding, reproductive behaviour, and parental behaviour
3. The neo-mammalian complex, or cerebral neocortex: responsible for language, abstraction, planning, and perception.

These three are, perhaps, less 'Levels of Organization,' more evolutionary overlays. The Triune Brain model has been partly overtaken by recent research, but it still offers a seductively straightforward way of regarding the human brain.

The Triune Brain does offer one potential model—that of layered processing. Instead of replacing redundant computing systems, with their accumulated wealth of software and data, perhaps these systems could be retained as a nucleus and surrounded with 'shells' of advanced processing permitting faster calculation, correlation, communication and control such that computing systems might interact with humans via speech and vision, rather than keyboards and mice...and might anticipate the human need for planning and presentation...

4 Layers in the Brain

The human brain is the most complex part of the human body, and although some may like to compare it with a computer, and/or with the software of a computer, such comparisons do not seem to offer useful models by which we might advance our manmade computing systems...so far.

The cerebral cortex is the outermost sheet of neural tissue of the cerebrum of the human brain. It is divided into left and right hemispheres, and plays a key role in memory, attention, perceptual awareness, thought, language and consciousness. It consists of up to six horizontal layers, each with a different composition in terms of neurons and connectivity.

4.1 Levels of Organization within the Brain

The brain is an exceedingly complex organ that is defying attempts to analyse and reduce its performance and behaviour. However, there have been attempts to show levels of organization within the brain, as part of the Human Brain Project (Markram, 2012) as follows:

- **Whole Organ**, comprising some 89 billion neurons and 100 trillion interconnections (making over 1000 connections per neuron on average)
- **Regions**, mutually interacting major neural substructures: amygdala (emotions), hippocampus (memory), frontal lobes (executive control)
- **Circuits**, neural interconnections among neighbouring cells and between different brain areas
- **Cellular**, neurons, non-neuronal glial cells, dendrites and axons
- **Molecular**, parts of a neuron and its transmission of electrical and chemical signals

On the other hand, there appear to be functional 'subsystems' within the brain, concerned with speech, vision *(q.v.)*, motor control, etc.

The Human Brain Project is undertaking an ambitious task, employing 'synthesis biology;' this is essentially simulating the human brain within a computer. As a test case, the project team built a unifying structure called a *cortical column:* this is described as analogous to putting a miniature apple corer through the cortex and pulling out a cylinder of tissue about one half millimetre in diameter and 1.5 millimetre in length; this would constitute a column.

The column penetrates the six vertical layers of the neocortex; the neural connections between it and the rest of the brain are organized differently in each layer. A few hundred neuron types reside in the column.

The team simulated the behaviour of a column from a new born rat, allowing the virtual neurons to connect up as real neurons would, eventually providing them with a static model of a column, as in a comatose brain. They then 'jolted' the column with a simulated electrical impulse: the neurons began to interact and intercommunicate. 'Spikes,' or action potentials, spread through the column as it began to work as an integrated circuit; this was spontaneous, not programmed behaviour. And the column stayed active after the stimulation stopped, briefly developing its own internal dynamics...

Observation. Here, then, is a fascinating, new way of looking at the way the human brain might work, and—although not its primary, medical purpose—it may afford potential for novel design of complex computing and information systems of the future.

5   Brain Cells for Concepts

Neuroscientists dispute how the brain stores memories—which it does remarkably well. Some view memories as somehow spread across the brain and interspersed with each other: others suspect that memories are stored in individual neurons, or groups of neurons. Research, particularly into the brains of those with debilitating epilepsy, is showing interesting results. (Quiroga, Fried and Koch, 2013)

Researchers have been able to insert fine probes into the brain, particularly into the hippocampus, and have discovered individual neurons firing when the subject is presented with a picture of a well-know film or TV star, or has seen their name written. An individual neuron fired for Halle Berry, again to her name written on a screen, and again to her name spoken by a synthesized voice. The same phenomenon occurred for Oprah Winfrey, and for Luke Skywalker; each appeared to have his or her own 'neuron.' The research team could not assert this, however, because they were able to sense only a few neurons at a time. Others could have been firing but undetected.

However, further research showed that individual neurons might fire for more than one such star. The neuron that fired for Jennifer Anniston also fired for Lisa Kudrow, her co-star on *Friends*. Again, it was sufficient to show any picture of either star, or the written name, or the name spoken with a synthesized voice. Perhaps the neuron was firing for blondes, or for the TV program *Friends?* The neuron that fired for Luke Skywalker also fired for Yoda: was there a Jedi neuron? Moreover, the neuron for e.g. Jennifer Anniston fired when shown only part of her, when wearing different clothing, etc. Such neurons, then, appeared to be firing in response to the 'concept' of Jennifer Anniston, or Yoda, or... how could this be?

The organization and structure of visual information in the brain is outlined in Figure 4. Top left is shown the eyeball, with the optic nerve leading to the primary visual cortex at the back of the head. Here a detailed picture is formed, such that, for every detail in the observed image, there is some correspondence in the primary visual cortex. One neuron firing does not indicate whether it is part of a tree, a wall, or a person, however, and the observer is interested in whether they are looking at an object, and if so, what object…

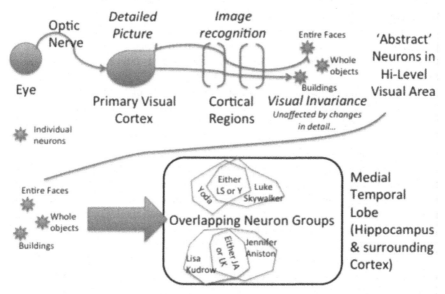

Neuron groups respond to 'concepts' about e.g. people. Group of neurons that respond to Luke Skywalker may also respond to Yoda, spoken, written, etc. There is overlap – both from *Star Wars* – but also some difference. No overlap with Jennifer Anniston from *Friends*

Fig. 4 Concept memory

Next the visual information goes through a series of cortical regions towards the front of the brain. Individual neurons in these higher visual areas respond to entire faces, or whole objects, and not to local detail. So, minor changes in the visual scene will not affect these neurons. This is 'visual invariance.'

Neurons in the higher visual areas send their information to the medial temporal lobe – the hippocampus and surrounding cortex – which is involved in memory functions and where the so-called Jennifer Aniston neurons were found. The response of neurons in the hippocampus is much more specific than in the higher visual cortex. Each neuron responds, not so much to an individual, as to a concept of some individual.

As the figure suggests, there may be a 'patch' of neurons, a relatively sparse grouping, which respond to, say, Luke Skywalker and another patch that responds to Yoda, and these two patches overlap, meaning that some of the neurons that fire are common to both memories. Similarly, neurons for Jennifer Anniston and for Lisa Kudrow may also overlap. However, there may be little or no commonality between the sets for Lisa and Jennifer from *Friends* and the sets for Luke and Yoda from *Star Wars*.

What does this all mean? Surgical removal of the hippocampus leaves the patient still able to recognize people and objects, and to remember events, but the patient can no longer make new, long lasting memories. It was as though the means of transferring from short to long-term memory had been removed, as if the 'memory folder index' had been erased.

So, the Jennifer Anniston neuron was not necessary to recognize the actress, or to remember who she was, but it was critical to bring her into awareness for forging new links and new memories about her, such as later remembering seeing her picture.

Memories are more than single isolated concepts. A full recollection of a single memory episode involving a person or thing – perhaps even a place – requires links between different but associated concepts. If two concepts are related, some of the neurons encoding one concept may also fire the other one. This hypothesis suggests how the neurons in the brain encode association.

"The tendency for cells to fire to related concepts may indeed be the basis for the creation of episodic memory (such as the particular sequence of events during an encounter) or the flow of consciousness moving spontaneously from one concept to another... A similar process may also create the links between aspects of the same concept stored in different cortical areas, bringing together the smell, shape, colour and texture of a rose." (Quiroga, Fried and Koch, 2013).

If the research is justified, an elusive aspect of the flow of human consciousness may have been explained. There are implications here for the memory systems, not only of humans, but also of autonomous machines, which will also need to associate concepts. An autonomous (robotic) peace officer,[1] for example, would need to establish concepts of people, places and things[2] on 'his' beat, together with episodic concepts of misdemeanours and crimes in progress. He would recognize individuals with past records and observe their behaviour, comparing it no doubt with models of acceptable behaviour, threatening behaviour, etc.

Observation. The above is cutting edge research, and holds out exciting prospects, not only for understanding how memory and perhaps even consciousness function, but also for conceiving advanced computing systems that learn about concepts and association between concepts for themselves.

6   Homeostasis

Homeostasis, dynamic equilibrium, is maintained in the body partly by negative feedback processes, and partly by dynamics in open systems. A biologist, Ludwig von Bertalanffy, developed general transport equations for open systems (Bertalanffy, 1968) as follows:

---

[1] Sometimes used as a hypothetical test case...

[2] As does a human peace officer.

$$\frac{\partial Q_i}{\partial t} = T_i + P_i$$

where:

Qi = is a measure of the ith element of a system

Ti = the velocity of transport of Qi at that point in space

Pi = the rate of production or destruction of Qi at a certain point in space.

A system so defined may have three types of solution: first there may be an unlimited growth in the system, Q; second, a time independent state may be reached; and third, there may be periodic solutions.

In the case where a time independent solution is reached:

$$T_i + P_i = 0$$

In these two simple equation can be seen both the conservation laws of physics and the open systems stability of organisms.

An everyday example of the general transport equation might consider how a person maintains his or her weight, by eating as much food energy as is expended in basal metabolism, activity and exercise. This is homeostasis.

6.1   Control through Feedback

While open system dynamics may be the principal means of establishing and maintaining homeostasis in natural open systems, many incremental feedback systems are also at work. Two well-known processes follow; one for maintaining blood sugar level, and the second for maintaining core body temperature. This ubiquitous 'opposing pairs' arrangement in natural systems is curiously at odds with engineering and cybernetic practices.

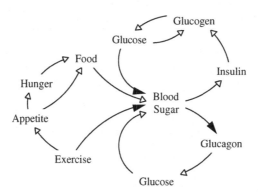

Fig. 5 Maintaining Blood-Sugar Level. Open headed arrows indicate positive, supporting relationship. Solid arrows indicate a negative, detracting relationship. E.g., increasing blood sugar level increases insulin secretion (open arrowhead); decreasing blood sugar level increases glucagon secretion (solid arrowhead).

Figure 5 presents a simplified causal loop model (CLM) showing how blood sugar level is usually regulated in the human body. The upper loop is well known: excess blood sugar caused, perhaps, by eating, stimulates the secretion of the hormone insulin principally by the pancreas, which allows glucose in the blood to be converted to glucogen (and possibly to fat) in the liver and muscles, so restoring the correct blood sugar level.

The lower loop works differently. Sustained physical exercise (or starvation) might cause the blood sugar level to drop. This is also detected in the pancreas, which secretes a different hormone, glucagon, which increases (restores the level of) glucose in the blood. Together, these elements within the body's endocrine system maintain a steady level of blood sugar in a healthy human, although that level may change throughout a typical day.

At left in the figure above is shown the relationship between exercise and food ingestion, which forms part of the energy transport equation for open organic systems. In this instance—and generally—open system dynamics operate in conjunction with the feedback controls to maintain homeostasis. The complex human system also has feedback mechanisms, not shown, which encourage eating when blood sugar falls, and when food is not needed but is expected as a result of regular, habitual feeding, as opposed to eating when hungry.

Figure 6 is a second causal loop model, showing a different pair of contra-acting causal loops. The upper loop shows the well-known effect of being in a hot environment: perspiration leads to evaporation cooling to maintain core body temperature. Less well appreciated, perhaps, is what happens when the body experiences a cold environment and the surface skin and flesh cool. As the lower loop indicates, blood capillaries near the surface contract, closing off the flow of warm blood near the skin surface, so that the blood is not cooled so much.

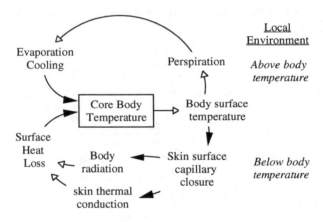

Fig. 6 Maintaining core body temperature

Flesh with the blood withdrawn (or not circulating as after death) forms an effective insulator, so that heat conduction through, and radiation from, the skin surface are also reduced.

The combined effect of these is to maintain essential body core temperature, at the expense of a cold surface skin. As in the previous figure, we see two contra-acting causal loops resulting in maintenance of a stable value of blood sugar and core body temperature respectively. Homeostasis is established and maintained with many such contra-acting feedback loops and with the open system dynamics presented earlier.

Observations. Natural systems such as the human body clearly find it advantageous to maintain open system homeostasis by maintaining an equable inflow/outflow regime of energy, material and information, but also by controlling short-term variations using contra-acting, incremental feedback control 'mechanisms.'

There appear to be advantages in this approach compared with the usual single-loop negative feedback technique used by engineers and cyberneticists. Equable inflow/outflow regimes allow the organism/platform to maintain homeostasis even when feedback control mechanisms are prejudiced, by changing behaviour and/or environment.

Some of these abilities are peculiarly human; we are the only members of the great apes to have subcutaneous fat, for example. This feature of human physiology may have evolved to sustain human hunters during long hunting trips where it was necessary to pursue prey over many miles under the heat of the African plains. It offers an explanation, too, for the naked ape's loss of hair covering; hair and subcutaneous fat together would have caused bodily overheating, so with fat being essential to sustain energy for the hunt, hair covering the whole body became prejudicial to survival. (Morris, 1967)

7   Social Insects, Social Organization

The social insects are often described as 'super-organisms.' Certainly, hives, colonies, armies, etc., do behave in a highly coordinated manner. But then, so do trained and disciplined human armies: would we describe them as super-organisms? Bertalanffy (1968) introduced the term "organismic analogy," suggesting that human systems (teams, groups, nations, even civilizations) are not organisms, but in many ways behave as though they were organisms. The organismic analogy works in both directions: human to insect; and, insect to human. So, a hive of honeybees is evidently not a single organism, but in many ways behaves as though it were...a human city is not a single organism, but in many ways it behaves as though it were.

7.1   Hymenoptera, Honey Bees

Honeybees include the well-known European honeybee (Apis *mellifera*), which often nest inside hollow trees. (Raina & Kimbu, 2005). They construct vertical wax combs with individual hexagonal cells for storing honey and rearing brood. Each hive is "ruled" by a single Queen whose only job is to lay eggs. Workers are adult females – daughters of the Queen who mates initially with between perhaps as many as fifty male drones, stores their sperm in her body and uses it throughout her life to fertilize eggs. There are, then, as many as fifty lines of workers in each hive. Their duties change upon the age of worker bees in the following order:

- Clean their own cell after eating through their capped brood cell
- Feed brood;
- Receive nectar;
- Clean hive;
- Guard hive;
- Forage.

Some workers engage in other specialized behaviours, such as:

- "Air conditioning," using their wings to move air through the hive to prevent overheating;
- "Undertaking," removing corpses from inside the hive.

Adult workers live for about six weeks during the summer, but Queens may live for several years. During cold winters, the bees cluster together, feeding on stored food reserves and sharing their body heat. As the weather gets colder, the 'ball' of bees tightens, and there is a continuous circulation of bees from the outside to the inside of the ball. This is analogous to the way chilled humans maintain core body temperature (q.v.).

The Queen secretes a pheromone to maintain calm and order in the hive. As the hive grows in numbers, there is progressively less of this pheromone to go around 'per bee.' At the same time, forager bees have to go further and further afield to fetch food for the growing numbers of bees in the hive. A mature, healthy colony may grow to as many as 80,000 workers, at which point foragers may be gathering barely enough nectar and pollen to feed, let alone expand, the hive. With so many worker bees, the Queen's pheromone becomes too diluted to maintain calm and order, and the hive may 'reproduce' by swarming, with the old Queen accompanied by many bees. A new Queen, reared within the hive, will hatch out, and restock the depleted colony.

The prime directive for the colony, or hive, is evidently propagation of the species through propagation of the hive. The survival strategy involves limiting the size of each colony to that which may be supported by local resources (flora), at which point they relocate to a new area... Hives are perennial, accumulating and storing food to survive winters in cooler climes; this appears to give the hive a fast start to the new season when the weather improves. Honeybees do not employ any technology as such, but their ability to secrete wax and to create hexagonal shaped honeycombs is surely equivalent. This 'technology' directly contributes to propagation, in being used to raise the next generation and to store food compactly for over-wintering.

Observations. European honeybee survival strategy contrasts with that of humans, who build ever-bigger and more densely populated cities. Humans create complex import-export systems to transport foodstuffs from far and wide, making their cities potentially vulnerable to breakdown in supply chains, as well as supporting populations beyond that which could reasonably be supported from locally resourced supplies. This is surely a recipe for unbounded population expansion (and eventual disaster), which is what humanity is experiencing, but which honeybees avoid by living within their neighbourhood resources, and by limiting hive population to that which the local flora will support.

7.2   Hymenoptera, Ants

Ants share a common ancestor with wasps, and there are over 8,800 species. (Höllobler & Wilson, 1990.) A typical nest contains at least one fertile egg-laying Queen, hundreds or thousands of adult female workers, a nursery for rearing eggs, larvae and pupae, a storage area for food reserves, and a disposal site for waste and dead bodies.

The size of most ant colonies is limited by their ability to forage sufficient food. Leaf cutter ants may establish lengthy supply chains to bring in their food source – although they do not eat the leaves they harvest, most of which are poisonous. Instead, they use them to feed a farm, which grows a fungus on which the ants live, an early example of symbiosis. The fungus farm is supplied with an antibiotic covering farmworker ant bodies, so preventing disease; these ants discovered antibiotics some 40M years ago.

Army ants are carnivorous nomadic predators that form bivouacs, but do not nest. Honeypot ants farm aphids, living on their excreted honeydew and moving them from plant to plant as a farmer might move cows between pastures.

Recently results from an interesting 6-year[3] study, publicized by Professor Laurent Keller, Director of Ecology and Evolution at the University of Lausanne, Switzerland, have shown that ants have a similar allocation of worker duties with age to honeybees:

- Young ants, about one third of the workers, were nurses;
- About a third were cleaners, having graduated from nursing;
- And about a third were foragers, collecting food outside the colony, having graduated from cleaning

---

[3] Read more at http://www.dailymail.co.uk/sciencetech/
article-2311688/ANTS-change-job-grow-older-scientists-
discover.html

Researchers found that ants, with lives lasting about a year, tended to socialize within their current 'profession,' e.g., nurses with nurses, foragers with foragers. It was not clear what promoted ants from one profession to the next, nor was promotion automatic: researchers found occasional old nurses and young foragers.

Observations. Between them, the many and various ant species display a wide variety of lifestyles and survival strategies. Some are carnivorous, some vegetarian. The biomass of all ant species in the world is reputed to exceed that of all humans[4]. Throughout, their prime directive is survival of their species; they limit their nest/colony size according to locally available resources, a repeated lesson, perhaps, for humans.

Recent research showing that ants progress through different 'professions' in a similar way to honeybees suggests a further instance of convergent evolution, this time of societal behaviour... Ant tendency to intra-profession socialization may suggest that these 'professions' are akin to classes or castes, although without any morphological distinctions.

7.3   Blattodea (Formerly Isoptera), Termites

Termites share a common ancestor with cockroaches, and are universally vegetarian. (Darlington & Bagine, 1999.) Although similar in appearance to ants, they are unrelated, another instance of convergent evolution solving the same social problem. There are some 2,600 known species of termite. They eat anything from lichen and fungus to wood.

In Africa, termites create enormous mounds, with funnels to create air conditioning, particularly for the Queen's chamber, which is held at 30°C and near to 100% humidity – not dissimilar to the breath of a mammal. These mounds can become very large, with some 35km of tunnels. The relatively huge Queen may live up to 30-45 years, supported and enabled by her king who remains virtually attached to her as she continues to lay eggs. There may be over one million termites in a mound, with the Queen as mother to all.

Like other eusocial insects, termites have castes:

- Reproductives
- Workers
- Soldiers

Periodically, members of the winged sexual caste take to the air, mate, and the newly fertilized young Queen seeks a new spot where she may start a new colony. So, new colonies are budded off the old, but—unlike honeybees—the old may stay put with its original Queen.

---

[4] See en.wikipedia.org/wiki/Biomass_(ecology), extracted April 2013).

Observations. Termites, at least African termites, may be more established than their ant and bee 'cousins,' and they build strong, deep nests, mostly underground. Queens 'entertain' their king, and together they are much longer lived. Like the Hymenoptera, they live very much within their neighbourhood, foraging over the surrounding surfaces, clearing up vegetable detritus, advantageously reducing the risk of forest fires in the locale.

8   Do We Have Anything to Learn from the Social Insects?

Insect societies are inherently different from human societies. We humans would find insect social behaviour intense, exceedingly tiring and beyond our ability or desire to sustain. Busy bees, for instance, literally work themselves to death. The social insects know instinctively what job they are intended to do, and they do it: no instructions, no supervision—although, failure to perform may result in summary execution.

Much of the social insects' apparent decision-making is democratic, with insects voting in preference for one option over others, as with the bees and the famous waggle dance, where the 'strongest waggle' wins. There is no control, as such. Neither is the Queen in charge, in any conventional sense; that is surely an anthropomorphic viewpoint.

There are, however, castes. While all eggs may be laid equal, the way in which they are fed may produce morphologically different results, castes:

- bees and ants have worker and reproductive castes (Queen and drone). Workers tend to progress from nursing, though cleaning to foraging. Some ant species have soldier castes where the soldiers are very much larger than the workers, and may have to be fed by those workers.
- termites have reproductive, soldier and worker castes (immature males and females (nymphs), with potential to moult into replacement soldiers or 'reproductives' if needed. Soldiers may differ from others morphologically, with 'hoses' on their heads to squirt irritating fluids at their enemies—often ants...

Castes may be compared with classes in human society, which have existed since the earliest times—the earliest probably being that of ancient Egypt. This isolated proto society was a 'pyramidal' (sic!) three-class system of a pharaoh and a few upper class nobles ('nomarchs'), a middle class of merchants, lawyers, priests, schoolteachers, doctors, scribes, etc., and a working class of very many who used their hands to till, fish, grow, tend, operate and make things. Despite efforts to eradicate, reclassify and rename class, it is still with us and seems to be as innate within human social order as it is within insect social order.

Class mobility, on the other hand, the ability to move between classes, is evident within the social insects, except where caste is morphologically determined. Within the social insects, workers progress through 'professions,' finally working outside of the colony/hive as foragers. Within human society, there is no such obvious progression…although as humans age in society they tend to accumulate knowledge and experience, (may) gain social stature, (may) own property, etc. It is not really the same…

The eusocial insect example suggests that a contemporary Western idea, that all people should be essentially equal[5], may prove difficult to achieve. People naturally fall into classes and castes: some are leaders, some followers, some develop quickly, others more slowly, and some develop degrees of social sophistication with age and experience. Attempts to destroy/discredit so-called upper classes (determined by inheritance, wealth, education and ownership) results in the creation of pseudo classes, such as celebrities—people celebrated more for ridiculous, disgraceful, socially unacceptable behaviour than for anything meritorious.

All of the eusocial insects create societies that are closely related genetically, to promote cooperation and coordination. Humans have no obvious equivalent, outside of the family group, nor would we want or accept one. But do we have any corresponding attributes that would similarly promote cooperation and cooperation? Perhaps it would be easier with human society to identify those phenomena that prevent, or detract from these 'desirable social traits.'

- **Culture.** People of different cultures, upbringing, educational backgrounds, political persuasions, religions, etc., may behave, express attitudes and formulate decisions that differ from each other, so constraining cooperation and coordination
- **Belief Systems.** Our view of the world, our *Weltanschauung*, and our beliefs—cultural and religious—may differ radically from others, so that we are unable to see things as they do, and vice versa.
- **Training.** Intense military-style training may overcome different belief systems, such that people who have trained together will cooperate and coordinate in their actions, decision-making and will even overcome their differing natural reactions under pressure.

So, we may reasonably expect human societies where people are the same/similar background, culture, education, upbringing and beliefs to form cooperative, coordinated societies. Within those societies we may expect to find different classes. And this is generally, although far from always, the case.

Alternatively, cooperative societies may form and self-sustain where culture encourages individuals to consider themselves part of a close-knit family group. (Toyota in Japan has operated in this way, with notable success. (Womack, 1990))

What price, then, multi-cultural society? If the eusocial insects are to offer any guidance, it might be that an effective, coherent multi-cultural society is unlikely, while a complex of different side-by-side discrete cultures is more likely, with the inevitable risk of friction at the cultural boundaries.

---

[5] Promoted by various political notions such as Political Correctness, feminism, equal opportunities, 'gay' equality and marriage, etc.

9 Conclusions

Biomimetics offers a wide range of models from individual organisms and from the social insects, which may be of benefit to systems, systems thinking and systems engineering. Comparing natural and human systems engineering suggests that human systems engineering parallels natural systems engineering to a surprising degree. Further that, in both instances, systems engineering appears fundamentally to be the integration of functional structures from one level of organization lower—although integration may be far from simple...

Natural systems employ a more sophisticated approach to homeostasis than do engineers and cyberneticists. Nature's methods generally involve contra-acting control loops providing *incremental regulation* of open system 'flow through' of energy, substance and information: they are robust, non-linear, fast, precise, effective and tested by time. There may be much to learn here.

On-going research into the human brain is funding intriguing concepts for advancing our ability to create autonomous machines that can think like humans, which have our phenomenal ability for concept associations, and with which we may be able to interact more comfortably, since they will seemingly think and behave like us, rather than like the archetypal clumsy, inarticulate robotic machine.

Social insect societies remind us that the prime directive of all animals is the propagation of species. For hives and colonies it is quite evident that their 'business' is to raise the next generation; that is what they are preoccupied with doing, and what their equivalent of technology is dedicated toward.

Many—most—human societies, on the other hand, seem to be unaware of this prime directive, leaving people to question 'the purpose of life,' to devote their lives to accumulating wealth, to hedonism, to being continually entertained by their governments, to devote themselves to war, conflict, sport, business, work, music, etc., to decide not to have children, and many, many more ways of denying, overcoming, or supressing their nature and ignoring the prime directive. Despite this—or, perhaps *because* of it—human population continues to rise...

The eusocial insects suggest that human societies should be able to manage population size much better than we are presently doing. The hive, the colony, etc., limit their population to that which can be supported by the immediately surrounding flora and fauna, according to species. If there is insufficient food, they do not continue to multiply, but instead they swarm, or bud off, another hive or colony in another area where the resources are adequate. Humans could do the same. (Howard 1898.)

The eusocial insect example does suggest that countries, or regions within countries, could easily establish whether or not they are under-populated or overpopulated.

- A place (village, town, city, nation) would be overpopulated if the existing population absorbed more resources than the immediate surrounding environment was *actually providing* for human consumption.

- Similarly, it is possible in principle to draw a circle around centres of population, with diameter sufficient to encompass the actual, productive farmland area needed to support that population: i.e., the "food resource footprint."

As a non-political observation, such concepts challenge globalization that would see populations growing unchecked, supported by global exports and imports—inevitably leading to further overpopulation... Social insects also differ from humans in maintaining their essentially agrarian lifestyle; many humans, on the other hand, have abandoned the land physically, socially and mentally, encouraged by the Industrial Revolution, which still rumbles around the planet. As human populations continue to grow and have to be fed, however, it seems possible that there may be, of necessity, a general return to a more agrarian lifestyle for many, if only to feed themselves and live in balance with, rather than progressively destroy, their supporting local environments.

Social insects also entertain class and caste, concepts that politicized western societies find objectionable. If human social history and the social insects are to be appreciated, class and caste are inevitable; attempts to demolish class will serve only to see class re-emerge under different headings. Certainly the present situation in some western societies, where it is somehow despicable to be of a high social status, wealthy and well educated, is inverted. Should we all not, rather, aspire to be like that than to denigrate it, supplant it with highly doubtful celebrity status, and degrade society in the process? Although few of us, perhaps, would want to emulate the social insect Queen, with nothing to look forward to but spending more years producing ever more eggs...

References

Bar-Cohen, Y.: Biomimetics: Nature-based Innovation. CRC Press (2011)

Benyus, J.M.: Biomimicry: Innovation Inspired by Nature. William Morrow (1997)

von Bertalanffy, L.: General System Theory: Foundations, Development, Applications. George Braziller, New York (1968) (revised edition 1976)

Darlington, J.P.E.C., Bagine, R.K.N.: Large termite nests in a moundfield on the Embakasi Plain, Kenya (Isoptera: Termitidae). Sociobiology 33, 215–225 (1999)

Höllobler, B., Wilson, E.O.: The Ants. Springer (1990)

Howard, E.: Garden Cities of Tomorrow. S. Sonnenschein & Co., London (1898)

Kazlev, M.A., et al.: The Triune Brain. KHEPER (October 19, 2003) (retrieved May 25, 2007)

Markram, H.: The Human Brain Project. Scientific American 306(6) (2012)

Morris, D.: The Naked Ape: A Zoologist's Study of the Human Animal. Jonathan Cape, UK (1967)

Quiroga, R.Q., Fried, I., Koch, C.: Brain Cells for Grandmother. Scientific American 308(2) (2013)

Raina, S.K., Kimbu, D.M.: Variations in races of the honeybee Apis mellifera (Hymenoptera: Apidae) in Kenya. International Journal of Tropical Insect Science 25(4), 281–291 (2005)

Womack, J.P., Jones, D.T., Roos, D.: The Machine that Changed the World. Rawson Associates, NY (1990)

Integrated Product Team in Large Scale and Complex Systems

Sorin Aungurenci and Aurel Chiriac

Abstract. The development process of Large and complex systems are a highly complex process involving different disciplines which shall take decision and develop interrelated subsystems. Integrated Product Team is a management tool used to improve system development performance. Among the potential benefits of implementing the integrated product team principles are reduced development time, reduced risk of failure, enhanced quality, flexibility and better knowledge sharing. This paper provides an overview of the integrated product team's principles and evaluates the lessons and guidance in existing literature applicable to develop large and complex systems. Recommendations based on own experience and lessons learned developing large and complex systems are presented here.

1   Why Do We Need Integrated Product Teams?

Because of the environment changes and the market needs, solutions are being more complex and develop large and complex systems are a fact.

Systems are entities that shall provide functionalities based on the customer needs and the result is the integration of well defined components.

Sorin Aungurenci
EADS Deutschland GmbH,
Landshuter Str. 26
D-85716 Unterschleissheim, Germany
e-mail: sorin@premium-soft.eu

Aurel Chiriac
ISBC Romania
Rahova 23
Bucharest, Romania
e-mail: aurel.chiriac@intersystems.ro

M. Aiguier et al. (eds.), *Complex Systems Design & Management 2013*,
DOI: 10.1007/978-3-319-02812-5_24, © Springer International Publishing Switzerland 2014

One of the main characteristic of the Large Scale and Complex System is the multi-dimension and the high number of parts that are involved to produce the system. Another characteristic is the large number of the complex system decomposition levels needed to manage it and to integrate it. For example a complex system can be decomposed as: systems; these systems are formed of sub-systems/system elements, which are formed of system components, which are formed of software and physical components. All this "parts", that shall work together to implement the complex system, typically are developed in many locations in the country or different countries. The multi-level sub-systems and systems integration and verification effort are required to support the complex systems development.

Large Scale and Complex Systems challenging characteristics:

- *High Complexity* – the complexity to develop large-scale systems is a considerable increase. It is not enough just to be able to manage the complexity, it is required to be able to reduce the complexity of the system, identify and manage the system components [2].
- *Time of implementation* – usually the large scale and complex systems take very long time, years in term of timescale, and the main reasons are that the scope is large and   complexity. At the first level of decomposition there are many subsystems that involve many development projects and a large integration effort. Program management and information coherence are factors that contribute to duration lengthening.
- *Customers are each of them different* – many times the contractual customer is not the same with the final user, they can request new, diverse and unpredictable scenarios. In this case it is very important to have clear defined in the contract the terms and conditions, otherwise the cost will increase.
- *Large number of stakeholders* [1]– the number of internal or external stakeholders is high and not always located under the same authority or hierarchy.
- *Geographic distribution* – different locations in the same country or different countries. The time synchronization between different locations shall be across several time-zone and even week days (for example Europe and Arabic countries).
- *Internationalization/Globalization* – different cultures and people mindsets have to be taken in consideration and have an important impact on system development.
- *Communication* – different languages, different dictionaries and understandings.
- *Design and Evolution* – because today, two year cycle needs for new technology to reach the market, for example in a project that takes 5 years long to be delivered is quite hard to provide the latest technologies available because the design was frozen 3 years ago.
- *Integration of the system components to a combined result* – how relationships between the components generate the collective behaviors of a system, and how the system interacts and forms relationships with the environment [3].
- *Monitoring and Assessment*

Integrated product team concepts represent a modern approach to address the complexity and technology associated with large and complex systems development.

Enterprises that are able to deliver Large Scale and Complex Systems have complex organization charts. The geographic dispersion of personnel and the number of the functional departments are factors that challenge the success of the systems delivery.

2   What Are the Main Features of Integrated Product Teams

Integrated Product Team (IPT) is a multidisciplinary group of people (i.e., design, marketing, production engineering, process planning, and support) who work together and are collectively responsible for delivering a defined product (system).

"Integrated Product Teams (IPTs) are composed of representatives from all appropriate functional disciplines working together to build successful programs and enabling decision-makers to make the right decisions at the right time." - DoD Acquisition Directive 5000.2-R. (Fig. 1).

The failure rate to deliver the large and complex system (delays, increased cost, no compliancy) is quite high, not because of technical reasons, is more often caused by organizational and human factors.

Because the cost of failure of the large and complex systems is too high, applying the principles of the integrated product teams to develop large and complex systems could reduce the risk of failure [6].

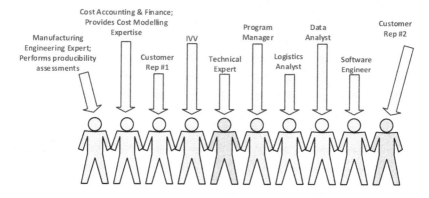

Fig. 1 Example of Integrated Product Team Composition

In the last years the almost all large-scale systems development efforts included the use of the Integrated Product Team approach as a project management tool.

Integrated Product Teams (IPT) could be used as an important tool for development of the large scale and complex systems, but they are generally not used as effectively as they are supposed to be used.

In an organization where enterprise activities are directed by functional managers in order to provide effective early involvement and parallel design, is quite hard to manage important changes (Fig. 2). The integrated product teams (IPT) offer the ability to breakdown the organizational complexity and the full hierarchy decision making process, and put together the required skills and resources to effectively develop a complex system/product.

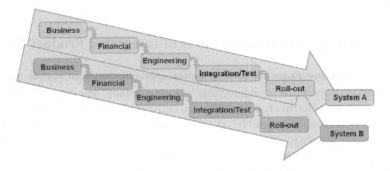

Fig. 2 Stovepipe Approach to System Development [13]

Using the Integrated Product Teams (Fig.3) to implement large scale and complex systems represents a transition from waterfall approach focus to deliver a system according with the contractual requirements, to a team work approach focus to deliver a system according with the customer expectations.

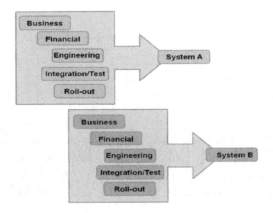

Fig. 3 IPT Approach for System Development[13]

The team members of the Integrated Product Teams are the personnel from different functional departments for support, design, development and transition needed to develop of a system. The IPTs facilitate earlier involvement of the key functions that are involved in the design, production and support of a product and provide a mechanism to manage earlier the needed changes.[11] This early involvement of different functional departments is intended to facilitate the design and production of a system on schedule and within budget, higher in quality, and more reliable and supportable.

Following the IPT approach, each team shall possess the knowledge and skills to work together, collaboratively identify problems and propose solutions, minimizing the amount of rework that has to be done. When this knowledge is accompanied by the authority to take key decisions, IPTs can make trade-offs between competing demands and implement design changes quicker, if necessary [9].

The Integrated Product Team concept promotes the collaborative work, open discussion and innovative thinking resulting in providing high quality of products, more efficient result of work and, not ultimately, a more satisfied customer. The focus of the team is to develop a system to satisfy the customer requirements, as well as the internal stakeholders requirements related to factors such as cost, supportability, testability, reducibility, etc.

Although IPTs require more resources early in the development phase, the result is reduced the number of iterations and the resources over the life cycle of development to production and less effort to correct identified deficiencies through the reviews and engineering changes[12].

The main feature of IPTs is "collaborative work", the teams shall have open discussion with no hidden thoughts and all open topics shall be discussed and clear for each member. Because is a multidisciplinary team, each member brings a unique experience to the team and contributes to the performance of the team [8]. Because of that expertise, each person's views are important in developing a successful system and these views need to be taken in consideration. Open discussion does not mean that the team must act on each view.

3   How Do Integrated Product Teams Achieve Their Goal?

In considering whether an IPT is the best way to develop the large and complex system, it is important to ask three questions [6]:

Integrated – Does the work to be done require input, analysis, and decision-making from a variety of skill sets, perspectives, and/or constituencies?

Product – Does the work produce a unique specific product within a specific timeframe?

Team – Does the work to be done require an environment where team members are accountable for outcomes, and consensus-building is essential?

IPT's are applied at various levels of the overall structure of an organization in the different stages of the project life cycle. The teams can be created, formed, and applied at all levels of the organization ranging from the overall structure of the organization to ad hoc teams that address specific problems.

To develop complex systems we have different levels of IPTs and the level of the decomposition of the system defines the level of hierarchy of IPTs. Usually a complex hierarchy imposes also a long process for decision- making. Usually we can assume to have three levels hierarchy of IPTs for multi-disciplinary integration and teamwork[6]:

- "Executive" IPT - equivalent to an executive steering committee and focus on strategic guidance, program assessment and resolve issues escalated by management.
- "Management" IPT - Manage complete scope, the resources and the risk of complex system.
- "Implementation" IPT – implement the system.

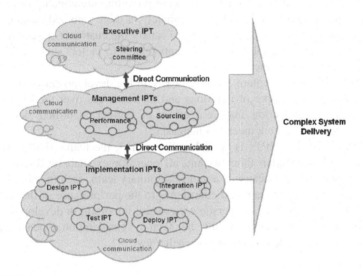

Fig. 4 IPT hierarchy

The Basic Ground Rules for Integrated Product Teams."(Department of Defense 1995, p.4)

- Team discussion are open with no secrets
- Qualified, empowered team members
- Team participation is consistent, success-oriented, and proactive
- Continuous "up-the-line communications"
- Team member disagreements must be reasoned:

    - Focused on alternative plans of action, not simply opposition to the proposed plan
    - (Formal or informal) Trade studies and other analyses are used to resolve issues
    - Complaints about the team are not voiced outside the team; conflicts are resolved internally

- Issues are raised and resolved early
- Design results must be communicated clearly, effectively and timely

- Design results must be compatible with initially defined requirements
- Each team member must be familiar with all system requirements

In order to determine whether a team is in fact an integrated product team, two elements are essential: authority to make cross-cutting decisions and to recognize problems by the knowledge and authority needed to know them. Authority is present when the team is responsible for making both day-to-day decisions and delivering the system. In the programs experiencing problems, the teams either did not have the authority or the right mix of expertise to be considered integrated product teams. Knowledge is sufficient when the team has the right mix of expertise to master the different facets of product development[14].

When it is defined a new IPT shall be clear why it is needed to define an integrated product team, the objectives of the IPT, how the IPT will perform and the skills and behavior of the team members.

4 What Are the Main Problems/Constraints on Their Success

The large and complex systems are implemented by complex organizations (big companies, consortium, join venture, etc.) with complex processes and imply large and complex teams (Management, Engineering, Roll-out, Operations).

Usually, the organizations that implement large and complex systems have the teams distributed across multiple geographical sites.

Distributed development seems to have become the standard rather than the exception[17]. The nature of highly distributed teams is that they

Why we have distributed teams?

- Location of projects (customer) – different places in the country, different countries around the world
- Lack of specialists at certain locations and hard to find a high number of skilled people in a specific geographical area
- Differences in cost of personnel and resources
- Mergers and acquisitions
- Organizations serving local markets

The collocated teams are supposed to be more productive that the distributed teams, but it is not all the time true. There are teams that work very well and are productive even that they are distributed in different physical places and teams that are collocated and they need to improve their productivity.

When having geographically distributed teams it is important time to time to meet and to try to know each other and establish trust. This trust then helps in bridging some of the communication gaps that come with distances and will help in decision making process [14].

Sometimes the culture and language differences are quite obvious, but to reduce the problems coming out of this, shall continuously educating the team members about the other culture and try to create a common team behavior.

To coordinate different teams at different locations, it has to be carefully defined for whom to deliver what and when. The decisions and information shall be communicated on time otherwise there is the possibility to not have the same information when it is needed to take the decision.

Relationships between people inside an organization and those considered outside (customers, suppliers, managers of collaborating organizations, other stakeholders) are becoming more important [12].

To be successful using distributed teams is recommended having a well-defined process and communication tools.

A clear understanding of the team's goals, responsibilities, and authority should be established.

5   What Have We Done to Ensure That IPTs Are Successful

The question becomes how to improve the development process and further how to implement IPTs concepts.

To improve the IPTs performance we ensure that teams have all available information and we have enhanced team decision-making at all levels.

The organization develops a positive culture oriented toward continuous improvement and team-based approaches. Management provides leadership and defines explicitly roles, and responsibilities of team members, functional managers, and program managers.

In order to change the organizational culture and continuously improving the team work a significant investment in training, includes team building training or workshops, cross-functional training, and trainings that provides minimal technical knowledge to allow the non-technical members to participate effectively to deliver the system.

The following practices have been applied to assist in ensuring success of IPTs:

5.1   Enhance the Team Empowerment

The top-down style of management, in which one manager has the control, while the team members just follow orders, can lose the creativity and initiative of every member of the team and is quite hard to gain the team buy-in. Accordingly with the IPTs principles to have experienced and skilled team members, when working with a team of professionals, empowering every team member to play a key role is likely to yield the most productive results. Even so, they need a leader that shall set the tone for the team. The top down style of management may be appropriate when team members are inexperienced and unskilled.

The team members of IPTs shall be empowered. They needed to know their limits, and they could and should act promptly on team matters for which they have experience and authority [6].

Empowerment of the team is linked to decision making process and can be defined as giving to team members' authority to reach their goals as they see fit or to make their own decisions and choices. Empowerment is linked to authority, responsibility, opportunity and motivation.

The ultimate goal of empowerment is developing an environment of *trust*, where every team member is accountable for their own actions.

Empowerment, in regards to IPTs, is the process whereby the team members contribute with more inputs into the operation of the team. In order to be more efficiently and to achieve the goals of the team, empowerment of team members is about the team members sharing decisions and responsibility [10].

How many times have you heard from the team members? "It is not my responsibility".

Is a matter of company philosophy or culture that shall be changed to have empowered teams.

One characteristics of the Large and Complex systems is that are developed by medium-large organizations with a predominant corporate culture which promotes top-down decision-making, extensive up-front planning, stove-piped functional organizations, and a directive style of leadership.

Even it is quite hard to change the organizational culture, some team members don't want to be empowered, identifying potential cultural issues and setting proper processes to empower the teams is the main task of the upper management. Sometimes added responsibilities, accountability and challenges are not always welcomed by all team members.

Some practices applied to create an empowered team:

- Ensure that all contributors are informed about team decision in time (build trust).
- Take the decisions in defined limits (responsibility and budget).
- Share the information and informing every team member of the true issues, concerns and obstacles. Everyone can then make decisions on an equal playing field. Even the information means power, the information shall be shared between them members to have the same level of understanding of the current status.
- Ensure buy-in of key stakeholders and build trustful relationships
- Train anyone who needs to improve their ability to make stronger and wiser decisions. The worst thing is to make an important decision without having the right skills, knowledge and to know the possible strategies.
- Develop self-confidence and motivate the team.
- Encourage team members to have "voice" and to make suggestions that will be evaluated by the team. The result of missing analyses can lead to a team not voting on the best choice, or even a misguided choice.

- Perform team and leadership training courses.
- Management support – without management support is quite hard to change the organization culture.
- The roles and responsibilities need to be well communicated to all teams members within an organization;
- Take analysis and lessons learnt (LL) after each decision as a source for improvement and team development

In an empowerment team, each team member feels responsible to the team and part of the decision-making process.

5.2 Improve Communication

The communication is critical in developing a team that shall perform in a complex environment.

Implementation of the distributed IPTs requires that we take a larger view about communications.

To communicate the team members and the teams can use the following types [5]:

Synchronous communications (identified as "same-time"), includes face-to-face meetings, phone calls (voice), video conferencing (audio – visual), Instant Messaging (text) or virtual environments (audio, visual, text). The delay between the time the sender initiate the communication, send message and the time to receive the answer is short ("real time" communication).

Asynchronous communications includes a significant delay between the time the sender sends a message/initiate the communication and the intended person receives the message/end communication. The asynchronous communication includes email, voice mail, text messaging (SMS), faxes, notification services (e.g. SharePoint, Doors), discussion forums, and regular mail.

The synchronous communication is much more important in building team relationships more quickly than asynchronous communication. The building of relationships between team members is a critical factor in developing "confidence" in what the team is trying to do [15].

While different communication methods and technologies such as phone, e-mail, video conferencing, etc., contribute to team building, the trust and the confidence contribute to the team's survival.

The lack of communication between the teams at different levels generates delays and increases the cost of system development. Without sharing the information on time and to be aware about the decisions taken by different teams, each team has the own approach to reach the same objective (deliver the system).

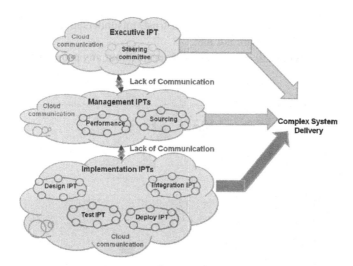

Fig. 5 Lack of communication

Collocation and physical proximity increases the quality and the frequency of communication, and decreases the cost of initiating communication and the time to take the decisions.

The members of collocated team, have the advantage that is easier for them to be aware of the status of the product being developed, by simply using their social abilities, being involved in uninformal meetings (coffee break discussion, lunch, etc.), and having a much greater opportunity to have access to different information earlier.

We have to choose the right type communication tool for different situations. For example, some teams may perform some tasks which team members can do on their own without interacting with anyone else on the team and in this case they can use regularly e-mails, and time to time to have conference calls or video conferences.

Sometimes because of the "disconnected" feeling, the team members can use over reporting as strategy to give people the feeling of knowing what's going on and the current status.

It is a common cause of information overload which can sometimes result in team members avoiding engaging in the communications which actually are important to the team.

The team needs to agree on a common communication strategy, and try to avoid this kind of situations. For exchanging information the teams shall define a clear process and make a commitment to sticking to agreements about when and what will be produced by each member of the team. Because the time is quite important in the communication, it is recommended to use "Pull" (you go get what you need when you need) instead "Push"(having it pushed automatically) to have the information. The distributed teams should agree on common guidelines for their communication.

Poor communication can lead to lack of trust, misunderstandings, wasted time, energy and team objectives are not met.

The team members shall communicate direct without any proxy or facilitator.

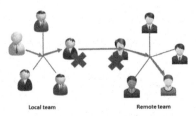

Fig. 6 Intermediate communication

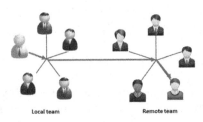

Fig. 7 Direct communication

The information can be altered or delayed. Also the quality of the information is important, in case of using distributed teams, both the amount and the quality of information decreases and the control of the process shall be increase.

6   Conclusions

Even the principles of Integrated Product Teams are simple, to change the company's culture is challenging and to implement the IPTs process and rules can successful achieved with a well-planned and managed effort. It is not enough for company management to only understand the concepts of IPTs, but also they shall apply the process of managing change within the organization. The responsibility for making these major changes in culture, organization, business process and technology cannot be delegated.

References

[1] Holmes, W.: Developing confidence in large scale and complex simulations with software systems (2001)
[2] Ericsson, M.: Developing Large-scale Systems with the Rational Unified Process (2000); Department of Defense. DOD Guide to Integrated Product and Process Development (Version 1.0), Office of the Under Secretary of Defense (Acquisition and Technology), Washington D.C. 20301-3000 (February 1996)

[3] Feiler, P., et al.: Ultra-Large-Scale Systems. The Software Challenge of the Future (2006)

[4] DSMC System Engineering Fundamentals publication, ch. 18 (January 2001)

[5] Farshchian, B.A.: Integrating geographically distributed development teams through increased product awareness (2001)

[6] Creekmore, R., Muscella, M., Petrun, C.: Integrated Project Team (IPT) start-up guide (2008)

[7] Jeffs, M., Douglas Hamelin, R.: Systematic Approach to the Development, Evolution, and Effectiveness of Integrated Product Development Teams (IPDTs), INCOSE (2011)

[8] Moulder, R.: Systems Engineering/Integrated Product and Process Development (IPPD) in Science & Technology (2005)

[9] Springsteen, B., et al.: Integrated Product and Process Development Case Study: Development of the F/A-18E/F (2006)

[10] DiTrapani, A., Geithner, J.D.: Getting the Most Out of Integrated Product Teams, IPTs (1996)

[11] SE Handbook Working Group, INCOSE Systems Engineering Handbook (2011)

[12] Crow, K.: Building effective Integrated Product Teams (1996)

[13] Department of Defense, DOD Teaming Practices Not Achieving Potential Results (2001)

[14] Katzenbach, J.R., Smith, D.K.: Harvard Business School Press (1993)

[15] Kuhrmann, M., Kalus, G.: Providing Integrated Development Processes for Distributed Development Environments (2009)

[16] Monk, B.: Integrated Product Team Effectiveness in the Department of Defence (2002)

[17] Farshchian, B.A.: Integrating geographically distributed development teams through increased product awareness (2001)

Author Index

Printed in the United States
By Bookmasters